T0292688

MODEL MANAGEMENT AND ANALYTICS FOR LARGE SCALE SYSTEMS

MODEL MANAGEMENT AND ANALYTICS FOR LARGE SCALE SYSTEMS

Edited by

BEDIR TEKINERDOGAN

ÖNDER BABUR

LOEK CLEOPHAS

MARK VAN DEN BRAND

MEHMET AKŞIT

ACADEMIC PRESS
An imprint of Elsevier

ELSEVIER

Academic Press is an imprint of Elsevier
125 London Wall, London EC2Y 5AS, United Kingdom
525 B Street, Suite 1650, San Diego, CA 92101, United States
50 Hampshire Street, 5th Floor, Cambridge, MA 02139, United States
The Boulevard, Langford Lane, Kidlington, Oxford OX5 1GB, United Kingdom

Notices

Knowledge and best practice in this field are constantly changing. As new research and experience broaden our
understanding, changes in research methods, professional practices, or medical treatment may become necessary.

Practitioners and researchers must always rely on their own experience and knowledge in evaluating and using any
information, methods, compounds, or experiments described herein. In using such information or methods they
should be mindful of their own safety and the safety of others, including parties for whom they have a professional
responsibility.

To the fullest extent of the law, neither the Publisher nor the authors, contributors, or editors, assume any liability
for any injury and/or damage to persons or property as a matter of products liability, negligence or otherwise, or
from any use or operation of any methods, products, instructions, or ideas contained in the material herein.

Library of Congress Cataloging-in-Publication Data
A catalog record for this book is available from the Library of Congress

British Library Cataloguing-in-Publication Data
A catalogue record for this book is available from the British Library

ISBN: 978-0-12-816649-9

For information on all Academic Press publications
visit our website at https://www.elsevier.com/books-and-journals

Publisher: Mara Conner
Acquisition Editor: Mara Conner
Editorial Project Manager: John Leonard
Production Project Manager: Punithavathy Govindaradjane
Designer: Greg Harris

Typeset by VTeX

Contents

Part 1. Concepts and challenges

Part 2.   Methods and tools

8. Delta-oriented development of model-based software product lines with DeltaEcore and SiPL: A comparison   **167**

Christopher Pietsch, Christoph Seidl, Michael Nieke, Timo Kehrer

9. OptML framework and its application to model optimization   **203**

Güner Orhan, Mehmet Akşit

Part 3.   Industrial applications

10. Reducing design time and promoting evolvability using Domain-Specific Languages in an industrial context   **245**

Benny Akesson, Jozef Hooman, Jack Sleuters, Adrian Yankov

11. Model analytics for industrial MDE ecosystems   273

Önder Babur, Aishwarya Suresh, Wilbert Alberts, Loek Cleophas, Ramon Schiffelers,
Mark van den Brand

Contributors

Antonin Abherve
Softeam, Paris, France

Benny Akesson
ESI (TNO), Eindhoven, The Netherlands

Mehmet Akşit
University of Twente, Computer Science, Formal Methods & Tools Group, Enschede, The Netherlands

Wilbert Alberts
ASML N.V., Veldhoven, The Netherlands

Önder Babur
Eindhoven University of Technology, Eindhoven, The Netherlands

Alessandra Bagnato
Softeam, Paris, France

Konstantinos Barmpis
University of York, York, United Kingdom

Cagatay Catal
Information Technology Group, Wageningen University, Wageningen, The Netherlands

Michel R.V. Chaudron
Chalmers University of Technology, Gothenburg, Sweden
Gothenburg University, Gothenburg, Sweden

Loek Cleophas
Eindhoven University of Technology, Eindhoven, The Netherlands
Stellenbosch University, Matieland, Republic of South Africa

Antonio García-Domínguez
Aston University, Birmingham, United Kingdom

Regina Hebig
Chalmers University of Technology, Gothenburg, Sweden
Gothenburg University, Gothenburg, Sweden

Jozef Hooman
ESI (TNO), Eindhoven, The Netherlands
Radboud University, Nijmegen, The Netherlands

Truong Ho-Quang
Chalmers University of Technology, Gothenburg, Sweden
Gothenburg University, Gothenburg, Sweden

Aydin Kaya
Department of Computer Engineering, Hacettepe University, Ankara, Turkey

Ali Seydi Keceli
Department of Computer Engineering, Hacettepe University, Ankara, Turkey

Timo Kehrer
Department of Computer Science, Humboldt-Universtität zu Berlin, Berlin, Germany

Michael Nieke
Institute of Software Engineering and Automotive Informatics, Technische Universität
Braunschweig, Braunschweig, Germany

Güner Orhan
University of Twente, Computer Science, Formal Methods & Tools Group, Enschede,
The Netherlands

Christopher Pietsch
Department of Electrical Engineering and Computer Science, University of Siegen, Siegen,
Germany

Eric J. Rapos
Miami University, Department of Computer Science and Software Engineering, Oxford, OH,
United States

Gregorio Robles
King Juan Carlos University, Madrid, Spain

Ramon Schiffelers
Eindhoven University of Technology, Eindhoven, The Netherlands
ASML N.V., Veldhoven, The Netherlands

Christoph Seidl
Institute of Software Engineering and Automotive Informatics, Technische Universität
Braunschweig, Braunschweig, Germany

Jack Sleuters
ESI (TNO), Eindhoven, The Netherlands

Matthew Stephan

Miami University, Department of Computer Science and Software Engineering, Oxford, OH, United States

Harald Störrle

QAware GmbH, München, Germany

Aishwarya Suresh

Eindhoven University of Technology, Eindhoven, The Netherlands

Bedir Tekinerdogan

Information Technology Group, Wageningen University, Wageningen, The Netherlands

Burak Uzun

Information Technology Group, Wageningen University, Wageningen, The Netherlands

Mark van den Brand

Eindhoven University of Technology, Eindhoven, The Netherlands

Adrian Yankov

Altran, Eindhoven, The Netherlands

Analysis in the large: A foreword

The sheer complexity of software nowadays and its pervasiveness in everyday life have escalated the importance of managing abstraction by means models, modeling notations, and in general more model-based techniques. However, the proliferation of heterogeneous collections of related models in large-scale software development and their inherent complexity introduced an accidental complexity that requires advanced techniques for model management as well as model analysis. For instance, locating relevant information in a model repository presents interesting human and technical challenges because it requires the artifacts in the repository to be precisely classified, something that can not always be done manually.

This book provides an important overview of the most relevant and current aspects of model management and related analytical methods. It covers a full spectrum of topics covering foundations, methodological and empirical studies, and industrial applications. Interestingly, there is an aspect that emerges throughout the book that is particularly distinctive; most of the contributions assume that nowadays model-based techniques cannot neglect complexity, whether it be the size of the models, the extension of the collection of models, or their structural intricacy. As a consequence, it is of crucial relevance to have powerful and robust management and analysis techniques that build on a rigorous and solid basis and that focus not only on individual models but also on entire collections of them.

This volume also has another merit that goes beyond its technical content: it is current. It establishes that the most critical selling argument offered by Model-Driven Engineering goes beyond code generation. In this new age of global interconnectivity and interdependence, it is necessary to provide community-based methods and techniques to extract and analyze useful and relevant information from collections of modeling artifacts with state of the art knowledge on model management and model analysis. This book is a significant step in that direction.

Alfonso Pierantonio
May 2019, L'Aquila

Preface

In the context of increasingly large and complex software-intensive systems, modeling and model-based approaches are widely used to tackle their development and maintenance. Cyber-physical and automotive systems are two examples of such domains, where models are central and indispensable artifacts for the corresponding software. The widespread and large-scale application of modeling, in turn, leads to a deluge of models and other related artifacts. This observation addresses the main concern of the book: how we can analyze and manage large collections of models in a scalable and efficient manner. Given that this subarea in model-driven and model-based software engineering research is in its infancy, the book aims to pioneer and promote model management and analytics for large software systems. As a focused collection of chapters on various aspects around the topic, we target researchers and practitioners, serving towards the cross-fertilization of new ideas and strengthening model management and analytics as an established research area.

Introduction

The trend of ever-increasing ubiquity of software is continuing, and even pacing up with the new advancements in areas including Industry 4.0, cloud computing, and Artificial Intelligence. Software plays a central role in those systems; as a result, growing systems and systems-of-systems naturally lead to larger and more complex software. This can be traced, for instance, in the evolution of the software in passenger cars: from hardly any software up until a few decades ago, to the current state, moving towards self-driving cars with hundreds of millions of lines of code, and an elaborate network of numerous software components in interplay.

Not just the development, but also the maintenance and in general engineering of those systems proves to be a challenge. Model-based and model-driven approaches have been introduced to tackle this challenge by employing models with higher abstraction levels, rather than the general low-level code. Indeed, model-* approaches have been successfully adopted in key industries, such as cyber-physical and automotive systems. Models corresponding to several domains and abstraction layers are used for a variety of purposes, ranging from conceptualization of those large systems to simulation, verification, and automatic generation of software.

As model-* approaches are being applied to larger problems and operational context, however, the complexity, size, and variety of the models and other related artifacts increase as well. The aspect of scalability with respect to (a few) large and complex models has been pointed out in the literature, along the lines of model comparison, merging, persistence, or transformation. On the other hand, the scalability with respect to model variety and multiplicity (i.e., dealing with a large number of possibly heterogeneous models) has so far remained mostly under the radar.

In this book, we advocate this aspect of scalability as a big challenge for the management, and therefore broader adoption, of model-⋆ approaches for large-scale systems. We can trace these challenges in open source. There are tens of thousands of model artifacts in public repositories. In GitHub, for instance, the literature reports tens of thousands of UML models and Eclipse Modeling Framework (EMF) metamodels, with numbers increasing rapidly over the years too. On the other hand, in industry, even within single organizations with a large-scale and long history of model-⋆ adoption, we observe a large (and increasing) number and heterogeneity of artifacts in model-⋆ ecosystems. These include Domain-Specific Languages, the corresponding models, and transformations; all spanning across multiple domains, technologies, and organizational units. Note that there are two orthogonal factors which complicate the situation even more. First of all, the various artifacts constantly change and (co-)evolve, confirming (at least several of) Lehman's laws of software evolution. Secondly, for systems with implicit or explicit (e.g., as a Software Product Line) variability, *variants* can be considered another amplifying factor besides *versions* for the total number of model-⋆ artifacts to manage. The scalability and management challenges are further elaborated with respect to various aspects in the individual chapters of the book. It is, however, evident that this highly upward trend of a *deluge* of model-⋆ artifacts is indeed continuing. We foresee this to speed up even more, along with recent developments such as low-code platforms, intelligent techniques such as model learning and process mining gaining more popularity in the model-⋆ world.

Having set the scene, we go on stating that analyzing models to derive relevant information using traditional model management approaches does not scale for the current situation. For other types of artifacts, such as text documents, webpages, and more recently source code, the community has earlier recognized the challenge and has for long developed efficient and scalable techniques for dealing with them. An important note is that model-⋆ artifacts have certain distinguishing features, such as underlying graph structure, various abstract and concrete syntaxes, and tool-specific representations. This might render it difficult to directly apply techniques from other domains to model-⋆ approaches; therefore, it should be investigated how and to what extent the relevant techniques can be transferred into the model-⋆ technical space.

We next give a nonexhaustive list of related areas, with inspirational purposes for the modeling domain. Statistical approaches, descriptive ones for empirical analyses for discovering characteristics, patterns, and distributions, as well as predictive ones along with advanced Machine Learning/data mining techniques, can help analyze, classify, and ultimately manage large sets of model-⋆ artifacts. The indexing, storing, searching, and retrieval in large repositories can be facilitated using information retrieval techniques adapted for the new type of artifacts. Visualization, in the broadest sense, could be another essential ingredient for understanding large and heterogeneous sets of model-* artifacts. Natural language processing would be necessary to deal with the (potentially erroneous and ambiguous) text content of those artifacts, which can be inevitable in real-world applications. Finally, the challenge with respect to the computational complexity and large data sizes can be remedied by successful utilization of distributed computing infrastructure and techniques, with an eye towards the Big Data implications in the future of model management and analytics.

Why a book on model management and analytics

There is a sizeable community of researchers and practitioners of model-* approaches. On the one hand, there are dedicated conferences such as the International Conference on Model Driven Engineering Languages and Systems (MODELS) and several ones covering different aspects of the topic under the federation Software Technologies: Applications and Foundations (STAF). There are working groups and initiatives to advance and promote model-* approaches; examples include the Model-Based Systems Engineering promoted by the International Council of Systems Engineering, and many others within the Object Management Group. However, there has been hardly any focused effort tackling the aforementioned problems of model-* approaches. As the topic is quite new, no books have yet been published on this topic. Moreover, only a handful of scientific papers and PhD theses have been published that are somewhat related to the topic, but these are too specialized for the broader public. A notable set of activities have been initiated with support from the editors of this book. We have held a local symposium in the Netherlands with high industry participation, and a dedicated international workshop, Analytics and Mining of Model Repositories (AMMoRe), co-located with MODELS in 2018. This book aims to follow up on those and to present a concentrated volume of knowledge, as well as set the agenda for novel approaches in large-scale model management and analytics.

A highly related topic is (big) data management and data analytics. The approaches of this book will be based on and partially enhance these general topics. The topic of model management and analytics however is unique and needs further elaboration, as models are very complex units of (typically graph) data with a lot of special technical and domain-specific characteristics.

We target both academics and practitioners who are interested in model-based development and the analytics of large-scale models. Academics working in the field of model-based development, data analytics, or both could gain novel insights in the topic of model analytics that goes beyond both model-based development and data analytics. The book specializes in model management and analytics and as such could be used in courses on model-based development, data analytics, and data management. Typically, the book could be used for courses in later years of the academic study, that is, the third or fourth year bachelor, Master, or doctorate study programmes. The book comprises both chapters that discuss experiences from industry and ones that are more research-oriented. Practitioners will benefit from the book by identifying the key problems, the solution approaches, and the tools that have been developed or are necessary for model management and analytics. Researchers will benefit from the book by identifying the basic theory and background, the current research topics, the related challenges, and the research directions for model management and analytics.

Given the novelty of the topic, and the fact that this book presents the first concentrated effort around it, we think that this is a pioneering book that will set the scene for the paradigm and as such will be referred to frequently by practitioners and researchers in this field.

Book outline

The book consists of three main parts: concepts and challenges in model management and analytics; methods and tools; and finally industrial applications. Chapter 1 gives a general introduction to model management and analytics, outlining the problem, challenges, and relations to certain established domains. First of all, a case is made where a deluge of model-$\star$ artifacts is explicitly observed in both industry and open source. Distinguishing those artifacts from other traditional types of data, the chapter lists several related domains for inspiration. This is performed within a reference architecture for a (big) data analytics framework. MMA is discussed with the reservation that direct applications of techniques from other domains into MMA might not be possible; certain adaptations and extensions might be necessary. This is demonstrated using the SAMOS framework as a real world-example.

Chapter 2 is a contribution by Truong Ho-Quang, Michel Chaudron, Regina Hebig, and Gregorio Robles: the team behind the seminal work of the Lindholmen UML model dataset and the analytics research around it. The authors present a reference architecture for a community infrastructure for evidence-based research in software architecture. Software architecture and design, which are often represented as models, have been popular research areas for decades, with the aim of gaining insights into, as well as improving, industrial software and system development. A point is made of the lack of large-scale empirical research on those areas while validation is performed merely based on industrial experiences and cases. The authors, based on their previous experience with model analytics and empirical research on models, discuss several aspects and challenges for the domain and propose a reference architecture as a community-wide infrastructure for evidence-based research in software architecture and design.

Model clone and pattern detection, which can be considered as prominent subdomains of MMA, are discussed in Chapter 3. Matthew Stephan and Eric J. Rapos present potentials and challenges of using model clone detectors as emergent pattern miners.

As model-based software engineering approaches gain traction, the size, complexity, and prevalence of models increase. In those approaches, model analysis, and emergent pattern extraction in particular, can be used by analysts to support the software engineering life cycle. Applications areas include ensuring standard compliance and quality assurance. The authors present the underlying concepts and feasibility of their approach along with a conceptual framework called MCPM. The framework is demonstrated on Simulink models using clone detector software, with remarks on the generalizability of their approach on different types of models and tools. The demonstration is followed by a discussion of open challenges and potential benefits of the approach from both researchers' and practitioners' points of view.

In the final chapter of the first part, Chapter 4, Burak Uzun and Bedir Tekinerdogan present a domain-driven analysis of architecture reconstruction methods. The chapter outlines different approaches in the literature for the reconstruction of system architecture from various sources such as documentation, logs, and code. The focus is given on automated approaches, excluding the various manual (and more costly and less scalable) reconstruction techniques. A systematic domain-driven survey on the automated approaches is conducted. As the major outcome, the authors construct a domain model of the area along with key concepts and terms, as well as the business process model and variability model addressing the domain from complementary points of view. The presented generic knowledge and models are supported by the accompanying method for deriving concrete architecture reconstruction methods, illustrated in two cases.

Part 2 consists of five chapters covering various methods and tools for MMA. Chapter 5, by Konstantinos Barmpis, Antonio García-Domínguez, Alessandra Bagnato, and Antonin Abherve, presents the Hawk toolset for large-scale monitoring and analysis of models. The need for large-scale analysis and tool support arises from modeling practices for large and complex systems, where traditional approaches based on single-file persistence do not suffice. The authors further emphasize that existing solutions to this, based on model persistence and operating on small model fragments, can lead to other challenges for performance and I/O. They introduce their heterogeneous model indexing framework, Hawk, which has been developed and applied on a variety of scenarios. In this chapter, the authors provide an overview and high-level design of the framework along with the next steps, both specifically for Hawk and for model analytics research in general. The chapter also includes an additional demonstration and evaluation of the framework in terms of its model querying capabilities and performance, applied on real-world industrial datasets.

Chapter 6, by Aydin Kaya, Ali Seydi Keceli, Cagatay Catal, and Bedir Tekinerdogan, addresses software defect prediction in early stages of software development, noting the advantages of such early detection. Defect prediction models can indeed lead to identifying error-prone components before the testing phase and can contribute to quality of the software. The authors apply Machine Learning techniques using design-level metrics and data sampling techniques as an effective means of defect prediction. The study demonstrates a strong correlation of design-level metrics with defect probabilities, while it applies advanced ensemble methods and data sampling to yield high-accuracy defect prediction models.

Haralt Störrle addresses the structuring and visualization of large models in Chapter 7. The premise of the chapter is based on the observation that large modeling projects increasingly need better internal structuring to organize their artifacts and that bad structuring can lower the efficiency and quality of modeling. The author aims to improve the state of the art for model structuring through providing templates, examples, and best practices. This is to be achieved via a practical visual notation to describe model structure. Based on two distinct case studies from academic and industrial modeling projects with diverse underlying technologies and conceptualization, a common notation called MONO is presented to cover both modeling structures, thus serving as a validation for the approach. The resulting process and set of artifacts, i.e., the templates and notation, demonstrate the effectiveness and genericness of the approach in the two case studies and provide a basis for further application in other modeling projects.

Chapter 8, by Christopher Pietsch, Christoph Seidl, Michael Nieke, and Timo Kehrer, contributes with a comparison of delta-oriented development of model-based software product lines using DeltaEcore and SiPL. The widespread use of model-based development in embedded systems calls for variability management where customized functionality needs to be delivered to the clients. Model-based software product line engineering addresses this by explicitly modeling and managing the variability. Delta modeling is one such technique in the domain with emphasis on a core model and delta modules and transformations on top of them. While there is limited tool support for the approach, it is realized in two major tool suites, i.e., DeltaEcore and SiPL. The authors provide an overview and comparison of the capabilities of these tools. The extensive comparison and demonstration aims to help both researchers and practitioners that are interested in the topic, in terms of the state of practice, research directions, and tool adoption and use.

The OptML framework and its application in model optimization are discussed in Chapter 9, by Guner Orhan and Mehmet Akşit. The authors argue that companies need to deal with larger model bases as their assets in widespread adoption of model-driven approaches. Besides the sheer size of model bases, software engineers may additionally need to use configurations of existing models, further expanding the design space with alternatives. The OptML framework addresses this problem with the capability of computing optimal models over a number of Ecore-based models based on user-defined criteria. The framework is implemented and validated on a set of models for image processing, but provides a first attempt in the literature as a generic framework for model optimization.

The final part of the book, Part 3, tackles MMA from an industrial application perspective with two extensive industrial studies. Chapter 10, by Benny Akesson, Jozef

Hooman, Jack Sleuters, and Adrian Yankov, discusses the improvement of design time performance and evolvability of systems using Domain-Specific Languages. The authors note the increasing complexity of cyber-physical systems and the arising need for variability and mass-customization along with adaptation of changing requirements and technologies. Model-Based Engineering based on Domain-Specific Languages can be used to remedy many of these challenges for developing such complex systems. The chapter elaborates several aspects of successful application of Model-Based Engineering around a real-world industry case study from the defence domain. The aspects include modularity and reuse, evolution, model quality, and generated artifacts.

Chapter 11, by Önder Babur, Aishwarya Suresh, Wilbert Alberts, Loek Cleophas, Ramon Schiffelers, and Mark van den Brand, is the final chapter of the book. The authors explore a wide set of model analytics techniques using the SAMOS framework in the context of real domain-specific model-driven engineering ecosystems in the lithography industry. Given the multidisciplinary and heterogeneous nature of the ecosystems, automated analyses are proposed as a means for managing those artifacts, in terms of getting an overview of or detecting duplication within the ecosystems. The case studies involve clone detection on data and control models within one of the ecosystems, cross-language conceptual analysis and language-level clone detection on three ecosystems, and finally architectural analysis and reconstruction on another ecosystem. The authors discuss how model analytics can be used to discover insights in model-driven engineering ecosystems (e.g., via model clone detection and architectural analysis) and opportunities such as refactoring to improve them.

Bedir Tekinerdogan[a]
Önder Babur[b]
Loek Cleophas[b,c]
Mark van den Brand[b]
Mehmet Akşit[d]

[a]Information Technology Group, Wageningen University, Wageningen, The Netherlands
[b]Eindhoven University of Technology, Eindhoven, The Netherlands
[c]Stellenbosch University, Matieland, Republic of South Africa
[d]University of Twente, Computer Science, Formal Methods & Tools Group, Enschede, The Netherlands

PART 1

Concepts and challenges

CHAPTER 1

Introduction to model management and analytics

Bedir Tekinerdogan[a], **Önder Babur**[b], **Loek Cleophas**[b,c], **Mark van den Brand**[b], **Mehmet Akşit**[d]

[a]Information Technology Group, Wageningen University, Wageningen, The Netherlands
[b]Eindhoven University of Technology, Eindhoven, The Netherlands
[c]Stellenbosch University, Matieland, Republic of South Africa
[d]University of Twente, Computer Science, Formal Methods & Tools Group, Enschede, The Netherlands

Contents

1.1. Introduction

The idea of creating business value from data has always been an important concern. Many businesses have extracted information from data to gain new insights and make smarter decisions. Together with the advancements of disruptive technologies such as Cloud Computing and Internet of Things, the ability to capture and store vast amounts of data has grown at an unprecedented rate and soon did not scale with traditional data management techniques. Yet, to cope with the rapidly increasing volume, variety, and velocity of the generated data we can now adopt the available novel technical capacity and the infrastructure to aggregate and analyze huge and variant sets of data. This situation has led to new and unforeseen opportunities for many organizations. Data science, and Big Data in particular, has now become a very important driver for innovation and growth for various industries, such as health, administration, agriculture, and education.

Big Data is usually characterized using four V's, that is, volume, variety, velocity, and veracity. Volume is one the key characteristics of Big Data, indicating the large amount of data that usually does not fit on one computer. Variety relates to the different data types, including structured data, semistructured data, and unstructured data. Velocity refers to the speed at which the data are generated as well as being processed. Finally, veracity refers to the trustworthiness of the data.

With Big Data, different types of data have been addressed, including text, audio, and video. A different type of data that did not get much attention yet is the whole set of models that is used in various engineering and science disciplines. A model is an abstract representation of selected parts of the considered domain. In science, models can relate to, for example, physical and chemical models. In engineering, models can relate to the intermediate artifacts for realizing the eventual system. In software engineering, models are, for example, the UML design artifacts.

In this context, model management and analytics (MMA) aims to use models and related artifacts to derive relevant information to support, in the most general sense, decision making processes of organizations. This can have several benefits, including the understanding, development, maintenance, and management of those artifacts. Various different models as well as analytics approaches could be identified. In this chapter we will borrow from existing domains, potentially to be adapted and exploited for MMA.

1.2. Data analytics concepts

Data analytics is the discovery, interpretation, and communication of meaningful patterns in data. Using data analytics, data are examined in order to gain insight, from which one can make decisions and take actions that lead to effective outcomes. Hence, the goal of analytics is usually to improve the business by gaining knowledge which can be used to make improvements or changes. Traditionally, data analytics predominantly refers to various set of applications, such as basic business intelligence (BI), online analytical processing (OLAP), and various forms of advanced analytics. Data analytics is related to the term business analytics, with the difference that the latter is focused on business uses, while data analytics has a broader focus.

Data analytics is important to support decision making and likewise help organizations better achieve their business goals. Different types of analytics can be distinguished, including:

- Descriptive analytics uses historical data to provide insight into what has happened.
- Diagnostics analytics again uses the historical data to answer why something has happened.
- Predictive analytics elaborates on the insight of the analyzed data and answers what can happen.
- Prescriptive analytics, which usually builds on and uses the other types of analytics, answers the question what to do.

Each of these types of analytics can be understood quite simply as using data to answer different types of questions. In the case of data analytics the source is the traditional data (e.g., text, audio, video, etc.). With model analytics we assume that the analytics takes as input a set of models to support the decision making process.

Model analytics can thus be considered as a subcategory of data analytics orienting particularly on model artifacts.

Based on the observations above, we thus advocate a perspective where model-⋆ (as an umbrella term for model-driven, model-based, and other related approaches) artifacts are treated holistically as data, processed, and analyzed with various scalable and efficient techniques, possibly inspired by related domains. Tackling large volumes of artifacts has been commonplace in other domains, such as text mining for natural language text [1] and repository mining for source code [2]. While we might not be able to apply those techniques as-is on model-⋆ artifacts, the general data analytics workflow appears to be applicable with the steps of data collection, cleaning, integration and transformation, feature engineering and selection, modeling (e.g., as statistical models or neural networks), and finally deployment, exploration, and visualization.

1.3. The inflation of modeling artifacts

The problem of the large number of models is not new in the model-⋆ world. Several efforts have indeed been initiated to store and manage large numbers of models and related artifacts [3,4]. Further efforts include mining public repositories for MDE-related items from GitHub, e.g., Eclipse-based MDE technologies [5] and UML models [6] (the Lindholmen dataset). In the latter, the number of UML models can reach more than 90k. We have previously demonstrated the increasing number of newly created Ecore metamodels, as well as a number of commits on them, on GitHub [7]. A part of the results is given in Fig. 1.1, depicting a strong upward trend for the number of commits and newly created files over the years. It is evident that there are increasingly more Ecore metamodels in GitHub. We have further evidence that other types of modeling artifacts and repositories follow similar trends [8].

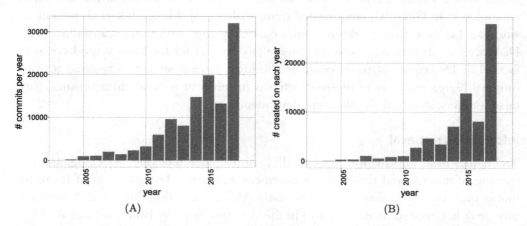

Figure 1.1 GitHub results on Ecore metamodels. (A) Number of commits per year. (B) New files per year.

Moreover, it appears that even within a single industry or organization, a similar situation emerges with the increasing adoption of model-★ approaches. Just a single company might need to manage an ecosystem consisting of dozens of metamodels, thousands of models, and tens of thousands LOCs of transformations. With the complete revision history, the total number of artifacts can reach tens of thousands. Along with conventional forward engineering approaches, we can observe an increasing trend with legacy software: automated migration into model-driven/-based engineering using process mining and model learning.

All the presented facts, from open source as well as industry, let us confirm the statement by Brambilla et al. [9] and Whittle et al. [10] that the adoption of model-★ approaches in (at least some parts of) the industry grows quite rapidly, and we conclude that tackling these artifacts, namely MMA, will be increasingly important in the future.

1.4. Relevant domains for MMA

Despite the different nature of models, as exemplified above, we can be inspired by techniques from other disciplines and try to adapt them for the problems in MMA. As a preliminary overview, in this section we list and discuss several such domains. While there is related model-★ research on some of the items on the list, we believe a conscious and integrated mindset would mitigate the challenges for scalable application of model-★ approaches.

Descriptive statistics

Several model-★ researchers have already performed empirical studies on model-★ artifacts with a statistical mindset. For instance, Kolovos et al. assess the use of Eclipse technologies in GitHub, giving related trend analyses [5]. Mengerink et al. present an automated analysis framework on version control systems with similar capabilities [11]. Di Rocco et al. perform a correlation analysis on metrics for various model-★ artifacts [12]. Descriptive statistics could in the most general sense be exploited to gain insight in large numbers of model-★ artifacts in terms of general characteristics, patterns, outliers, statistical distributions, dependence, etc.

Information retrieval

Techniques from information retrieval (IR) can facilitate indexing, searching, and retrieving of models, and thus their management and reuse. The adoption of IR techniques on source code dates back to the early 2000s, and within the model-★ community there has been some recent effort in this direction (e.g., by Bislimovska et al. [13]). Further IR-based techniques can be found in [14,15] involving repository management and model searching scenarios.

Natural language processing

Accurate natural language processing (NLP) is needed to handle realistic models with noisy text content, compound words, and synonymy/polysemy. In our experience, it is very problematic to blindly use NLP tools on models, e.g., just WordNet synonym checking without proper part-of-speech tagging and word sense disambiguation. More research is needed to find the right chain of NLP tools applicable for models (in contrast to source code and documentation), and reporting accuracies and disagreements between tools (along the lines of the recent report in [16] for repository mining). Note that NLP offers further advanced tools, such as language modeling, which are still to be investigated for model-★ approaches.

Data mining

Following the perspective of approaching model-★ artifacts as data, we need scalable techniques to extract relevant units of information from models (*features* in data mining jargon) and to discover patterns, including clusters, outliers/noise, and clones. Several example applications can be found in [14,15,17] for domain clustering EMF metamodels, and in [18] for classifying forward vs. reverse engineered UML models. To analyze, explore, and eventually make sense of the large model-★ datasets (e.g., the Lindholmen dataset [6]), we can investigate what can be borrowed from comparable approaches in data mining for structured data.

Graph databases and graph-based methods

Given that quite some commonly used models, such as UML, are based on an underlying graph, graph databases can be used to store, query, and reason about models. There has already been some effort using graph databases, such as Neo4EMF [19] as a persistence layer for (potentially very big) models, and Mogwaï [20] as a fast and complex querying mechanism for models. Another related idea is presented by Clarisó and Cabot [21], who advocate using graph kernel-based methods for several model-★ tasks such as model searching and clustering.

Machine learning

The increasing availability of large amounts of model-★ data can be exploited, via Machine Learning, to automatically infer certain qualities and predictor functions (e.g., performance). There has been a thrust of research in this direction for source code (e.g., for fault prediction [23]), and it would be noteworthy to investigate the emerging needs of the model-★ communities and feasibility of such learning techniques for model-★ approaches. The approaches in [24] for learning model transformations by examples and [25] for automatic model repair using reinforcement learning are some of the few pieces of such work in model-* approaches.

Visualization

We propose visualization and visual analytics techniques to inspect a whole dataset of artifacts (e.g., cluster visualizations in [15], in contrast with visualizing a single big model in [26]) using various features such as metrics and cross-artifact relationships. The goals could range from exploring a repository to analyzing a model-* ecosystem holistically and even studying the (co-)evolution of model-* artifacts.

Distributed/parallel computing

With the growing amount of data to be processed, employing distributed and parallel algorithms in model-* approaches is very relevant. There are conceptually related approaches worthwhile investigating, e.g., distributed model transformations for very large models [27,28] or model-driven data analytics [29]. Yet we wish to draw attention here to performing computationally heavy data mining or Machine Learning tasks for large model-* datasets in an efficient way.

(Big) data analytics

One of the key domains that is related to and required for managing large collections of models is obviously Big Data. Based on the literature, we have derived the family feature model for Big Data systems, as shown in Fig. 1.2. This has been largely adopted from our earlier work on a feature-based analysis of Big Data systems [22]. The feature diagram contains features representing the essential characteristics or externally visible properties of the system. Features may be mandatory or alternative/optional and may have subfeatures which as such can lead to a hierarchical tree. Besides the conceptual model, MMA can get inspiration from the different Big Data reference architectures in the literature. Notable ones would include the lambda architecture, which is a three-layer architecture aimed at a robust and fault-tolerant system [30], and a functional architecture with six key modules representing phases from data extraction and loading to interfacing and visualization [31]. MMA could be seen from the Big Data perspective. In that case, the MMA approaches can be represented using the features of the family feature diagram of Fig. 1.2.

The above represents a nonexhaustive list of domains as a preliminary exploitation guideline for MMA. Although the aforementioned domains themselves are quite mature on their own, it should be investigated to what extent results and approaches can be transferred into the model-* technical space, particularly for MMA. The chapters in this book approach the topic from these and other different perspectives, and likewise provide valuable insight in the MMA domain.

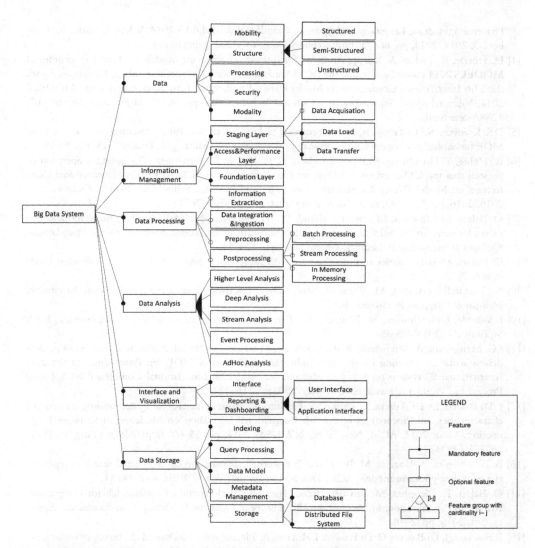

Figure 1.2 Top-level feature model for Big Data systems (adapted from [22]).

References

[1] A. Hotho, A. Nürnberger, G. Paass, A brief survey of text mining, LDV Forum 20 (2005) 19–62.

[2] H. Kagdi, M.L. Collard, J.I. Maletic, A survey and taxonomy of approaches for mining software repositories in the context of software evolution, Journal of Software Maintenance and Evolution 19 (2007) 77–131.

[3] F. Basciani, J.D. Rocco, D.D. Ruscio, A.D. Salle, L. Iovino, A. Pierantonio, Mdeforge: an extensible web-based modeling platform, in: Proceedings of the 2nd International Workshop on Model-Driven Engineering on and for the Cloud Co-Located with the 17th International Conference on Model

Driven Engineering Languages and Systems, CloudMDE@MoDELS 2014, Valencia, Spain, September 30, 2014, 2014, pp. 66–75, http://ceur-ws.org/Vol-1242/paper10.pdf.

[4] H. Störrle, R. Hebig, A. Knapp, An index for software engineering models, in: Joint Proceedings of MODELS 2014 Poster Session and the ACM Student Research Competition (SRC) Co-Located with the 17th International Conference on Model Driven Engineering Languages and Systems, MODELS 2014, Valencia, Spain, September 28 – October 3, 2014, 2014, pp. 36–40, http://ceur-ws.org/Vol-1258/poster8.pdf.

[5] D.S. Kolovos, N.D. Matragkas, I. Korkontzelos, S. Ananiadou, R.F. Paige, Assessing the use of eclipse MDE technologies in open-source software projects, in: OSS4MDE@MoDELS, 2015, pp. 20–29.

[6] R. Hebig, T. Ho-Quang, M.R.V. Chaudron, G. Robles, M.A. Fernández, The quest for open source projects that use UML: mining GitHub, in: Proceedings of the ACM/IEEE 19th International Conference on Model Driven Engineering Languages and Systems, Saint-Malo, France, October 2–7, 2016, 2016, pp. 173–183, http://dl.acm.org/citation.cfm?id=2976778.

[7] Ö. Babur, L. Cleophas, M. van den Brand, B. Tekinerdogan, M. Aksit, Models, more models, and then a lot more, in: M. Seidl, S. Zschaler (Eds.), Software Technologies: Applications and Foundations, Springer International Publishing, Cham, 2018, pp. 129–135.

[8] Ö. Babur, Model analytics and management, IPA Dissertation Series, Technische Universiteit Eindhoven, 2019.

[9] M. Brambilla, J. Cabot, M. Wimmer, Model-Driven Software Engineering in Practice, 1st edition, Morgan & Claypool Publishers, 2012.

[10] J. Whittle, J. Hutchinson, M. Rouncefield, The state of practice in model-driven engineering, IEEE Software 31 (2014) 79–85.

[11] J.G. Mengerink, A. Serebrenik, R.R. Schiffelers, M.G. van den Brand, Automated analyses of model-driven artifacts: obtaining insights into industrial application of MDE, in: Proceedings of the 27th International Workshop on Software Measurement and 12th International Conference on Software Process and Product Measurement, ACM, 2017, pp. 116–121.

[12] J. Di Rocco, D. Di Ruscio, L. Iovino, A. Pierantonio, Mining metrics for understanding metamodel characteristics, in: Proceedings of the 6th International Workshop on Modeling in Software Engineering, MiSE 2014, ACM, New York, NY, USA, 2014, pp. 55–60, http://doi.acm.org/10.1145/2593770.2593774.

[13] B. Bislimovska, A. Bozzon, M. Brambilla, P. Fraternali, Textual and content-based search in repositories of web application models, ACM Transactions on the Web (TWEB) 8 (2014) 11.

[14] Ö. Babur, L. Cleophas, M. van den Brand, Hierarchical clustering of metamodels for comparative analysis and visualization, in: Proc. of the 12th European Conf. on Modelling Foundations and Applications, 2016, pp. 2–18.

[15] F. Basciani, J. Di Rocco, D. Di Ruscio, L. Iovino, A. Pierantonio, Automated clustering of metamodel repositories, in: International Conference on Advanced Information Systems Engineering, Springer, 2016, pp. 342–358.

[16] F.N.A.A. Omran, C. Treude, Choosing an NLP library for analyzing software documentation: a systematic literature review and a series of experiments, in: Proceedings of the 14th International Conference on Mining Software Repositories, MSR 2017, Buenos Aires, Argentina, May 20–28, 2017, 2017, pp. 187–197, https://doi.org/10.1109/MSR.2017.42.

[17] Ö. Babur, Statistical analysis of large sets of models, in: Proceedings of the 31st IEEE/ACM International Conference on Automated Software Engineering, ASE 2016, ACM, New York, NY, USA, 2016, pp. 888–891.

[18] M.H. Osman, T. Ho-Quang, M. Chaudron, An automated approach for classifying reverse-engineered and forward-engineered UML class diagrams, in: 2018 44th Euromicro Conference on Software Engineering and Advanced Applications, SEAA, IEEE, 2018, pp. 396–399.

[19] A. Benelallam, A. Gómez, G. Sunyé, M. Tisi, D. Launay, Neo4EMF, a scalable persistence layer for EMF models, in: European Conference on Modelling Foundations and Applications, Springer, 2014, pp. 230–241.

[20] G. Daniel, G. Sunyé, J. Cabot, Mogwaï: a framework to handle complex queries on large models, in: Research Challenges in Information Science (RCIS), 2016 IEEE Tenth International Conference on, IEEE, 2016, pp. 1–12.

[21] R. Clarisó, J. Cabot, Applying graph kernels to model-driven engineering problems, in: Proceedings of the 1st International Workshop on Machine Learning and Software Engineering in Symbiosis, MASES 2018, 2018, pp. 1–5.

[22] C.A. Salma, B. Tekinerdogan, I.N. Athanasiadis, Feature driven survey of big data systems, in: IoTBD, 2016, pp. 348–355.

[23] C. Catal, B. Diri, A systematic review of software fault prediction studies, Expert Systems with Applications 36 (2009) 7346–7354.

[24] I. Baki, H.A. Sahraoui, Multi-step learning and adaptive search for learning complex model transformations from examples, ACM Transactions on Software Engineering and Methodology 25 (2016) 20.

[25] A. Barriga, A. Rutle, R. Heldal, Automatic model repair using reinforcement learning, in: Proc. of MODELS 2018 Workshops, Co-Located with ACM/IEEE 21st Int. Conf. on Model Driven Engineering Languages and Systems, MODELS 2018, Copenhagen, Denmark, October, 14, 2018, 2018, pp. 781–786.

[26] D.S. Kolovos, L.M. Rose, N. Matragkas, R.F. Paige, E. Guerra, J.S. Cuadrado, J. De Lara, I. Ráth, D. Varró, M. Tisi, J. Cabot, A research roadmap towards achieving scalability in model driven engineering, in: Proceedings of the Workshop on Scalability in Model Driven Engineering, BigMDE '13, ACM, New York, NY, USA, 2013, 2, http://doi.acm.org/10.1145/2487766.2487768.

[27] A. Benelallam, A. Gómez, M. Tisi, J. Cabot, Distributed model-to-model transformation with ATL on MapReduce, in: Proceedings of the 2015 ACM SIGPLAN International Conference on Software Language Engineering, ACM, 2015, pp. 37–48.

[28] L. Burgueño, M. Wimmer, A. Vallecillo, Towards distributed model transformations with LinTra, 2016.

[29] T. Hartmann, A. Moawad, F. Fouquet, G. Nain, J. Klein, Y.L. Traon, J.-M. Jezequel, Model-driven analytics: connecting data, domain knowledge, and learning, preprint, arXiv:1704.01320, 2017.

[30] N. Marz, J. Warren, Big Data: Principles and Best Practices of Scalable Real-Time Data Systems, Manning Publications Co., New York, 2015.

[31] D. Chappelle, Big data & analytics reference architecture, An Oracle White Paper, 2013.

CHAPTER 2

Challenges and directions for a community infrastructure for Big Data-driven research in software architecture

Truong Ho-Quang[a,b]**, Michel R.V. Chaudron**[a,b]**, Regina Hebig**[a,b]**, Gregorio Robles**[c]
[a]Chalmers University of Technology, Gothenburg, Sweden
[b]Gothenburg University, Gothenburg, Sweden
[c]King Juan Carlos University, Madrid, Spain

Contents

2.1. Introduction

Research into software architecture and design blossomed in the 1990s. At that point in time, many organizations were experiencing exponential increase in the size of their software systems. Almost at the same time, projects were struggling with the increasing amount of changes to the software that needed to be handled. The practices of software architecting were proposed as one of the main tools for addressing the challenges of both scale and evolvability. Academic research in software architecture has focused on several areas, including architecture description through views and architecture description languages, and on methods for evaluating architectural designs. While much of the contribution of research in software architecture was inspired by industrial experiences, little of the research was validated beyond individual case studies. Many scientific disciplines are currently harvesting fruits from large-scale data collection about their subjects of study. Indeed, such "Big Data" promises insights by finding patterns by analyzing large datasets. Therefore, this chapter contributes a discussion of challenges and directions for Big Data-driven studies of software architecture. We discuss lessons from various projects that focus on particular questions that are building blocks in the overall landscape of Big Data for empirical software architecture research. Based on these lessons, we synthesize a proposal for a reference architecture for a community-wide infrastructure for evidence-based research in software architecture and design.

The structure of this chapter is as follows. In Section 2.2, we discuss existing work related to our research topic. Then, we present our experiences on building, maintaining, and sharing a big corpus of models (Section 2.3). This is followed by a discussion on the challenges when conducting empirical studies in software architecture (Section 2.4). The discussion reflects our observations from research in the field as well as our experience building the Lindholmen dataset of UML software designs. In Section 2.5, we list nine requirements for building such an infrastructure. Lastly, we propose a reference architecture for such an infrastructure (called **CoSARI**) and present our on-going efforts on building this (Section 2.6).

2.2. Related work

In this section, we discuss works that are in various ways related to the topic of this chapter. Empirical data of software architecture serve as a basis for any evidence-based research in the field. Therefore, at first, we summarize existing corpora of software architecture artifacts. The software architecture artifacts/documentation (SAD) can be split into software modeling artifacts (such as UML models, DSLs, etc.) and textual-based artifacts (such as software architecture specification, etc.).

The desired infrastructure should ultimately support researchers with not only empirical data on software architecture but also means for analyzing the data and sharing the analyses. Therefore, we discuss existing work on discovering architecture knowledge and review some scientific workflow systems as a reference for building the infrastructure.

2.2.1 Existing corpora of software modeling artifacts

Störrle et al. introduced the Software Engineering Model Index (SEMI), which contains a list of contemporary model repositories [1]. We take this as a starting point for our search of software modeling corpora. In fact, three our of eight corpora to be reviewed in this section are listed in SEMI. In the paper, the authors also outline four main challenges when building a successful model repository: (i) archiving ("how to archive data with very high reliability, for a very long time, yet readily accessible and economically viable"), (ii) Access Support ("how to search for models"), (iii) Intellectual Property ("how to manage intellectual property such as models"), and (iv) Incentives ("how to motivate researchers/practitioners to publish their models").

The *BPM Academic Initiative (BPM AI)* is a platform where business process models are shared for teaching purposes [2]. A business process model is defined as a set of business activities and execution constraints between these activities [3]. It can be used to describe complex interactions between business partners and to indicate related business requirements on an abstract level. Currently, BPM AI claims to host 29,285 business process models in various machine-readable formats. The dataset has however not been updated since 2012. The process of collecting models is not clearly mentioned; apparently, most of the models in the dataset derive from students as part of modeling assignments.

The *Repository for Model-Driven Development (ReMoDD)* is created to support researchers and practitioners in sharing exemplar models and other modeling practices [4]. Currently, it contains around 90 modeling artifacts, including models in different modeling languages and artifacts of some MDD conferences. Models are stored in various formats, mostly PDF but also some in XMI.

The *Open Models Initiative (OMI)*,[1] similar to ReMoDD, offers a platform that allows researchers and practitioners to share models. It is currently hosting around 70 models stored mostly in image formats. There is no report on whether the models are derived from industrial or academic contexts.

Karasneh et al. used a crawling approach to automatically fill an online repository with so far more than 700 model images[2] from Google Image Search [5]. Registration is not required in order to get access to the repository. The repository also provides a comprehensive search which could be used to form and share subsets of the data.

Mengerink et al. collected a dataset of 9,188 OCL expressions derived from 504 EMF metamodels in 245 GitHub repositories [6]. To this end, the authors firstly performed a couple of GitHub searches, then downloaded all *.ecore* and *.ocl* files in the result list, then removed all duplicated files, and finally parsed all the unique files to extract OCL expressions.

---

[1] http://openmodels.org.
[2] http://models-db.com/.

Basciani et al. built MDEForge as a web-based modeling platform which aims at fostering a community-based modeling repository [7]. The number of metamodels hosted in this platform is not available.

GenMyModel[3] is a web-based online tool that supports collaborative modeling for UML, BPMN, RDS, and flowcharts [8]. At the time of writing, GenMyModels claims to host about 777,000 diagrams. However, it is not clear how many of these diagrams are open to public access and how many are private.

The Lindholmen dataset[4] contains more than 93,000 UML models from more than 24,000 GitHub repositories [9]. Different from the above-mentioned corpora, the Lindholmen dataset also includes metadata of the projects where UML models are used. This enables researcher to study the use of UML models in their context, e.g., how frequent and in which phase of the project the models are updated. The UML models are collected from GitHub using complex settings of tools and technologies (such as image processing). The models are provided in various formats, mostly in .uml, .xmi, and image files. This is currently the biggest dataset of UML models.

2.2.2 Other software architecture collections

Ding et al. present the retrieval and analysis of a collection of SADs obtained from 108 open source projects [10]. We have reviewed this document and have run into two issues. Firstly, many SAD links expired, resulting in 404-browser errors. This means that either the document was moved or deleted, so the URL pointing to it was faulty, or for the SAD documents that we could find, we found they were of "mixed quality": some documents were identified as SAD but are almost empty or contain only little text.

The well-known book "The Architecture of Open Source Applications" is a collection of SADs of 48 open source applications [11]. In particular, each chapter describes the architecture of an open source application, i.e., how it is structured, how its parts interact, etc. It is noted that there is no uniform representation of the architectures. In fact, every chapter in the book has its own structure and uses different kinds of diagramming notations.

2.2.3 Mining architectural knowledge

Another relevant area of empirical studies in software architecture focuses on the notion of "mining" artifacts for architectural knowledge.

The work by Soliman et al. aims to mine architectural knowledge from natural language sources, in particular from StackOverflow [12]. They apply advanced natural language processing and Machine Learning algorithms for the recognition and classification of sentences.

---

[3] https://www.genmymodel.com/.
[4] http://oss.models-db.com/.

Musil et al. describe a novel architecture knowledge management approach with similar objectives: use information retrieval and natural language processing to extract architectural knowledge about systems from all documents available in project repositories [13]. Their Continuous Architectural Knowledge Integration (CAKI) approach consists of four stages:

- Information acquisition – where relevant data are to be collected.
- Architecture knowledge synthesis – where architecture knowledge is automatically synthesized from raw data.
- Architecture knowledge dissemination – the stage at which architecture knowledge is represented to relevant stakeholders in various ways.
- Feedback – where inputs from user and experts are used to improve the learning model.

CAKI utilizes ontology-based models as a basis for personalization mechanisms and exploratory search. However, in the paper, the ontology models are not presented. This work involved industrial collaboration with Siemens.

Lin et al. propose a system (called IntelliDE) in which software Big Data could be aggregated, mined, and analyzed to provide meaningful assistance for developers across the software development processes [14]. Similar to CAKI, IntelliDE follows a three-stage knowledge discovery approach, including Data Aggregation, Knowledge Acquisition, and Intelligent Assistant. For each stage, a number of key research issues and challenges are listed. Unlike CAKI, IntelliDE uses a so-called Software Knowledge Graph for "storing knowledge in software domains, projects, and systems." In order to construct such graphs from various knowledge sources, a process of parsing and extracting knowledge entities is proposed. However, no extra review or feedback steps are undertaken in order to validate the knowledge to be added to the graph.

2.2.4 Scientific workflow systems

In various scientific disciplines, software tools have emerged for automating sequences of (typically data-intensive) steps of a scientific analysis or experimental procedures. Such system are called "scientific workflow systems" and these build on the approaches of general workflow management tools. Automation of workflows enables the reuse, replication, and incremental improvement by making changes to the details of the study. In a survey of existing scientific workflow technology, Barker et al. list 14 frameworks from the business and scientific domain [15]. Among the listed frameworks, some specialize in particular fields of science, while others aim to be more generic. In the paper, the authors suggest six key factors to consider when developing scientific workflow systems, i.e., (i) collaboration is key (to avoid overlapping requirements and reinvention of any wheels); (ii) use a conventional scripting language; (iii) reuse existing workflow language; (iv) research your domain (before building the systems); (v) stick to standards; and (vi) have a portal-based access.

In the Software Engineering domain, we find eSEE – a novel framework to support large-scale experimentation and scientific knowledge management in Software Engineering proposed by Travassos et al. [16] – the most relevant. Four main requirements of eSEE are: (i) having integrated experimentation support tools; (ii) being a Web System; (iii) using an e-services-based paradigm; and (iv) providing knowledge management mechanisms. The architecture of eSEE consists of three distributed macrocomponents, i.e., Meta-Configurator (MC), Instantiation Environment (IE), and Execution Environment (EE). The framework, however, does not support automated execution and data analysis from the experiment specification for technology-oriented experiments.

2.3. Experiences in creating & sharing a collection of UML software design models

In this section, we summarize our experiences in building, curating, and sharing the Lindholmen dataset, which is an extensive dataset of more than 93,000 UML models from more than 24,000 GitHub projects [9]. This dataset also contains metadata of the projects (such as commits, commit messages, and committers) which enables researchers to study UML models in their usage context.

2.3.1 Models extraction from GitHub

The data extraction process is described in detail in [17]. In this section, we provide a brief summary of the extraction steps and highlight the challenges when extracting the dataset. In general, the data extraction comprises four steps, i.e., (i) retrieving the file list of all GitHub repositories, (ii) identifying potential UML files, (iii) examining (and manually evaluating) the existence of UML notation in the obtained files, and finally (iv) collecting the metadata of the repositories where a UML file has been identified.

In steps (i) and (iv), we used the GitHub API[5] to retrieve the list of files from GitHub projects as well as to query metadata from the projects that contain UML models, respectively. At these stages, we faced two Big Data retrieval challenges. Firstly, the GitHub API limitation (of 5000 requests/hour) could hugely affect the crawling time. In particular, with up to three GitHub calls for each repository, given the limit of 5000 requests/hour, it would take around 14 months to perform the retrieval of data in step (i). We worked around this by downloading the JSON files in parallel with over 20 active GitHub accounts, which were donated from fellow colleagues and students during this process. This reduced the time span to approximately one month. Secondly, while GitHub is a dynamic environment where projects change over time, we could only work on a static snapshot (and thus, a somehow outdated version) of it (captured by GHTorrent). In particular, many repositories might have been removed or made

---

[5] https://developer.github.com/v3/git/trees/#get-a-tree.

private in the time that passed from GHTorrent obtaining its data (which is before 1 February 2016) and our request to the GitHub API (during Summer of 2017). This resulted in a huge number of repositories (around 3 million) where we obtained an empty JSON file or an error message from the GitHub API.

In steps (ii) and (iii), the main aim was to identify those files that actually contain UML models. There were two main challenges. Firstly, browsing through the enormous amount of files stored on GitHub is a challenge itself. Secondly, the file formats in which UML models are stored are diverse, making it difficult to develop a systematic searching approach. For example, UML models are often stored as images, for which a simple textual searching approach does not work – it required image processing technology. In fact, we split the files into textual-format and image-format files and developed different technologies to treat them separately. Details about the technologies that were used can be found in [9].

Last but not least, since some steps are carried out manually and therefore expensive, we could not provide updates on the dataset in a frequent and automatic manner. This results in a number of threats to the availability of models in the dataset, such as models becoming unavailable/inaccessible and missing mid-flight projects in which UML models were introduced later than the time of analysis.

2.3.2 Data curation

Having the dataset collected, we moved on to more in-depth studies about the use of UML models in the context of open source projects. As these studies require a set of projects and models with specific characteristics, the dataset had to be curated. For example, when studying the practices and perception of UML use, we were interested in the projects where we could observe long-term use of UML models and collaboration between contributors [18]. To obtain these projects, we applied some filters on the number of contributors, number of commits, and active time span. With that, we are willing to accept false rejections (e.g., "serious" projects that use UML might be rejected) in favor of having no (very few) false positives.

Successful curation can also be achieved by adding extra knowledge to the existing dataset. In particular, we performed a number of classifications, i.e., (a) classifying types of UML models (which was done manually) and (b) automatically classifying reverse engineering diagrams and forward design diagrams [19]. The classification results were then added/annotated in the dataset.

2.3.3 Sharing the Lindholmen dataset

The availability of the Lindholmen dataset has attracted researchers in the field to use and study the dataset. For example, El Ahmar et al. used more than 3500 diagrams from the dataset to study the use of visual variables (such as size, brightness, and texture/grain) in UML models in open source projects [20]. Schulze et al. used 50 sequence diagrams from the Lindholmen dataset for evaluating their automatic layout and label management [21]. Unfortunately, the results of these studies have never been integrated/annotated into the Lindholmen dataset because of two reasons. Firstly, these investigations have been conducted with small subsets of the dataset, making it hard to generalize the research result to the whole dataset. Secondly, there has been no systematic and convenient way for researchers to integrate their findings to the dataset.

2.4. Challenges for Big Data-driven empirical studies in software architecture

In this section, we discuss the challenges (**C**) for conducting Big Data-driven empirical studies on software architectures. The discussion reflects our observations from research in the field as well as our experience in building the Lindholmen dataset.

C1: **Finding a common representation for software architectures.** Source code is always represented as some type of text file that conforms to some formal grammar. For example, object-oriented source code consists of classes, methods, and interfaces. Indeed the aim to be "compilable" enforces that the source code conforms to a formal grammar. Notwithstanding the existence of standards for software architecture and UML, there is a very high diversity in the representation of software architecture across different projects. Software architecture documents may be represented in formats as diverse as Word (doc(x)), PDF, HTML, PowerPoint and (ppt), among others. The content of software architecture documents is a mix of natural language, images, and sometimes tables and diagrams. Indeed the content is a mix of descriptions of the system architecture, sometimes including design principles, design rationale, and even source code examples. This complicates the definition of a common representation (data model) of which information to represent for each architecture.

C2: **Capturing relevant context information.** Source code has as main purpose to represent the implementation in a manner that is compilable and executable by a computer. Software architecture on the other hand, serves different purposes to different consumers over time: in early stages of projects, architecture documentation is typically used to create a shared understanding among architects. Later on such documenting happens after (or in concert with) making the implementation, and serves to align architecture and implementation. Moreover, the documentation serves as a reference for developers to record which parts of the system have been implemented and stabilized. Also, testers of the software draw on information from the software architecture, e.g., to understand quality objectives as well as scoping decisions. In open source repositories, we can observe the production of architecture, but not its use/purpose/aim(s). The way an architecture is used is key to analyzing the benefits that can be harvested from it. This includes processes

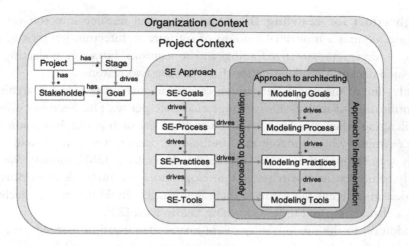

Figure 2.1 Impacts of contexts to software modeling approach.

and practices of the project (such as quality assurance, processes for monitoring conformance of implementation, or the way in which architecture is used in producing implementation).

Indeed, the representation, completeness, and level of abstraction of the description of an architecture depend on the stage of the project it is used in: at the start of a project, architectures may not be crystallized very much, hence little of the system is represented by an explicit architecture representation. For mature projects, architecture documentation usually focuses on high-level views of the system (so as to be able to provide one overview of the system), especially in large software projects. As a consequence, the representation will need to leave out many details. In summary, when we want to understand the role of architecture in a project, we need to consider as well various contextual factors, such as the stage of development, project size, and geographical distribution of the development team. Fig. 2.1 generalizes the complex nested contexts that influence the goals of architecture and thereby the various processes, practices, and tools used (generalized from [22]). The figure illustrates the empirical finding that there is a hierarchy of contexts that influence how software practices are used. There are organizational and project factors that include the goals of the stakeholders. For example, these may prioritize delivery date over quality of the software. Such priorities in turn affect the ways in which architecting is done. In particular, they will affect the goals of doing architecting and via this also the processes, practices, and tools used for architecting. Indeed, for a true understanding of the value of architecture practices, all these context factors would need to be understood. However, these contextual data are typically not obtainable via "artifact mining" approaches.

C3:   **High effort for crawling Big Data.** Empirical research into software architecture requires a nontrivial amount of software architecture (empirical) data in order to draw representative findings and conclusions. However, collecting/building such dataset is challenging for the following two reasons.

Firstly, due to the vast variety in representation and use of software architecture, identification of such SADs is a huge challenge per se. This becomes even more challenging when searching for SADs in Big Data such as GitHub, SourceForge. For example, when building the Lindholmen dataset, it was impossible to manually scan through the whole GitHub data to look for UML models. We had to apply some heuristic searches and develop automated methods to identify UML models in different file formats, including images. Building up such technology was a challenging and time consuming task in itself [23].

In addition to the unavailability of automatic identification methods, it is worth noting that the limited (human and machine) resources could hugely affect the amount and the quality of software architecture data to be collected. In particular, studies that involve the identification of SADs often target a small amount of SADs because of limited human resources within the research team (for identifying, verifying, and maintaining the data), thus running the risk of data not being representative. Moreover, to many studies that use the GitHub API (such as [24]), the limitation of a maximum of 5000 requests per hour is a technical challenge that limits the speed and scope of SAD search.

C4:   **High effort for curation.** Collecting software architecture artifacts from open source requires a lot of curation. Firstly, public repositories are frequently very "noisy": they do not only contain software development projects, but also, e.g., student projects and course material [25]. Secondly, as argued in the previous section, SADs exist in a very wide variety. Studies that aim to employ "Big Data"/Machine Learning techniques must realize that there are "many different animals in the SAD-zoo" that share very little commonality. One way to understand the zoo of SADs is to enable community/crowdsourcing curation, e.g., through annotation and classification. We elaborate on the need for curation in the next section. Another recommendation is to set up mechanisms as early as possible to monitor and improve the quality of the dataset. Given that typically large volumes of data are involved, this must be automated as much as possible. This is complicated by the fact that each "entry"/data point for one software architecture is very rich in many different types of attributes and context factors.

C5:   **Collaborating in empirical software architecture research.** Collaboration has become a common practice in doing Big Data-driven empirical research. This is due to the fact that such type of research often requires a huge amount of efforts, of which a single researcher might not be able to cover all parts alone. For example, in order to build the Lindholmen dataset, collaboration between researchers who are specialized in specific fields was necessary – some researchers were more specialized in mining Big Data from GitHub, others were responsible for developing techniques for detecting UML content in arbitrary files. Prior to forming the working team, it was important to establish the research intent and look for potential collaborators via the researcher's own network. When analyzing data, the researchers needed to communicate with each other on the steps and progress of data analysis as well as the preliminary results. Team effort was also needed in developing a community around the research. This included creating a website, communicating with relevant research groups at various conferences/workshops, and responding to (extra-feature) requests from the research community. However, the level of (tool) support for the collaborative empirical research activities was far from sufficient. The authors of the Lindholmen dataset were not aware of any tool that supports collaboration for all the above-mentioned activities.

2.5. Directions for a community infrastructure for Big Data-driven empirical research in software architecture

Given the challenges for Big Data-driven empirical research in software architecture, a solution could be to build a community-based infrastructure that enables researchers to share and reuse software architecture artifacts as well as to collaborate in their empirical studies. In this section, we discuss the main requirements (**R**) for building such an infrastructure (called "the infrastructure" hereafter).

R1: **Be able to host big & heterogeneous data of SADs.** As mentioned in the previous section, to many empirical studies identifying and collecting software architecture artifacts is a big challenge. Therefore, the first and foremost requirement is that the infrastructure should be built upon and be able to host (an) enormous corpus (ora) of software architecture artifacts. Since the representation of architecture artifacts is highly heterogeneous, the hosting solution might need to be flexible enough to integrate and handle both structured and unstructured data (such as design documentations, requirements).

R2: **Share not only data but also software architecture knowledge and analysis.** Having access to a huge corpus of software architecture artifacts is a big advantage, but it is not enough. Often, researchers and practitioners collect data for their studies with a set of criteria in mind. In many cases, the criteria concern not only the data themselves, but also additional knowledge about and existing analyses on the data. For example, El Ahmar et al. browsed the Lindholmen dataset manually to be able to collect 3500 UML diagrams with different visual variables for their study [20]. In the study, the authors find many UML diagrams where *color variable* was misused, which could thus have had a negative impact on the communication where the diagrams were used as an intermediate. The additional knowledge and analysis (hypotheses, method, and results) could potentially serve as an input for other researchers to make a more knowledgeable selection of data or even to make a follow-up study.

R3: **Enable links to contextual data.** Mining studies have given a huge boost to empirical studies on software development, and source code-related studies in particular. Indeed, various important open questions related to software architecture require studying the relation between the architecture and the source code, e.g., how software architecture affects the quality of the source code or how architecture affects the evolution/maintainability of a system.

R4: **Support evolving artifacts and architectural knowledge.** Software development is an evolving process in which software artifacts, including architecture artifacts, are constantly updated during a project's life time. Accordingly, software architecture and corresponding architectural knowledge are subject to change as the project progresses. In order to facilitate research about the evolution of software architecture, the infrastructure should provide means for managing versions of not only software architecture artifacts, but also corresponding architecture knowledge. While there have been numerous solutions for managing versions of software artifacts, collecting, organizing, and administrating versions of architectural knowledge remain challenging. This is mainly due to the fact that the process of discovering architecture knowledge is not always fully automated, making it hard to collect multiple versions in a frequent and systematic manner. Indeed, the infrastructure should help to mitigate this issue by allowing users to define their scientific workflow and enabling reuse of analysis and computational services.

R5: **Keep annotations separately.** Ongoing research efforts require the enriching of empirical data by annotations, for example, annotations delineating the location of features in source code (such as in [26]) or annotations indicating whether a piece of source code (or architecture) is related to security. Performing annotations invasively by adding annotations into the artifact itself is not scalable. In fact, this will become very messy when multiple types of annotations need to be combined. Indeed it should be a requirement that one artifact from a software project can be annotated by multiple parties. Hence, we should implement exogenous ways of annotating the artifacts found in software projects, i.e., the annotations should exist outside the actual artifact. This triggers the question on how to refer to particular parts of an artifact (source code or document) in order to link an annotation to a fragment of an artifact.

R6: **Crowdsource annotation.** Annotating large systems, especially from multiple perspectives, is a colossal task. In some cases, such annotations can possibly be done automatically, either using external sources or through Machine Learning algorithms. For those cases where automation is impossible, this task is probably best addressed as a community effort. For supporting such efforts, the tooling infrastructure should support some way of crowdsourcing. When opening a system up for annotation and crowdsourcing, one would need to also introduce mechanisms for (i) authentication, (ii) quality assurance, and (iii) traceability of annotations.

R7: **Enable comparison & aggregation of research findings.** Empirical studies in software architecture sometimes contradict in their findings. One of the main reasons is that different studies base their analyses and findings on different samples of the population, and thus observe different phenomena. This infrastructure provides a way for different research to study the same empirical process and possibly the same data sources, and therefore it should also enable comparison and aggregation between research findings from different studies.

R8: **Encourage discussion and peer-review.** A benefit of joining a community is to get early feedback and support. This benefit should therefore be considered as a core value when building the infrastructure. As sharing interests in studying software architecture, researchers in the community should indeed be able to report their experience and give comments/questions/feedback on other studies. The infrastructure should learn from scholarly social networks such as ResearchGate[6] to encourage members to interact and exchange knowledge, for example with bonus points and badges for reviewers/commenters.

R9: **Promote collaborative research and enrich a collaboration network.** Many initiatives in creating a corpus for software architecture artifacts have become inactive or have been left outdated. Only five out of twelve reported corpora in the SEMI index [1] are active at the time of writing this chapter – none of them have been updated since 2014. It seems these corpora have not been successful in growing a research community around them. A reason could be that inadequate effort has been spent on promoting the corpora as well as maintaining the research networks around the corpora. This should be taken as a lesson when building the infrastructure. For example, the following methods are successfully used in ResearchGate: (i) suggest researchers with related studies, analyses, and questions/answers; (ii) allow researchers to invite other researchers (that might be interested) to review/visit their studies; and (iii) support announcing research plans and calls for joint efforts (and then team composition).

While the above list of requirements and desiderata is by no means complete, it should be a starting point to capture many of the core characteristics. As the infrastructure aims at facilitating collaboration within a scientific domain, requirements for building a scientific workflow management system are relevant. In particular, together with the above-mentioned nine requirements, we would recommend to consider an addition of nine requirements mentioned by Ludäscher et al. [27] when building the infrastructure (Table 2.1 summarizes these requirements).

---

[6] https://www.researchgate.net/.

Table 2.1 Nine requirements for building a scientific workflow management (by Ludäscher et al. [27]).

Requirement	Content
Seamless access to resources and services	Using web services for remote service execution and remote database access.
Service composition & reuse and workflow design	Web services should be constructed in a way that can be combined to perform complex tasks. This is similar to the idea of using microservices (in SOA architecture).
Scalability	Should support data-intensive and compute-intensive workflows.
Detached execution	Long running workflows require an execution mode that allows the workflow control engine to run in the background on a remote server, without necessarily staying connected to a user's client application that has started and is controlling workflow execution.
Reliability and fault tolerance	To make a workflow more resilient in an inherently unreliable environment, contingency actions must be specifiable, e.g., fail-over strategies with alternate web services.
User interaction	Allows users to inspect intermediate results and select and rerank them before feeding them to subsequent steps. Allows user to reconnect to the running instance and make a decision before the paused (sub)workflow can resume.
"Smart" reruns	A "smart" rerun would not execute the workflow from scratch, but only those parts that are affected by the parameter change.
"Smart" (semantic) links	Assists users in workflow design and data binding phases by suggesting which actor components might possibly fit together, or by indicating which datasets might be fed to which actors or workflows.
Data provenance	The results of a conventional experiment should be reproducible, computational experiments and runs of scientific workflows should be reproducible and indicate which specific data products and tools have been used to create a derived data product.

2.6. Overview of CoSARI

In this section, we present our proposal for the novel framework CoSARI (**Co**llaborative **S**oftware **A**rchitecture **R**esearch **I**nfrastructure) and our ongoing efforts to construct it.

2.6.1 Overview of the CoSARI framework

Fig. 2.2 illustrates the architecture of CoSARI, which is composed of three layers: a Data Storage layer, a Business SaaS layer, and a Presentation layer.

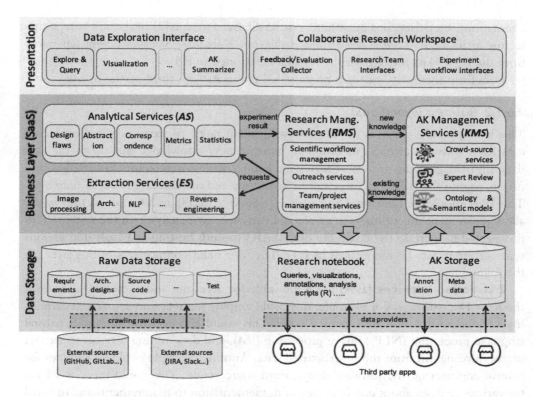

Figure 2.2 Architecture for architecture research framework.

2.6.1.1 Data access layer

This layer provides access to the data sources of the system, which consists of the following:

Raw Data Storage stores SADs as well as contextual data related to the project that the SAD belongs to (**R1, R3**). The contextual data include, amongst others, source code, requirements, and testing documents. Data in this storage are crawled from common code sharing systems (such as GitHub, GitLab), issue tracking systems (such as JIRA), and developer communication channels (such as Slack).

Research Notebook is the component that stores research profiles of all empirical studies conducted on the system, thus enabling sharing of scientific analysis on software architecture (**R2, R4**). The research profile might consist of hypotheses, analyses, queries, progress, result, etc., of a specific empirical study. This component also keeps track of the human records of empirical research such as research individuals and teams which are essential for promoting collaborative research network (**R8, R9**).

Architecture Knowledge Storage (AK Storage) is dedicated to storing all architecture knowledge generated on top of the raw data. Data in this database can be annotations, metadata, and quantitative or qualitative assessments of the software architecture artifacts. *Research Notebook* and *Architecture Knowledge Storage* are kept separately from the Raw Data Storage for the reasons mentioned in **R5**.

Data from these databases can be provided via an API for third parties to use.

2.6.1.2 Business layer (SaaS)

This layer provides services for the following main tasks: (i) extracting and analyzing data from software artifacts in order to generate new knowledge on top of the raw data; (ii) managing the process of reviewing and retrieving architecture knowledge, and (iii) managing the collaborative scientific workflow and maintaining research networks. This layer should be built on a Software-as-a-Service (SaaS) architecture as this would allow researchers/practitioners to create and run their own analyses on the system. This is expected to increase the flexibility of CoSARI towards hosting and analyzing various types of input data (as mentioned in **R1**). The four main components of this layer are the following:

Extraction Services (ESs) are responsible for accessing and extracting data from the Raw Data Storage. For example, this could employ services for parsing and extracting various information from source code. This could also contain numerous natural language processing (NLP), image processing (IM), and data mining services to extract useful information from the unstructured data. Another example could be services for reverse engineering the software design from source code. This could serve as a basis for various analyses about conformance of implementation to requirements and original design, etc.

Analytical Services (ASs) employ methods for analyzing data collected from the ESs to provide qualitative or quantitative results for answering research questions about the software architecture. An example of such a service is a service that computes the correspondence between software design and implementation (source code). For this, the AS service needs ES to provide a list of class names from source code and a list of class names from the design documentation of the project. Then, the AS service is responsible for checking the similarity between the names and calculating the correspondence rate (as the naming convention used when designing and coding might be very different, this is a challenging task itself). Other services that might fall in this layer could be calculating design flaws (from source code and architecture documents), analyzing the role of software components in the design of a system, establishing mappings and traces between software artifacts, building statistical analysis, etc.

Collaborative Research Management Services (RMSs) provide core services for managing the collaborative workflow for software architecture research (**R9**). This should look similar to existing scientific workflow management systems such as Kepler [27] with additional support for teamwork and outreach of the research. In particular, using the services, research teams should be able to discuss, modify, and experiment with various experimental settings. RMS also allows research teams/individuals to search and compare research profiles and findings from other empirical studies (**R7**). Outreach services aim at promoting the research within and outside of CoSARI by various activities (depending on the stage of the projects), for example: (i) they support creating a project and looking for potential co-researchers in the network; (ii) they support announcing research results/events/milestones within and outside of CoSARI; (iii) they allow other researchers to follow/subscribe to specific studies; and (iv) they manage a rewarding system to encourage researchers to contribute more to projects.

RMS interacts with the Research Notebook in order to store and update research profile. Besides, RMS delegates management of research findings (and software architecture knowledge) to the Architecture Knowledge Management Services (KMSs).

Architecture Knowledge Management Services (KMSs) are services that manage architecture knowledge within CoSARI. In particular, the KMS have two main tasks: (i) collecting and storing new architecture knowledge and (ii) generating meaningful answers to questions regarding software architecture of a specific system.

Regarding the first task, new knowledge is derived from two main sources, including human intelligence (e.g., experts and crowdsourcing assessments) and results from empirical studies. The request to store/update architecture knowledge is directed from RMS or the crowdsource services to KMS (**R6**). The new knowledge might need to be reviewed before being stored/updated at the AK Storage.

Regarding the second task, the KMS makes use of an ontology which describes generic concepts used in software architecture and the relationship between the concepts. This allows KMS to be able to generate answers to both predefined questions and user-articulated questions (**R2**). The answers will then be returned to the RMS or to the *Presentation* layer. Feedback from experts and users to the answers can be used to improve the vocabulary of the ontology.

2.6.1.3 Presentation layer

This layer provides interfaces for end-users to explore and start a new empirical research project on CoSARI. The interfaces can be split into the two following categories:

Data Exploration Interface, which supports users to search, identify, and understand the architecture artifacts to be used in their future research. Example applications in this layer include textual summarization or graphical visualization of software architecture, dashboards that show architecture quality, etc.

Collaborative Research Workspace, which supports users with everything to create and corporate on a collaborative research project. This also provides application forms for collecting feedback from users and experts. Section 2.6.2 shows the use of applications in this layer with more details.

2.6.2 Main use cases of CoSARI

In this section, we provide two main use cases of CoSARI: (i) identifying a set of SADs to study and (ii) starting and managing an empirical research project on CoSARI.

Use case 1: *Identifying a set of SADs to study*.

A researcher wants to search for an architecture dataset in CoSARI for his study. The researcher can either browse the existing (public) datasets and studies hosted by CoSARI or perform a search to filter relevant data. In particular, the researcher can choose to apply different filters on different properties of the data. For example, he can search across all UML diagrams that contain a specific "search term" and are large-sized (e.g., have more than 25 classes).

Maybe the researcher wants to filter on a property that is not yet provided. In that case, he can implement and run his own extraction services and analytical services to extract the property from the Data Storage. Maybe the researcher wants to further understand a specific dataset or project; he can then consult visualizations or summaries of the architecture data.

After this step, the researcher should have identified a dataset to start with his experimental study.

Use case 2: *Starting and managing an empirical research project*.

After identifying a dataset, the researcher can start an empirical research project by creating a work space for the project. This includes (i) creating a research profile and a work plan; (ii) defining a scientific workflow; and (iii) inviting fellow researchers/practitioners to join the project. CoSARI can suggest researchers that might be interested in joining based on matching of their research profiles to the chosen dataset and research topic; creating the experimental workflow consists of small steps of data extraction & analysis, data visualization and statistical tests, etc.

As the project runs, CoSARI supports teamwork by allowing team members to cooperate in creating, configuring, and executing analytical services. A research team can also use outreach services provided in CoSARI to call for joint efforts as well as to get feedback about the project's approach. Feedback to a research can be given via comments and discussions. The system records update on the experimental approach and result in the project notebook (profile). As a result, the history of the research approach can be traced.

2.6.3 Ongoing and future work

In this section, we present our on-going efforts and future work on building CoSARI.

Creating raw data storage & querying interface. The Lindholmen dataset of UML designs is the initial dataset to be part of the core database for CoSARI. We are constantly working on curating this dataset, for example by labeling UML diagrams with their type (Class Diagram, Sequence Diagram, etc.) and by identifying reverse engineered and forward-design diagrams [19]. We provide access to the Lindholmen dataset

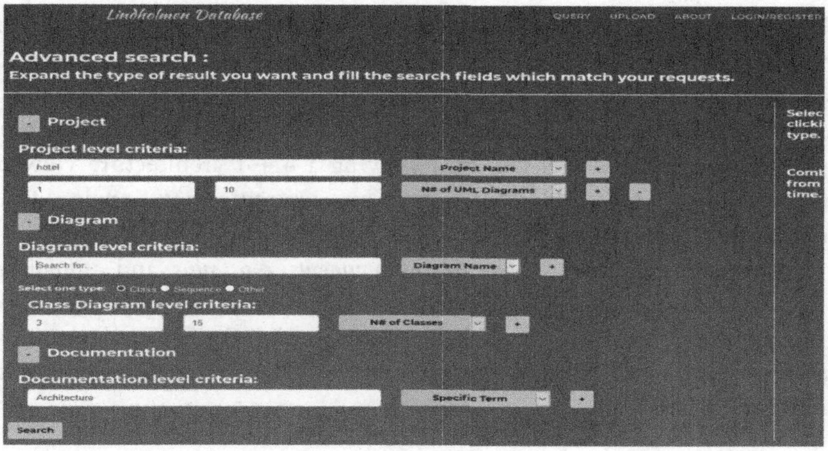

(A)

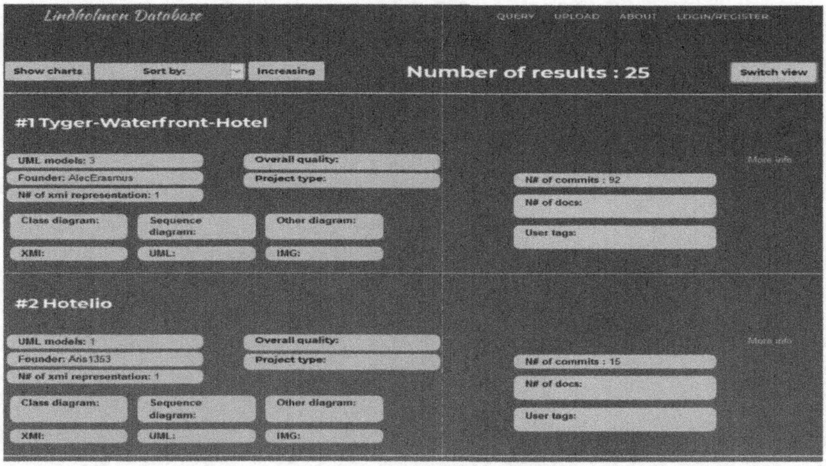

(B)

Figure 2.3 GUI for sharing Lindholmen DB. (A) Advanced search user interface. (B) Search result UI – "card" view.

via a website and a REST API (these are currently under testing and not yet available for use). In particular, the website would allow users to search for UML diagrams with multiple filters at both project level (such as project name, founder, number of commits/stars/issues, etc.) and model level (such as model name, type of UML model, number of elements inside, etc.). Fig. 2.3A shows the advanced search interface of the website when searching for GitHub projects that (i) have a project name containing the word "hotel," (ii) have at least one and maximum ten UML diagrams, and (iii) have class diagrams that are small-/medium-sized, i.e., the diagrams contain three to fifteen classes. Fig. 2.3B shows the search result, i.e., 25 projects that match the advanced search.

Building an ontology of software architecture knowledge. As mentioned, such an ontology is an essential part for the KMS of CoSARI. The design of our ontology was originally inspired by the ontology demonstrated by De Graaf et al. [28]. Some modifications were made for the ontology to be able to capture different perspectives of software architecture, i.e., by following a 4 + 1 architecture view model. Moreover, the ontology is also able to reason about requirements, rationale, and implementation regarding software architecture. For example, the rationale behind a *design option* of software architecture is expressed by *arguments*, *constrains*, and *assumptions* and results in specific technologies (e.g., libraries, languages, frameworks). Fig. 2.4 shows a part of the ontology (a full-size version is available online[7]). Tao et al. demonstrated how to use the ontology to compose answers to various questions that are commonly asked by developers [29].

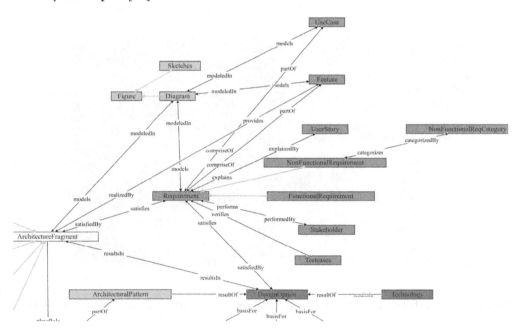

Figure 2.4 A part of the architecture knowledge ontology.

Future work. CoSARI is as of now a reference architecture and there are lots of work to be done in order to implement such architecture. For example, at the Data Storage Layer, data collected from different external sources are heterogeneous and thus need to be standardized/harmonized (e.g., popularity metrics are different between GitHub and Bit Bucket). Thus, CoSARI should be equipped with an Extract-Transform-Load

---

[7] http://bit.ly/software-explanation-composer.

(ETL) tool for harmonizing the data. At the Business Layer, as of now, the RMS has not yet been implemented. A first step toward implementing it would be to evaluate how much can be reused from existing scientific workflow management systems [15].

As CoSARI aims at supporting the research community of software architecture, getting early feedback from the community when building it is a must. We plan to build a prototype of CoSARI and conduct a user study to assess its usability (e.g., by using the System Usability Scale (SUS) [30] and improving it accordingly.

2.7. Summary and conclusions

In this chapter we explored the topic of doing Big Data-driven empirical studies in software architecture, and we in particular reflected on what an infrastructure that could support a community to do collaborative research on this topic would look like. We pointed out that we can build on earlier works in the area of scientific workflow systems and software knowledge discovery. Also, we discussed several existing and ongoing efforts in creating large collections of software architecture representations (either models or complete design documents). We discussed some lessons learned as well as remaining challenges and future directions. Building on these experiences, we proposed a reference architecture, CoSARI, that can serve as a starting point for our community towards building a common infrastructure for performing collaborative empirical research in software architecture.

References

[1] H. Störrle, R. Hebig, A. Knapp, An index for software engineering models, in: International Conference on Model Driven Engineering Languages and Systems, MoDELS 2014, 2014, pp. 36–40.
[2] M. Kunze, P. Berger, M. Weske, BPM academic initiative-fostering empirical research, 2012.
[3] M. Weske, Business Process Management: Concepts, Languages, Architectures, 2nd edition, Springer-Verlag, Berlin, Heidelberg, 2012.
[4] R. France, J. Bieman, B.H. Cheng, Repository for model driven development (ReMoDD), in: International Conference on Model Driven Engineering Languages and Systems, MODELS 2006, Springer, 2006, pp. 311–317.
[5] B. Karasneh, M.R.V. Chaudron, Online Img2UML repository: an online repository for UML models, in: EESSMOD@ MoDELS, 2013, pp. 61–66.
[6] J. Noten, J.G. Mengerink, A. Serebrenik, A data set of OCL expressions on GitHub, in: Proceedings of the 14th International Conference on Mining Software Repositories, IEEE Press, 2017, pp. 531–534.
[7] F. Bascianidi, J. Di Rocco, D. Di Ruscio, A. Di Salle, L. Iovino, A. Pierantonio, MDEForge: an extensible web-based modeling platform, 2014.
[8] M. Dirix, A. Muller, V. Aranega, Genmymodel: an online UML case tool, in: ECOOP, 2013.
[9] R. Hebig, T.H. Quang, M.R.V. Chaudron, G. Robles, M.A. Fernandez, The quest for open source projects that use UML: mining GitHub, in: Proceedings of the ACM/IEEE 19th International Conference on Model Driven Engineering Languages and Systems, ACM, 2016, pp. 173–183.
[10] W. Ding, P. Liang, A. Tang, H. Van Vliet, M. Shahin, How do open source communities document software architecture: an exploratory survey, in: Engineering of Complex Computer Systems (ICECCS), 2014 19th International Conference on, IEEE, 2014, pp. 136–145.

[11] A. Brown, G. Wilson, The Architecture of Open Source Applications, volume I&II, Ebook, May 2011.

[12] M. Soliman, A.R. Salama, M. Galster, O. Zimmermann, M. Riebisch, Improving the search for architecture knowledge in online developer communities, in: IEEE International Conference on Software Architecture, ICSA 2018, Seattle, WA, USA, April 30 – May 4, 2018, 2018, pp. 186–195.

[13] J. Musil, F.J. Ekaputra, M. Sabou, T. Ionescu, D. Schall, A. Musil, S. Biffl, Continuous architectural knowledge integration: making heterogeneous architectural knowledge available in large-scale organizations, in: Software Architecture (ICSA), 2017 IEEE International Conference on, IEEE, 2017, pp. 189–192.

[14] Z.-Q. Lin, B. Xie, Y.-Z. Zou, J.-F. Zhao, X.-D. Li, J. Wei, H.-L. Sun, G. Yin, Intelligent development environment and software knowledge graph, Journal of Computer Science and Technology 32 (2) (2017) 242–249.

[15] A. Barker, J. Van Hemert, Scientific workflow: a survey and research directions, in: International Conference on Parallel Processing and Applied Mathematics, Springer, 2007, pp. 746–753.

[16] G.H. Travassos, P.S.M. dos Santos, P.G. Mian, A.C.D. Neto, J. Biolchini, An environment to support large scale experimentation in software engineering, in: Engineering of Complex Computer Systems, 2008, ICECCS 2008, 13th IEEE International Conference on, IEEE, 2008, pp. 193–202.

[17] G. Robles, T. Ho-Quang, R. Hebig, M.R.V. Chaudron, M.A. Fernandez, An extensive dataset of UML models in GitHub, in: Proceedings of the 14th International Conference on Mining Software Repositories, IEEE Press, 2017, pp. 519–522.

[18] T. Ho-Quang, R. Hebig, G. Robles, M.R.V. Chaudron, M.A. Fernandez, Practices and perceptions of UML use in open source projects, in: Software Engineering: Software Engineering in Practice Track (ICSE-SEIP), 2017 IEEE/ACM 39th International Conference on, IEEE, 2017, pp. 203–212.

[19] M.H. Osman, T. Ho-Quang, M.R.V. Chaudron, An automated approach for classifying reverse-engineered and forward-engineered UML class diagrams, in: 2018 44th Euromicro Conference on Software Engineering and Advanced Applications, SEAA, IEEE, 2018, pp. 396–399.

[20] Y. El Ahmar, X. Le Pallec, S. Gérard, T. Ho-Quang, Visual variables in UML: a first empirical assessment, in: Human Factors in Modeling, 2017.

[21] C.D. Schulze, G. Hoops, R. von Hanxleden, Automatic layout and label management for compact UML sequence diagrams, in: 2018 IEEE Symposium on Visual Languages and Human-Centric Computing, VL/HCC, IEEE, 2018, pp. 187–191.

[22] A.M. Fernández-Sáez, M.R.V. Chaudron, M. Genero, An industrial case study on the use of UML in software maintenance and its perceived benefits and hurdles, Empirical Software Engineering 23 (6) (2018) 3281–3345.

[23] T. Ho-Quang, M.R.V. Chaudron, I. Samúelsson, J. Hjaltason, B. Karasneh, H. Osman, Automatic classification of UML class diagrams from images, in: Software Engineering Conference (APSEC), 2014 21st Asia-Pacific, vol. 1, IEEE, 2014, pp. 399–406.

[24] D. Rusk, Y. Coady, Location-based analysis of developers and technologies on GitHub, in: 2014 28th International Conference on Advanced Information Networking and Applications Workshops, WAINA, IEEE, 2014, pp. 681–685.

[25] E. Kalliamvakou, G. Gousios, K. Blincoe, L. Singer, D.M. German, D. Damian, The promises and perils of mining GitHub, in: Proceedings of the 11th Working Conference on Mining Software Repositories, ACM, 2014, pp. 92–101.

[26] J. Krüger, M. Mukelabai, W. Gu, H. Shen, R. Hebig, T. Berger, Where is my feature and what is it about? A case study on recovering feature facets, Journal of Systems and Software 152 (2019) 239–253.

[27] B. Ludäscher, I. Altintas, C. Berkley, D. Higgins, E. Jaeger, M. Jones, E.A. Lee, J. Tao, Y. Zhao, Scientific workflow management and the Kepler system, Concurrency and Computation: Practice and Experience 18 (10) (2006) 1039–1065.

[28] K.A. De Graaf, A. Tang, P. Liang, H. Van Vliet, Ontology-based software architecture documentation, in: Software Architecture (WICSA) and European Conference on Software Architecture (ECSA), 2012 Joint Working IEEE/IFIP Conference on, IEEE, 2012, pp. 121–130.

[29] A. Tao, M. Roodbari, Towards Automatically Generating Explanations of a Software Systems, Master's thesis, Chalmers University of Technology, Gothenburg, Sweden, 2018.

[30] J. Brooke, et al., SUS - a quick and dirty usability scale, Usability Evaluation in Industry 189 (194) (1996) 4–7.

CHAPTER 3

Model clone detection and its role in emergent model pattern mining

Towards using model clone detectors as emergent pattern miners – Potential and challenges

Matthew Stephan, Eric J. Rapos
Miami University, Department of Computer Science and Software Engineering, Oxford, OH, United States

Contents

3.1. Introduction

As the proliferation of model-driven engineering continues [76], there is an emergence and continuous flow of software models. This includes model repositories, for example MDEForge [10] and the Lindholmen dataset [32]; common open source repositories that contain models, such as GitHub and SourceForge; and organizations' internal repositories. Source code-based techniques are not applicable nor suitable for model-specific analytics. While this, and the increasing scale and intricacy of software models, presents interesting challenges in repository management, model visualization, and other areas we address in this book, it gives rise to more advanced and learning-based opportunities for analytics.

One such opportunity is software pattern extraction based on a corpus of software models. Mining traditional software source code in general is a mature and growing field [30]. However, the mining of model-driven engineering artifacts is only recently gaining traction as model-based approaches become more widespread, and the complexity and size of the artifacts has grown. For this chapter, we are interested solely in *emergent patterns*, which are recurring software solutions that occur organically within software projects that are not known to analysts beforehand. Such a facility can help in areas such as standards enforcement, quality assurance, software development and refactoring, and maintenance. Sharma et al., from NEC Laboratories America, identify pattern matching as a key challenge in the field of cyber-physical systems (CPSs) [58], which are often developed exclusively or predominately as software models [20].

One tool for model-driven engineering analysis is model clone detection [19]. Model clone detection is a form of model comparison [66] that involves analyzing models to identify identical and/or similar models with respect to some measure of similarity. Similarity analysis can include both structural and semantic aspects. Clone detection in software has many uses, including estimation of maintenance costs, fault prediction, and refactoring [35]. In this chapter, we describe using model clone detection as an emergent model pattern miner within a conceptual Model Clone detection Pattern Mining (MCPM) framework. We describe model clone detection's role in the MCPM, including its potential and challenges; how existing work fits into the requirements of the MCPM framework; and examples. Our goal is to improve model analytics by helping pave the way for model clone detection to be used as an analytical tool that discovers emergent patterns.

We begin in Section 3.2 by providing the background material necessary to understand this chapter. In Section 3.3, we define our MCPM concept with a detailed description of the framework, its substeps, examples, and a review of existing tools and research for each of its phases. We identify open challenges and future work related to the MCPM framework in Section 3.4, and conclude in Section 3.5.

3.2. Background material

In this section, we describe background material on software patterns in general with a focus on pattern mining for source code and models, and model clone detection. For the latter, we describe the different types of model clones, as that is an important consideration when describing our framework's requirements for MCPM.

3.2.1 Software patterns

Software patterns are an integral part of software engineering. They can be thought of as successful, commonly employed, and validated approaches to solving a software engineering problem [57]. They are an abstraction that occurs in "nonarbitrary" and recurring contexts [51]. One of the most popular examples, both in education and practice, are design patterns [15], which describe established software development solutions to frequent design problems. The opposite of design patterns are antipatterns, which are commonly occurring incorrect/problematic ways that people use to solve problems using software [13]. The term "patterns" can be used as an umbrella term for both, and we do so in this chapter unless we otherwise use the explicit "design" or "anti" descriptors. Software patterns exist also for many other domains and contexts, for example, software security patterns [24], agent-oriented domains [41], architectural patterns [29,60], and agile development [44]. Software patterns have many uses, including software evolution [80], architecture evaluation [81], fault detection [21], and establishing traceability links [36]. Traditionally, software patterns have a posteriori been formulated manually by industrial and academic experts based on real-life development experiences and struggles [74]. However, it also is possible to have unknown patterns detected through analysis of existing systems through the process of pattern mining. For this chapter, we are interested specifically in *emergent patterns*, which occur naturally within software projects and are not known to analysts before system evaluation.

3.2.1.1 Source code pattern mining and detection

Analysis of software systems to discover patterns allows analysts to identify and codify commonly occurring solutions and implementations within software systems. This can involve mining patterns from software revision histories [42], source code [8,79], execution traces [54,72], architectures [45,81], and more. A common application of source code pattern mining is to detect the presence of known design patters. Dong et al. [22] identified seven different categories for design pattern mining through a survey of the literature: structural aspect mining; behavioral aspect mining; structural and behavioral aspect mining; structural and semantic mining; structural, behavioral, and semantic aspect mining; and pattern composition. Structural aspects refer to design relationships, such as aggregation and generalization. Balany and Ferenc [8] similarly mine design patterns, but do so by building an abstract semantic graph from C++ code. Shi and

Olsson [59] detect design patterns in Java code by focusing on structure-driven and behavior-driven patterns from the popular Gang of Four design pattern list [15]. Tsantalis et al. reverse engineer source code into matrices in order to detect design patterns by looking at similarity scores between the design patterns and source code [73]. This work on design pattern detection is different from what we focus on in this chapter, as we are concerned with emergent patterns that are not known before mining takes place, whereas design pattern mining looks explicitly for those known patterns. Additionally, all these approaches require and work explicitly on source code, which is not always present in model-based development or model-driven approaches.

3.2.1.2 Model pattern mining

Model pattern mining involves analyzing software models explicitly, instead of source code. This is beneficial and relevant for software organizations and domains that rely heavily, or exclusively, on model-based or model-driven ideologies. This often includes automotive, aerospace, telecommunications, and other formal and safety-critical domains. This can include modeling languages such as UML, SysML, and Simulink. Model pattern mining is considered challenging as it often involves the general subgraph isomorphism problem, which is NP complete [26]. Thus, model pattern mining techniques must account for this through heuristics, considering alternative forms of analysis, and other means. Paakki et al. analyze UML models using constraint satisfaction approaches to detect known architectural patterns, both good and bad [45]. Gupta et al. employ the state space representation of graph matching to find instances of existing UML patterns, represented as graphs, in other UML models [27]. Pande et al. [46] look for design patterns in UML diagrams of graphical information systems using graph distances. Bergenti and Poggi [11] detect the Gang of Four Patterns in UML models by using rule-based techniques, as do Ballis et al. [9]. Liutel et al. [43] use answer set programming, a declarative language, to detect patterns and antipatterns using both facts and rules on those facts. Fourati et al. use metrics to detect antipatterns on the UML level by looking at the amount of coupling between objects, the number of methods that can be invoked by a class, cohesion attributes, and complexity attributes such as imported interfaces [25]. Wenzel performs a fuzzy evaluation of UML models to determine if a model is exhibiting structural properties indicative of a failed or incomplete application of a design pattern [75]. Past research conducted by us involved analyzing EMF models to detect Java EE antipatterns [61]. We also later developed SIMAID to detect Simulink antipatterns [64,67] by looking for the intersection of clones of known antipatterns and target systems. All of these research ideas and tools are focused on detecting known/existing design patterns and/or antipatterns. The problem we focus on in this chapter is the detection of emergent patterns.

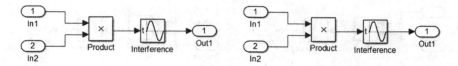

Figure 3.1 Example of an exact/identical, type 1, Simulink model clone.

3.2.2 Model clone detection

Software clone detection involves finding identical or similar sets of software artifacts, which are termed "clones." Similarity is established by a variety of measures for different tools. Tools often have similarity threshold criteria that must be met in order for two or more artifacts to be considered similar enough to be identified as a clone. Software clones can exist due to a variety of reasons, including a rush to deliver, bad reuse practices, and unfamiliarity with cloning [39]. Software clones are not necessarily an indicator of poor software quality [37], and are important to identify regardless [35]. The majority of research and tooling involving software clone detection has been focused explicitly on source code clones [50,52]. Only since the 2010s has model clone detection research begun to materialize [19]. Model clone detection is a form of model comparison [66] that discovers clones in software modeling artifacts. As we will be discussing model clone detection extensively in this chapter, it is necessary to define model clone types and overview existing model clone detection approaches.

3.2.2.1 Model clone types

Just as code clones can be categorized into different types based on their nature [52], so can model clones. It is generally accepted that there are four types of model clones, although the specific categorizations vary [4,68,69]. We now overview these as they are important concepts in understanding model clone detection's role and potential for model pattern mining. We describe them both textually and with contrived examples using Matlab Simulink[1] models.

3.2.2.1.1 Type 1 – identical/exact model clones

Identical, or "exact," type 1 model clones are those that are structurally identical to one another. They are identical except for aesthetic aspects such as layout/positioning, color, and formatting. That is, it ignores those aspects in its comparison.

We present an example of type 1 identical Simulink model clone in Fig. 3.1. Here we see two models that are completely identical, including labels, relationships, and block types. These models both take in two inputs, multiple them, apply a sine wave, and output the result.

---

[1] https://www.mathworks.com/products/simulink.html.

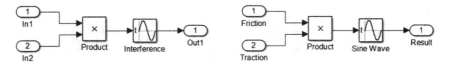

Figure 3.2 Example of a renamed, type 2, Simulink model clone.

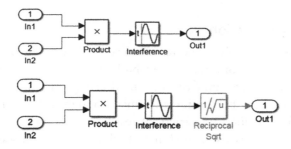

Figure 3.3 Example of a near-miss, type 3, Simulink model clone.

3.2.2.1.2 Type 2 – renamed model clones

Renamed, type 2 model, clones are those that are identical to each other as defined for type 1 identical model clones, except that type 2 model clones allow for differences in names and/or labels, attributes and/or values, and types and/or parts [4,69].

Fig. 3.2 presents a contrived example of a type 2 renamed model clone. These models are identical to one another, except for that we see four of the five blocks have been renamed by the modeler. Specifically, the two inputs are renamed, the Interference block is now named Sine Wave, and the Output block is now called Result.

3.2.2.1.3 Type 3 – near-miss model clones

A near-miss, type 3, model clone is a model fragment that is similar to another model fragment allowing for the same differences as type 2 model clones and additional structural modifications, such as adding or removing blocks/parts. Detectors typically use some form of threshold to indicate how "similar" two model fragments must be in order to be considered a clone, for example allowing a percentage difference of 20%, that is, model clones that are 80% similar to one another.

Fig. 3.3 demonstrates a sample type 3, near-miss, model clone. Here we illustrate two models that are similar to one another. In this case, however, the modeler made a slight modification compared to the original model at the top of the figure by adding an additional, Reciprocal Sqrt, block. Thus, a new block and additional line are present in the bottom model, causing them to be similar but not identical, near-miss clones. Their similarity percentage would be less than 100% but likely above a model clone similarity threshold.

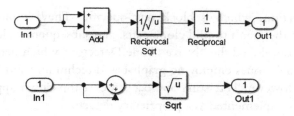

Figure 3.4 Example of a semantically equivalent, type 4, Simulink model clone.

3.2.2.1.4 Type 4 – semantically equivalent clones

A type 4, semantically equivalent clone, is one that is structurally different enough that it may not be similar enough to be identified as a type 3 model clone but is semantically and behaviorally equivalent [1]. There are a variety of reasons these may exist in software systems, including refactoring, coincidence, and language constraints [1,69].

We present a semantically equivalent, type 4, model clone in Fig. 3.4. In this case, we have two models that are semantically equivalent, but potentially different enough from one another that a model clone detector may not find them to be within its structural similarity threshold. Specifically, we see a contrast in number of blocks and types of blocks. Both of these models add the input from In1 to itself, and take the square root of that number. The upper model does so using an addition block and (naively) takes the reciprocal of the reciprocal square root. The bottom model uses Simulink's sum block to add the input from In1 to itself and takes the square root of that number directly.

3.2.2.2 Model clone detectors

Model clone detectors exist that employ varying means of detection and have the ability to detect model clones in a variety of modeling languages. Simulink model clone detector research is the most mature of all modeling languages [3,18,47,48]. However, techniques are emerging for other modeling languages, including Stateflow models [14, 40] and UML models [6,69,70]. Techniques are also being devised by researchers to detect clones in metamodels, such as EMF models [7], and for model transformation languages [71].

We now discuss notable model clone detection approaches, which we categorize based on how they view, and/or the form they consider for comparing, modeling elements.

3.2.2.2.1 Graph-matching model clone detection

One of the first model clone detectors was ConQAT [19]. It detects Simulink models by considering their underlying graphical relationships. ConQAT calculates a graph label that encapsulates the modeling elements' information deemed important by them for similarity detection, which is block type-specific. They employ a breadth-first search approach on the graph to find clone pairs, and subsequently cluster their clones. Similarly, Peterson developed the "Naive Clone Detector," which detects type 1, identical, Simulink model clones employing graph-based techniques and Simulink domain knowledge [47]. However, Peterson's approach uses a top-down approach instead of breadth-first, and is implemented as a proprietary detector.

Pham et al. developed the eScan and aScan model clone detectors for detecting exact and approximate model clones, respectively [48]. Their identical model clones are detected by eScan through a consideration of size and graph node labels representing topology. To facilitate approximate, type 3, model clones, they allow for differences in those labels. Both of these tools are no longer available nor supported.

3.2.2.2.2 Text-based model clone detection

The MQlone tool, developed by Störrle, detects UML model clones. MQlone works with a slightly different classification of model clone types, using the class types A, B, C, and D, and is capable of representing renamed clones. In their approach, they convert the XMI underlying representation of UML CASE models into Prolog representations.[2] In doing so, they are able to detect model clones using similarity metrics and static identifiers. Examples of metrics that MQlone considers are name distance, containment relationships of models, and size.

The Simone Simulink model clone detector detects Simulink model clones of types 1, 2, and 3 [3]. To do so, they focus on the underlying textual representations of Simulink models, and adapt the tried and tested source code clone detector, Nicad [53]. It is adapted in such a way that it is model-sensitive and considers Simulink concepts including systems, lines, blocks, and whole models. Based on industrial feedback and systems being the main unit of organization within Simulink, Simone detects system-level model clones. It has since been extended to work with Stateflow models [14] and UML behavioral models [6].

3.3. MCPM – a conceptual framework for using model clone detection for pattern mining

In this section, we discuss model clone detection's potential and possible role in the model analytics task of emergent pattern mining. Based on their work with industrial partners, Cordy et al. describe their exploratory research extracting emergent patterns in Simulink models [2,16]. They split up their project into three phases [16]: discovery, formalization, and application. They do not go into great detail about the phases

---

[2] www.swi-prolog.org.

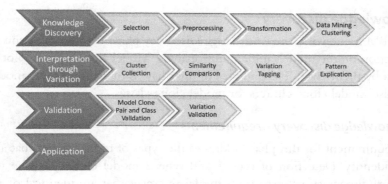

Figure 3.5 Overview of the conceptual MCPM framework.

themselves, but present their work in realizing them for their specific project. In this section, we formally refine/transform these phases to use accepted data mining terminology [23], which we use as the foundation for the MCPM conceptual framework with the intention that others can employ model clone detection for emergent pattern mining. To do this, we generalize these phases as steps within the framework, define the requirements for each of these steps, and discuss how existing research and tooling satisfies these requirements including some examples using Simulink models and our past experiences developing and employing a Simulink model clone detector, Simone [4]. That is, in each of the respective "Existing Research and Tooling" subsections for each step we are summarizing how existing work either meets or does not meet the respective requirements of the respective step.

We present the overall process within our conceptual MCPM framework in Fig. 3.5. In contrast to Cordy's work [16], we use terms that are more consistent with those in the data mining and analytics literature [23]. The first phase in the process is Knowledge Discovery, which we break up into four substeps: selection, preprocessing, transformation, and data mining through clustering. This involves employing model clone detection to identify models that are similar and cluster them together. The next phase in MCPM is Interpretation, which corresponds to Cordy's "formalization" phase. It also contains four substeps. Interpretation can sometimes be considered by analysts as part of the Knowledge Discovery process. However, in MCPM we consider it more of a postprocessing and visualization step whereby clusters are analyzed by variation detection tools and techniques to find variation points, and then visualized to analysts. The next phase, Validation, involves validating the results of both the Knowledge Discovery and the Interpretation/variation identification. The Application phase is the final phase, which focuses on applying the results of the model pattern mining in a form that is useful for analysts and end–users.

3.3.1 Knowledge discovery

In the MCPM framework, the Knowledge Discovery phase involves performing model clone detection on the models undergoing analysis. Model clone pairs are not useful in this context by themselves. The end goal of this phase is to yield not only models clone pairs, but also model clone clusters, or model clone classes.

3.3.1.1 Knowledge discovery – requirements

The first requirement for this phase addresses the types of model clones that a detector is able to identify. Detection of type 1 and type 2 model clones may be useful for identifying rudimentary patterns from model instances that are identical or renamed. However, truly emergent and not readily apparent model patterns would be discovered through detection of type 3, near-miss, model clones only, as verified through our experiences working with industrial partners [3,16]. Type 4, semantically equivalent, model clones may also prove interesting in discovering patterns, but would likely be more applicable in refactoring scenarios, and would be challenging to parameterize as required in the next phase of our framework. Thus, a model clone detector would need to be able to detect near-miss, type 3, model clones.

Since model clone pairs themselves are not useful in establishing an emergent pattern, a higher-level categorization must take place in the mining phase, specifically, a categorization of model clone clusters or classes that identify an abstraction of similar model clones beyond pairs. Not all code [52] or model clone detectors [63] are able to do this. However, model clone clustering and/or classification is a requirement for a model clone detector to be used within the MCPM framework.

Scalability is always a concern when it comes to data mining [28]. From a model clone detection perspective, the more models that can be analyzed and mined, the more informative and "correct" the results will be. With the increasing size and complexity of software models being a forefront concern of model analytics and management, especially for large scale systems, this too is a key nonfunctional requirement when evaluating a model clone detectors' appropriateness for this task.

3.3.1.2 Knowledge discovery – substeps

We break down this phase into four substeps based on data mining practices [23]. We outline and summarize the substeps in Fig. 3.6, which we describe herein.

3.3.1.2.1 Selection

The selection step involves choosing what data will be analyzed [23]. For MCPM, this requires analysts to decide what models should be selected for emergent pattern mining. Much of this is context-dependent. For example, if a specific organization is trying to codify and explicate emergent patterns based solely on their organization's

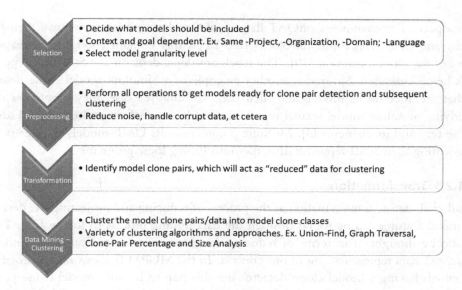

Selection
- Decide what models should be included
- Context and goal dependent. Ex. Same -Project, -Organization, -Domain; -Language
- Select model granularity level

Preprocessing
- Perform all operations to get models ready for clone pair detection and subsequent clustering
- Reduce noise, handle corrupt data, et cetera

Transformation
- Identify model clone pairs, which will act as "reduced" data for clustering

Data Mining – Clustering
- Cluster the model clone pairs/data into model clone classes
- Variety of clustering algorithms and approaches. Ex. Union-Find, Graph Traversal, Clone-Pair Percentage and Size Analysis

Figure 3.6 Summary of MCPM Knowledge Discovery phase.

past development for internal future reference, then they can select their own models. They may additionally, however, choose to include also models from the same domain, for example, an automotive company may choose to use their own models and all other automotive models to which they have access, such as open source models or models/software they have acquired through acquisition agreements. For more general analysis, for example by an academic or consultant, open source models from Github and Sourceforge and model-specific repositories such as the Lindholmen dataset [32] or MDEForge [10] can be included in the set of data. Additionally, this selection step is the time for analysts to decide on any subsets (models to include, in the case of MCPM) or variables within the data [23]. From a model perspective, this is dependent on the modeling language under consideration. For Simulink, for example, this can manifest itself into a question of granularity. Do analysts want to discover emergent patterns at the model level, the system level, or another level? For UML class models, this may be a question of package-level patterns, relationship-level patterns, or interface-level patterns.

3.3.1.2.2 Preprocessing

The preprocessing step in MCPM is for any models, or their data, that need to be manipulated and changed before analysis can occur. Each model clone detector must have the proper preprocessing in place so that model clone pairs can not only be discovered, but also clustered. This can include noise reduction, deciding on what information is necessary to consider or ignore for mining, how to handle corrupt/missing data, and other aspects. For example, ConQAT flattens all Simulink models to remove their hierarchy and transforms Simulink blocks into labels that contain enough information necessary for clone detection [19]. The labels are dependent on the Simulink type of block being evaluated. Simone is another example of a Simulink model clone detector that preprocesses its models and their data. In Simone's case, they preprocess the underlying Simulink model textual representations by normalizing, filtering, and sorting the text and its elements [3]. MQlone preprocesses its UML models of interest by transforming their XMI representation files into Prolog logic programs.

3.3.1.2.3 Transformation

Fayyad et al. define transformation as the process of reducing and projecting the data so that useful features can be found through the data mining process to follow [23]. This can also be thought of in terms of reducing variables under consideration or finding "invariant" data representations in our context. In the MCPM framework, transformation entails having a model clone detector use this step to identify model clone pairs, or some other immediate form, to facilitate model clone clustering/model clone class identification.

3.3.1.2.4 Data mining – clustering

The final step in the Knowledge Discovery phase involves clustering the model clones, or other reduced model clone data, into "similar" groups or classes. This can be accomplished through a variety of clustering algorithms and approaches, for example those discussed by Bishop [12]. In ConQAT's case, they discover model clone classes by devising a graph that represents clone pairs and having edges represent potential cloning relationships [19]. They then employ a combination of union-find structure analysis and graph traversal algorithms to cluster their clones. Simone generates model clone classes by clustering via the percentage differences among clone pairs and their sizes and uses connected component analysis. They then select the largest clone within the model clone class to use as a demonstration/representation of the class. MQlone does not perform clustering.

3.3.1.3 Knowledge discovery – existing research and tooling

When it comes to existing research and tooling, there are different model clone detectors to consider. ConQAT is currently unable to detect type 3, near-miss, model clones [62,68]. So, although they do perform model clone clustering, they do not fulfill the MCPM requirements of detecting the necessary clone types. MQlone detects type 3, near-miss, model clones for UML models. While we did not find any indications that they currently cluster or create model clone classes [69,70], there is no evidence suggesting MQlone cannot be extended to do so. With respect to MQlone's scalability,

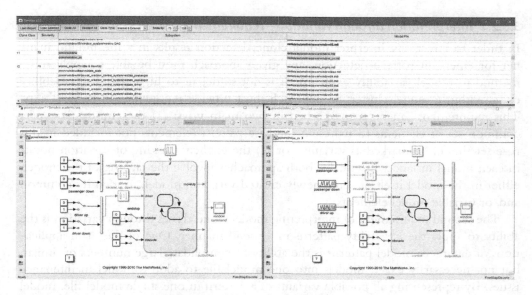

Figure 3.7 SimNav showing a detected clone class with two instances.

they have demonstrated its performance and ability to handle many clones, noting it is "mildly polynomial" with respect to model sizes [70]. Simone is another viable candidate for the MCPM framework. It is able to detect near-miss Simulink model clone pairs, and also cluster them into model clone classes. Simone has been shown to have higher precision and recall than other Simulink model clone detectors [68]. From a scalability perspective, large model sets including industrial automotive sets and open source systems with hundreds of Simulink systems were processed in minutes [3,68].

3.3.2 Interpretation through variation

While these model clone classes can be thought of as direct representations of emergent model patterns, it still lacks applicability in that the patterns have not been explicated nor parameterized [16]. That is, we have examples of the pattern, but not its general form or how it can be used.

One illustration of this shortcoming is shown by the clone class navigation and visualization tool SimNav [49]. SimNav is a tool that is capable of interpreting the clone detection results of Simone and visualizing them directly in the Simulink environment for developers to view and interact with the identified classes. However, it still lacks a direct representation of variability. Fig. 3.7 demonstrates SimNav's ability to select and display a model clone class to users, who are still left with the specific task of determining if a pattern exists, and, if so, how it can be used or generalized.

3.3.2.1 Interpretation through variation – requirements

In order to effectively interpret model clone detection results in a meaningful manner for emergent pattern detection, any given approach must be able to apply some method to model variability explicitly. It is not sufficient to highlight that variability exists within a model cluster; an approach must provide representation of variability. This requirement can be accomplished in two main ways: (1) the construction of a common set of elements to represent the base pattern (elements not included in this base set should be marked as variants), or (2) the explicit marking of variation points in each model instance. Essentially, both approaches involve the tagging of the model, either in the model itself or some newly created variation model, to identify common and/or variable model elements.

The second requirement of interpreting model clone classes through variation is the ability to apply the variability patterns in a useful manner. One of the main applications of emergent model patterns is the ability to consolidate large numbers of similar, but not necessarily exact, models into one model file to address model maintenance issues. By representing all possible variants of a pattern in one single model file, model maintenance teams are required to maintain one model only now as opposed to many different instances of an observed pattern, which may be spread over many different files at different levels of hierarchies within the models.

The third requirement relates to scalability. Any appropriate approach suitable for the MCPM framework should generalize to both large models and large sets of models. A pattern between a given clone pair does not sufficiently improve interpretation. In contrast, a large set of instances in a model clone class, summarized by a single pattern, can drastically improve the maintenance and understanding of a large-scale system through model instance replacements.

In summary, the following are the requirements we have conceived of for the Interpretation Through Validation phase of the MCPM framework. We propose that any candidate must demonstrate the ability to

1. model variability explicitly;
2. apply variability patterns in a useful manner;
3. scale to large model sets.

3.3.2.2 Interpretation through variation – substeps

Based on these requirements, we further break down this phase of MCPM into four distinct steps, as shown in Fig. 3.8, which we further describe in this section. We use a simple example using Simulink models to demonstrate these steps.

3.3.2.2.1 Cluster collection

The first step is to collect the model clone clusters in such a manner that each model instance is grouped along with all related instances, and all model instances are available

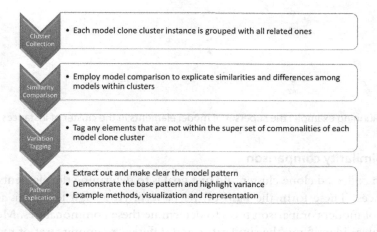

Figure 3.8 The four substeps to the Interpretation Through Variation phase of MCPM.

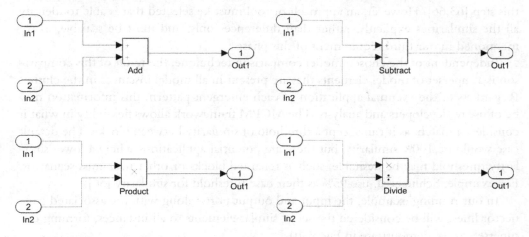

Figure 3.9 Running example: four model instances in the same model clone cluster.

for further analysis and manipulation. Since this phase may involve modifications to the models themselves, it may be prudent and/or necessary to work on copies of the original models to avoid unwanted overwriting of original artifacts. As we mentioned in requirement 3 of this phase, the automatic clustering technique and tooling must be scalable enough for large model sets.

The running example we employ to illustrate this phase consists of four Simulink models. Each model takes in two numbers as input, performs some calculation (the variance), and produces an output. The pattern and variance are fairly obvious in the example but the process and techniques apply generally to any complexity of models. We introduce the four models in Fig. 3.9.

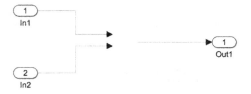

Figure 3.10 Running example: the superset of model elements in the cluster of instances.

3.3.2.2.2 Similarity comparison

Within each collected clone cluster, the first goal is to determine the elements common to all instances. These form the basis of the pattern. At a high level, this is a simple application of model comparison tools to determine these commonalities. Model comparison involves identifying the similarities and differences among a set of models [38]. There are many approaches and tools for automatic model comparison to employ in this step [63,66]. However, an approach or tool must be selected that is able to identify all the similarities explicitly, rather than differences only, and must be scalable, as we mentioned in our third requirement of this phase.

Independent of the chosen model comparison technique, the result of this comparison is a superset of model elements that are present in all model instances in the cluster. Regardless of the eventual application of each emergent pattern, this information may be of use to developers and analysts. The MCPM framework allows flexibility in what it considers a match, as it can accept a threshold of similarity between blocks. The default case would be 100% similarity, but there are potential applications where a lower similarity threshold may be desirable, such as renamed blocks or other near-miss scenarios. For example, Schlie et al. use 95% as their base threshold for similarity [56].

In our running example, the input and output ports, along with the associated connector lines, will be considered the set of similar elements to all instances, forming our superset, as we demonstrate in Fig. 3.10.

3.3.2.2.3 Variation tagging

In any particular instance model within a model clone cluster, any element that is not part of the previously identified superset of commonalities by the MCPM framework is a candidate for variation, which must be tagged through some method. This relates to requirement 1, being able to model variability explicitly. Any element that is not common to all instances represents one particular configuration of the pattern as observed by the respective model clone detector employed in the previous phase.

There are multiple forms we recommend for use within the MCPM framework. It can either be (1) based on the presence or absence of a model element or (2) based on a relative similarity. If the tagging is based on presence or absence, each different element would present a new variant of the pattern. If analysis reveals there is similarity (below

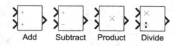

Figure 3.11 Running example: the variable model elements across the cluster of instances.

the threshold chosen for inclusion in the superset of similar elements) between elements in two or more instances, these can be tagged by the framework implementation as alternatives rather than new variants. In both cases, variation tagging would, ideally, be a fully, but potentially semi, automatic process in order for it to scale to large model sets.

In our running example, the four mathematical operators, addition, subtraction, multiplication, and division, are the only elements not tagged as part of the similarity superset. Thus, they are the logical candidates for variation tagging. At a minimum each is tagged as a new variation of the model pattern due to being present. However, after additional analysis of the properties of the blocks, mainly number of inputs and outputs in this example, it can be reasonably concluded that each of them is similar enough to be considered variants of the same operation. Thus we identify four variants, as we exhibit in Fig. 3.11.

3.3.2.2.4 Pattern explication

The final substep in this phase involves automatically explicating the variation/pattern in a meaningful way. This relates to the second requirement we identified for this phase. While this can take many forms in practice, each model pattern instance is an explication of the pattern, which as a minimum must demonstrate the existence of a base pattern and highlight the variance in some manner. The two main methods of explication within the MCPM framework are visualization and representation.

The first method is to create a visualization of the model pattern that identifies both the common elements and its variation points. In its simplest form, this is a merge of all model elements into a single model with duplicates of common elements removed. For our running example, a visualization of the pattern may appear, as we show in Fig. 3.12. This Simulink model contains all four variations of the pattern's instances. This abstraction is very explicit and, if implemented within the Simulink environment, can allow modeler interaction with the pattern by selecting a specific instance.

The second method of explicating model patterns, which is more pragmatic, is to represent the pattern in some new format from which it can be validated and applied by analysts. In the next section we will discuss some existing implementations for doing this, but, at a high level, the goal is the creation of a single model file capable of representing the pattern and its variance. One such approach, which we apply to the running example, was introduced by us in our past work [5] in which we use Simulink variant

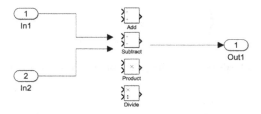

Figure 3.12 Running example: a sample visualization of variance.

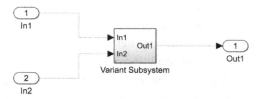

Figure 3.13 Running example: a sample representation of variance using Simulink variant subsystem blocks.

subsystem blocks.[3] Essentially all variants are contained in a special Simulink variant block that will connect the desired variant based on an environment variable selection only. For the running example, we present the resulting model in Fig. 3.13. The variant subsystem block will contain four variants that can be selected by a modeler. Essentially, and in general, the model pattern is the whole model, including the commonalities, represented by all the static blocks, and the variants, represented by anything stored inside variant subsystem blocks.

3.3.2.3 Interpretation through variation – existing research and tooling

There exist several approaches that go beyond the identification of patterns shown in the Knowledge Discovery phase. These approaches make use of clustering and differencing techniques to explicitly represent variability in some form. It is from these approaches that we draw motivation for the Interpretation phase of MCPM.

Wille et al. apply variability mining to generic block-based modeling languages [78]. Specifically, they focus their work on Simulink and state charts. This approach is preliminary work that was later expanded upon by Schlie et al. in their technique to represent variability in Simulink models [56]. Their method proposes the use of a family model [33], wherein model elements are tagged as "mandatory" if they appear in all variants, "alternative" if they represent a variation point, and "optional" if they appear in one variant with no counterpart in the other. This is very similar to feature models in product line engineering [17]. Their Family Mining Framework consists of

---

[3] https://www.mathworks.com/help/simulink/examples/variant-subsystems.html.

three distinct phases: compare, match, and merge. During the *compare* phase, models are compared using a comparison technique to generate a list of Compare Elements (CEs), which are essentially a pair of similar elements with a normalized similarity value. During the *match* phase, a subset of all CEs is chosen by the algorithm as those that feature the *best matches*. Finally, during the *merge* phase, the models are merged into a single model based on a user-specified mapping function which determines for each element if it is mandatory, alternative, or optional. Schlie et al. also present an expansion of this work where they apply the Family Mining Framework to Reverse Signal Propagation Analysis (RSPA) [55]. RSPA consists of three phases: signal set generation, comparing and clustering, and cluster optimization. During *signal set generation*, their approach traverses each hierarchical layer of the model to generate a set of outgoing signals present on that layer. These signal sets are then used as a basis for *comparing and clustering*, in which blocks associated with varying signals are clustered into sets of preliminary clusters. These preliminary clusters may contain some intersections, and thus *cluster optimization* is used to improve the clustering iteratively until no more intersections exist. Essentially, rather than relying on other similarity metrics to cluster blocks together, RSPA uses the unique signals and propagates them through the models at various levels to determine similarity and clustering, which is then used to determine variation points. Willie et al. [77] also devised an approach for configurable detection of variability relations for model variants at the block level. Their work would be helpful in an MCPM framework by providing guidelines for variability mining interpretation and preprocessing.

Our past work involved developing a method of using detected clusters of clones (clone classes) to model variation points in Simulink models [5]. Our approach asserts that a clone class with limited variability between each individual model is best represented by a single model using built-in Simulink variant subsystem blocks for each variation point. Each original version of the model can then be recreated by a modeler by setting environment variables to configure each choice to match the original subsystem, while allowing maintenance to be performed on one model file instead of the number of models originally contained in the detected clone class. This approach accounts for five different types of variability: block, input/output, function, layout, and subsystem naming, with each being handled in a similar manner. This technique can be applied to the model clone class detected and shown in Fig. 3.7 to create a single model file capable of representing both models. The original variance between the two instances can be seen on the left of the models; each has its own method of producing input signals. We present the resulting variability model in Fig. 3.14, with each expanded subsystem expanded in detail. One of the main advantages of this technique is its ability to scale to large-scale systems, both in terms of model complexity and size of model sets. The other techniques we discussed were applied mainly to pairwise variants, whereas this technique can combine arbitrarily large sets of model variants into a single variability model.

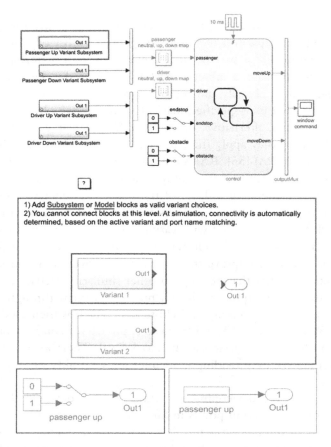

Figure 3.14 Applying Alalfi et al.'s technique [5] to represent the variants in the model clone class shown in Fig. 3.7.

3.3.3 Validation

This phase of the conceptual MCPM framework involves framework implementers validating the results of the previous two phases. Specifically, it is necessary to validate the results of the model clone detection and clustering, and the pattern interpretation of those clusters. Problems in mining can sometimes produce results that seem to be predictive and useful, but actually are misleading and cannot be reproduced [31]. This can occur at various phases within pattern mining, so we thus evaluate each phase.

3.3.3.1 Validation of the model clone pair and model clone clusters

As model clone detection tooling and research is still emerging, there are limited techniques for evaluating and validating model clone detection results. In our past work, we developed a framework, MuMonDE, for evaluating model clone detectors that employs automatic mutation analysis to ascertain correctness and recall [68]. It mutates the models using a predefined mutation taxonomy [65] to find out if the model clones that should be detected are detected. Currently it is employed by researchers to evaluate Simulink model clone detectors only. Störrle has performed some model clone evaluation experiments of their own to evaluate MQlone by manually seeding clones within existing model bases and seeing if they are detected [70]. Regardless of how validation is performed by an MCPM framework implementer, the general idea is to ensure the model clone detector is exhibiting high precision and recall in their model clone detection executions.

3.3.3.2 Validation of the interpretation through variation

With respect to validating results obtained by interpreting the observed variations within the MCPM framework, there are two main approaches that can be taken: a test-driven approach and a reverse engineering approach, both of which we describe in detail.

The test-driven approach applies mainly to any application of representation of variation, that is, any time a variant model is created regardless of technologies used. The general idea is that if a set of tests exist for the original corpus of models prior to model clone detection, clustering, and variant modeling, the newly created variant model should be capable of being configured in such a way so as to successfully pass each of the original test cases. This is analogous to the idea of finding semantically equivalent source code via testing [34]. Consider our running example. Each model would have its own set of unique test cases to ensure it functions as required, and would have been able to independently pass all of the required tests. Table 3.1 provides a small set of example tests. After the creation of a variant model, for example the one we presented in Fig. 3.13, to validate the correctness of the variant model, it must be able to be configured by a tester to be able to pass all four sets of tests. While this is not a guarantee of correctness of the newly created variant model, it provides a high level of confidence in its ability to represent the original behavior of all original pattern elements.

The second method of validating the results of Interpretation Through Variation within the MCPM is through the use of reverse engineering techniques. The goal of this type of validation is to be able to accurately recreate the original models that were used to create the resulting pattern, for example by means of model transformations to recreate the original models. This can result in a high level of confidence in the representation of the pattern. The added benefit of using reverse engineering to validate pattern representation is that it applies to both representations as a variant model, as well as a more simplified visualization of variance. The reverse engineering technique works by applying, in reverse order, the same steps we recommended to create the representation or visualization. If it is possible to return to the original models that form

Table 3.1 Running example: sample test case values for four math function models.

Addition test cases			Subtraction test cases		
Input 1	Input 2	Output	Input 1	Input 2	Output
0	1	1	0	0	0
1	1	2	1	0	1
2	1	3	2	0	2
...			...		

Multiplication test cases			Division test cases		
Input 1	Input 2	Output	Input 1	Input 2	Output
2	1	2	1	1	1
2	2	4	2	1	2
2	3	6	3	1	3
...			...		

the pattern, it becomes evident the new model pattern representation/visualization has not lost any of its original semantics. However, if it is not possible to obtain the original models from the resulting visualization or representation, this means there might be some loss of information, thus reducing the overall confidence in the interpretation of the pattern.

While there may exist other application- and domain-specific methods of validating variation results, the two techniques we discussed here present methods that generalize to most cases. Further, these techniques are sufficiently scalable to apply to large-scale systems without additional effort, provided the required inputs (test files or list of applied steps) exist previously.

3.3.3.3 Validation of pattern quality

A potential third area of validation is the evaluation of the quality of any detected patterns. The previous validations deal largely with correctness and the ability to express patterns. However, the "quality" of the patterns is still unknown to analysts. In its current state, we do not consider this aspect of validation within scope for the MCPM conceptual framework as the MCPM is currently concerned mainly with identifying emergent patterns and expressing them in a form conducive to evaluation and application. Quality evaluation and filtering, sorting, and/or ranking of patterns are interesting from a postprocessing perspective, but not part of the core MCPM.

3.3.4 Application

While the identification of valid emergent model patterns through model clone detection presents useful information, the MCPM framework posits that a particular application of the obtained knowledge is important to solidify its contributions. As such, this section presents our recommendations on potential applications of findings in industrial settings, including integration into workflows and tool chains.

The first potential application, and perhaps the simplest, is the incorporation of pattern visualization into the workflow of model development and model-driven engineering. Being aware of the existence of model patterns in a collection of models yields its own benefits in that developers are able to draw reasonable conclusions about their presence and adapt usual working conditions to make changes. While this seems abstract, it fits the goals of the MCPM framework and allows the domain experts to use the knowledge to improve their models as appropriate. Regarding pattern usefulness, the objective of MCPM is solely to identify emergent patterns. Thus, evaluation of the usefulness of the patterns that emerge is beyond the scope of the framework. Analysts can perform filtering, ranking, and sorting of the emergent patters depending on their intended use cases. One of the primary ways this information is applied is via the identification of potential faults. In some situations, an expert can be left asking the question, "Is this variance point intentional, or have we found a fault?" By applying the MCPM framework to a project periodically, relevant information can be revealed to improve overall software quality.

The second application combats the model management problem discussed earlier, and that is one of the focuses of this book. Determining the existence of model patterns in model sets provides candidates for model merging and retaining variance through metrics, allowing analysts to maintain only one model rather than the large number of original models. This process takes numerous forms as we discussed, but generally presents the outcome of a reduced number of models that are still capable of expressing the original functionality of the full set of models. Within Simulink, for example, this can be done through the use of subsystem variant blocks [5] or by creating tagged models with aspects similar to software product lines [56]. Regardless of approach, the introduction of variant models decreases the total number of models in the set, allowing for improved maintainability.

3.4. Summary of challenges and future directions

The main challenges faced by the MCPM framework are the full automation of variant model creation, effective visualization and explication of model patterns, postvalidation of emergent pattern quality, and the practitioner adoption issue. Each issue presents an opportunity for future work, which we discuss further.

3.4.1 Automatic variant model creation

We previously developed an approach [5] that is only semiautomated as a proof of concept. However, there is still no current method of creating a full set of variability models that work directly within Simulink. While the approach presents meaningful results, the manual interaction creates opportunity for error and does not scale well to large-scale systems, where the approach is most needed.

Schlie et al. present an approach [56] that deals with tagging elements in a Family model as mandatory, alternative, or optional. This does not translate directly to a Simulink model that can be used explicitly. Rather, it is merely an additional representation to present variability. While this approach scales better than the previous one, the separation from the application domain by representing the models outside of Simulink poses a pragmatic problem.

Both of these approaches demonstrate promise in variant modeling; however, the development of a fully automated approach capable of scaling to large-scale systems is still necessary and an open challenge to advance the application of the MCPM framework.

3.4.2 Effective visualization/explication

In order to use the results of model clone detection to find model patterns, an effective way of explicating these model patterns is needed. When working with experts in graphical modeling, this likely will take the form of visualization in order to maintain the higher-level abstractions afforded by modeling. However, there is no strong forerunner in the visualization of explicit model patterns. While we presented SimNav [49] as a method of visualizing the model clone classes, the approach focuses too specifically on clone classes, and not the underlying patterns.

In order to advance the application of the MCPM framework, further work in the effective visualization and explication of model patterns is required, specifically with some form of automated tool support.

3.4.3 Validating pattern quality

Currently, the MCPM framework does not address the quality of the observed patterns. Within the framework, detected patterns are always included in the resulting pattern set, as they are emergent. However, this may not always be the best course of action. In order to improve the effectiveness of applying patterns, it may be necessary to include a postprocessing method for validating pattern quality as part of the MCPM framework in its future implementations.

3.4.4 Practitioner adoption

The final challenge faced currently by the MCPM framework concept is the potential reluctance of practitioners to adopt elements of the MCPM framework. While this is a problem that is faced in many domains, and possibly related to the solutions to the previous two issues, it is still worth noting as an ongoing challenge in the field. Industrial developers of models need to be on board with the concept of applying model clone detection to observe and make use of model patterns for the framework to advance and gain traction. While we do not have any specific approaches to overcome this hurdle, it is a challenge to the community to work closely with practitioners to ensure their buy-in and adoption of research technologies and tooling. One aspect of this challenge would be some sort of ranking or feedback system associated with any discovered emergent patterns to ensure quality, that is, a way of having practitioners review and rate the patterns to help with "good" and "bad" emergent pattern identification.

3.5. Conclusion

In this chapter, we have investigated the possibility of employing model clone detection as a tool for extracting emergent model patterns from large model repositories. We defined a conceptual framework, MCPM, that can be realized and used by analysts and researchers to follow in order to use model clone detection for this purpose. We break down the details of each of MCPM's four phases, provide examples to follow, and describe how current research and tooling fit its requirements. We additionally outline current challenges to help set the path for researchers interested in this area. It is our intention for this chapter to help pave the way for more advanced model analytics allowing for the identification of emergent model patterns in large software systems, thus improving model-driven engineering on the whole.

References

[1] Bakr Al-Batran, Bernhard Schätz, Benjamin Hummel, Semantic clone detection for model-based development of embedded systems, in: International Conference on Model Driven Engineering Languages and Systems, Springer, 2011, pp. 258–272.

[2] Manar H. Alalfi, James R. Cordy, Thomas R. Dean, Analysis and clustering of model clones: an automotive industrial experience, in: Software Maintenance, Reengineering and Reverse Engineering (CSMR-WCRE), 2014 Software Evolution Week-IEEE Conference on, IEEE, 2014, pp. 375–378.

[3] Manar H. Alalfi, James R. Cordy, Thomas R. Dean, Matthew Stephan, Andrew Stevenson, Models are code too: near-miss clone detection for Simulink models, in: ICSM, 2012, pp. 295–304.

[4] Manar H. Alalfi, James R. Cordy, Thomas R. Dean, Matthew Stephan, Andrew Stevenson, Near-miss model clone detection for Simulink models, in: Proceedings of the 6th International Workshop on Software Clones, IEEE Press, 2012, pp. 78–79.

[5] Manar H. Alalfi, Eric J. Rapos, Andrew Stevenson, Matthew Stephan, Thomas R. Dean, James R. Cordy, Semi-automatic identification and representation of subsystem variability in Simulink models, in: Software Maintenance and Evolution (ICSME), 2014 IEEE International Conference on, IEEE, 2014, pp. 486–490.

[6] Elizabeth Antony, Manar H. Alalfi, James R. Cordy, An approach to clone detection in behavioural models, in: International Working Conference in Reverse Engineering, 2013, pp. 472–476.

[7] Ö. Babur, Clone detection for ecore metamodels using N-grams, in: International Conference on Model-Driven Engineering and Software Development, 2018.

[8] Zsolt Balanyi, Rudolf Ferenc, Mining design patterns from C++ source code, in: Software Maintenance, 2003. ICSM 2003. Proceedings. International Conference on, IEEE, 2003, pp. 305–314.

[9] Demis Ballis, Andrea Baruzzo, Marco Comini, A rule-based method to match software patterns against UML models, Electronic Notes in Theoretical Computer Science 219 (2008) 51–66.

[10] Francesco Basciani, Juri Di Rocco, Davide Di Ruscio, Amleto Di Salle, Ludovico Iovino, Alfonso Pierantonio, MDEForge: an extensible web-based modeling platform, in: International Workshop on Model-Driven Engineering on and for the Cloud, 2014, pp. 66–75.

[11] Federico Bergenti, Agostino Poggi Idea, A design assistant based on automatic design pattern detection, in: Proceedings of the 12th International Conference on Software Engineering and Knowledge Engineering, Springer-Verlag, 2000, pp. 336–343.

[12] Christopher Bishop, Pattern Recognition and Machine Learning, Springer, 2006.

[13] William H. Brown, Raphael C. Malveau, Hays W. McCormick, Thomas J. Mowbray, AntiPatterns: Refactoring Software, Architectures, and Projects in Crisis, John Wiley & Sons, Inc., 1998.

[14] Jian Chen, Thomas R. Dean, Manar H. Alalfi, Clone detection in Matlab stateflow models, Software Quality Journal 24 (4) (2016) 917–946.

[15] James O. Coplien, Software design patterns: common questions and answers, in: The Patterns Handbook: Techniques, Strategies, and Applications, 1998, pp. 311–320.

[16] James R. Cordy, Submodel pattern extraction for Simulink models, in: International Software Product Line Conference, 2013, pp. 7–10.

[17] Krzysztof Czarnecki, Simon Helsen, Ulrich Eisenecker, Staged configuration using feature models, in: International Conference on Software Product Lines, Springer, 2004, pp. 266–283.

[18] F. Deissenboeck, B. Hummel, E. Juergens, B. Schaetz, S. Wagner, J.-F. Girard, S. Teuchart, Clone detection in automotive model-based development, in: ICSE, 2009, pp. 603–612.

[19] Florian Deissenboeck, Benjamin Hummel, Elmar Jürgens, Bernhard Schätz, Stefan Wagner, Jean-François Girard, Stefan Teuchert, Clone detection in automotive model-based development, in: 30th International Conference on Software Engineering, ACM, 2008, pp. 603–612.

[20] Patricia Derler, Edward A. Lee, Alberto Sangiovanni Vincentelli, Modeling cyber–physical systems, Proceedings of the IEEE 100 (1) (2012) 13–28.

[21] Giuseppe Di Fatta, Stefan Leue, Evghenia Stegantova, Discriminative pattern mining in software fault detection, in: Proceedings of the 3rd International Workshop on Software Quality Assurance, ACM, 2006, pp. 62–69.

[22] Jing Dong, Yajing Zhao, Tu Peng, A review of design pattern mining techniques, International Journal of Software Engineering and Knowledge Engineering 19 (06) (2009) 823–855.

[23] Usama M. Fayyad, Gregory Piatetsky-Shapiro, Padhraic Smyth, et al., Knowledge discovery and data mining: towards a unifying framework, in: KDD, vol. 96, 1996, pp. 82–88.

[24] Eduardo Fernandez-Buglioni, Security Patterns in Practice: Designing Secure Architectures Using Software Patterns, John Wiley & Sons, 2013.

[25] Rahma Fourati, Nadia Bouassida, Hanêne Ben Abdallah, A metric-based approach for anti-pattern detection in uml designs, in: Computer and Information Science 2011, Springer, 2011, pp. 17–33.

[26] Michael R. Garey, David S. Johnson, Computers and Intractability: A Guide to the Theory of NP-Completeness, W.H. Freeman & Co Ltd, 1979.

[27] Manjari Gupta, Rajwant Singh Rao, Akshara Pande, A.K. Tripathi, Design pattern mining using state space representation of graph matching, in: International Conference on Computer Science and Information Technology, Springer, 2011, pp. 318–328.

[28] Jiawei Han, Jian Pei, Micheline Kamber, Data Mining: Concepts and Techniques, Elsevier, 2011.

[29] Neil B. Harrison, Paris Avgeriou, Uwe Zdun, Using patterns to capture architectural decisions, IEEE Software 24 (4) (2007).

[30] Ahmed E. Hassan, The road ahead for mining software repositories, in: Frontiers of Software Maintenance, 2008, FoSM 2008, IEEE, 2008, pp. 48–57.

[31] Douglas M. Hawkins, The problem of overfitting, Journal of Chemical Information and Computer Sciences 44 (1) (2004) 1–12.

[32] Regina Hebig, Truong Ho Quang, Michel R.V. Chaudron, Gregorio Robles, Miguel Angel Fernan-
     dez, The quest for open source projects that use UML: mining GitHub, in: International Conference
     on Model Driven Engineering Languages and Systems, ACM, 2016, pp. 173–183.
[33] Sönke Holthusen, David Wille, Christoph Legat, Simon Beddig, Ina Schaefer, Birgit Vogel-Heuser,
     Family model mining for function block diagrams in automation software, in: Proceedings of the 18th
     International Software Product Line Conference: Companion Volume for Workshops, Demonstrations
     and Tools, vol. 2, 2014, pp. 36–43.
[34] Lingxiao Jiang, Zhendong Su, Automatic mining of functionally equivalent code fragments via random
     testing, in: Proceedings of the Eighteenth International Symposium on Software Testing and Analysis,
     ACM, 2009, pp. 81–92.
[35] Elmar Juergens, Florian Deissenboeck, Benjamin Hummel, Stefan Wagner, Do code clones matter?,
     in: 31st International Conference on Software Engineering, IEEE, 2009, pp. 485–495.
[36] Huzefa Kagdi, Jonathan I. Maletic, Bonita Sharif, Mining software repositories for traceability links,
     in: Program Comprehension, 2007. ICPC'07. 15th IEEE International Conference on, IEEE, 2007,
     pp. 145–154.
[37] Cory Kapser, Michael W. Godfrey, "Cloning considered harmful" considered harmful, in: Reverse
     Engineering, 2006. WCRE'06. 13th Working Conference on, Citeseer, 2006, pp. 19–28.
[38] D.S. Kolovos, R.F. Paige, F.A.C. Polack, Model comparison: a foundation for model composition and
     model transformation testing, in: Proceedings of the International Workshop on Global Integrated
     Model Management, ACM, 2006, pp. 13–20.
[39] R. Koschke, Survey of research on software clones, in: Duplication, Redundancy, and Similarity in
     Software, 2006, pp. 1–24.
[40] Mythili Aravida Kumar, Efficient weight assignment method for detection of clones in state flow
     diagrams, Journal of Software Engineering Research and Practices 4 (2) (2014) 12–16.
[41] Jürgen Lind, Patterns in agent-oriented software engineering, in: International Workshop on Agent-
     Oriented Software Engineering, Springer, 2002, pp. 47–58.
[42] Benjamin Livshits, Thomas Zimmermann, Dynamine: finding common error patterns by min-
     ing software revision histories, in: ACM SIGSOFT Software Engineering Notes, vol. 30(5), 2005,
     pp. 296–305.
[43] Gaurab Luitel, Matthew Stephan, Daniela Inclezan, Model level design pattern instance detection
     using answer set programming, in: International Workshop on Modeling in Software Engineering
     (MISE), MiSE '16, ACM, New York, NY, USA, 2016, pp. 13–19.
[44] Robert C. Martin, Agile Software Development: Principles, Patterns, and Practices, Prentice Hall,
     2002.
[45] Jukka Paakki, Anssi Karhinen, Juha Gustafsson, Lilli Nenonen, A. Inkeri Verkamo, Software metrics
     by architectural pattern mining, in: Proceedings of the International Conference on Software: Theory
     and Practice, 16th IFIP World Computer Congress, Kluwer, Beijing, China, 2000, pp. 325–332.
[46] Akshara Pande, Manjari Gupta, A.K. Tripathi, Design pattern mining for GIS application using graph
     matching techniques, in: Computer Science and Information Technology (ICCSIT), 2010. 3rd IEEE
     International Conference on, vol. 3, IEEE, 2010, pp. 477–482.
[47] Hauke Petersen, Clone Detection in Matlab Simulink Models, Master's thesis, Technical University of
     Denmark, 2012, iMM-M. Sc.-2012-02.
[48] N.H. Pham, H.A. Nguyen, T.T. Nguyen, J.M. Al-Kofahi, T.N. Nguyen, Complete and accurate clone
     detection in graph-based models, in: International Conference on Software Engineering, ICSE, 2009,
     pp. 276–286.
[49] E.J. Rapos, A. Stevenson, M.H. Alalfi, J.R. Cordy Simnav, Simulink navigation of model clone classes,
     in: 2015 IEEE 15th International Working Conference on Source Code Analysis and Manipulation,
     SCAM, 2015, pp. 241–246.
[50] Dhavleesh Rattan, Rajesh Bhatia, Maninder Singh, Software clone detection: a systematic review,
     Information and Software Technology 55 (7) (2013) 1165–1199.

[51] Dirk Riehle, Heinz Züllighoven, Understanding and using patterns in software development, Theory and Practice of Object Systems 2 (1) (1996) 3–13.

[52] Chanchal Kumar Roy, James R. Cordy, A survey on software clone detection research, Queen's School of Computing TR 541 (115) (2007) 64–68.

[53] C.K. Roy, J.R. Cordy, NICAD: accurate detection of near-miss intentional clones using flexible pretty-printing and code normalization, in: ICPC, 2008, pp. 172–181.

[54] H. Safyallah, K. Sartipi, Dynamic analysis of software systems using execution pattern mining, in: 14th IEEE International Conference on Program Comprehension, vol. 1, ICPC, June 2006, pp. 84–88.

[55] Alexander Schlie, David Wille, Loek Cleophas, Ina Schaefer, Clustering variation points in Matlab/Simulink models using reverse signal propagation analysis, in: International Conference on Software Reuse, Springer, 2017, pp. 77–94.

[56] Alexander Schlie, David Wille, Sandro Schulze, Loek Cleophas, Ina Schaefer, Detecting variability in Matlab/Simulink models: an industry-inspired technique and its evaluation, in: Proceedings of the 21st International Systems and Software Product Line Conference-Volume A, ACM, 2017, pp. 215–224.

[57] Douglas C. Schmidt, Mohamed Fayad, Ralph E. Johnson, Software patterns, Communications of the ACM 39 (10) (1996) 37–39.

[58] Abhishek B. Sharma, Franjo Ivančić, Alexandru Niculescu-Mizil, Haifeng Chen, Guofei Jiang, Modeling and analytics for cyber-physical systems in the age of big data, ACM SIGMETRICS Performance Evaluation Review 41 (4) (2014) 74–77.

[59] Nija Shi, Ronald A. Olsson, Reverse engineering of design patterns from Java source code, in: Automated Software Engineering, 2006. ASE'06. 21st IEEE/ACM International Conference on, IEEE, 2006, pp. 123–134.

[60] Michael Stal, Using architectural patterns and blueprints for service-oriented architecture, IEEE Software 23 (2) (2006) 54–61.

[61] M. Stephan, Detection of Java EE EJB Antipattern Instances Using Framework-Specific Models, Master's thesis, University of Waterloo, 2009.

[62] M. Stephan, M.H. Alafi, A. Stevenson, J.R. Cordy, Towards qualitative comparison of Simulink model clone detection approaches, in: International Workshop on Software Clones, IWSC, 2012, pp. 84–85.

[63] M. Stephan, J.R. Cordy, A Survey of Methods and Applications of Model Comparison, Technical Report 2011-582 Rev. 3, Queen's University, 2012.

[64] M. Stephan, J.R. Cordy, Identification of Simulink model antipattern instances using model clone detection, in: International Conference on Model Driven Engineering Languages and Systems, MODELS, Sept 2015, pp. 276–285.

[65] Matthew Stephan, Manar Alalfi, James R. Cordy, Towards a taxonomy for Simulink model mutations, in: International Conference on Software Testing, Verification, and Validation 2014 (ICST) – Mutation Workshop, 2014, pp. 206–215.

[66] Matthew Stephan, James R. Cordy, A survey of model comparison approaches and applications, in: Modelsward, 2013, pp. 265–277.

[67] Matthew Stephan, James R. Cordy, Identifying instances of model design patterns and antipatterns using model clone detection, in: International Workshop on Modelling in Software Engineering, 2015, pp. 48–53.

[68] Matthew Stephan, James R. Cordy, Mumonde: a framework for evaluating model clone detectors using model mutation analysis, Software Testing, Verification & Reliability (2018) e1669.

[69] Harald Störrle, Towards clone detection in uml domain models, Software & Systems Modeling 12 (2) (2013) 307–329.

[70] Harald Störrle, Effective and efficient model clone detection, in: Software, Services, and Systems, Springer, 2015, pp. 440–457.

[71] Daniel Strüber, Vlad Acreţoaie, Jennifer Plöger, Model clone detection for rule-based model transformation languages, Software & Systems Modeling (2017) 1–22.

[72] Paolo Tonella, Mariano Ceccato, Aspect mining through the formal concept analysis of execution traces, in: 11th Working Conference on Reverse Engineering, IEEE, 2004, pp. 112–121.

[73] Nikolaos Tsantalis, Alexander Chatzigeorgiou, George Stephanides, Spyros T. Halkidis, Design pattern detection using similarity scoring, IEEE Transactions on Software Engineering 32 (11) (2006).

[74] John Vlissides, Richard Helm, Ralph Johnson, Erich Gamma, Design Patterns: Elements of Reusable Object-Oriented Software, vol. 49(120), Addison–Wesley, Reading, 1995, p. 11.

[75] Sven Wenzel, Automatic detection of incomplete instances of structural patterns in uml class diagrams, Nordic Journal of Computing 12 (4) (2005) 379.

[76] Jon Whittle, John Hutchinson, Mark Rouncefield, The state of practice in model-driven engineering, IEEE Software 31 (3) (2014) 79–85.

[77] David Wille, Önder Babur, Loek Cleophas, Christoph Seidl, Mark van den Brand, Ina Schaefer, Improving custom-tailored variability mining using outlier and cluster detection, Science of Computer Programming 163 (2018) 62–84.

[78] David Wille, Sandro Schulze, Christoph Seidl, Ina Schaefer, Custom-tailored variability mining for block-based languages, in: Software Analysis, Evolution, and Reengineering (SANER), 2016 IEEE 23rd International Conference on, vol. 1, IEEE, 2016, pp. 271–282.

[79] Tao Xie, Jian Pei Mapo, Mining API usages from open source repositories, in: Proceedings of the 2006 International Workshop on Mining Software Repositories, ACM, 2006, pp. 54–57.

[80] Andy Zaidman, Bart Van Rompaey, Serge Demeyer, Arie Van Deursen, Mining software repositories to study co-evolution of production & test code, in: Software Testing, Verification, and Validation, 2008 1st International Conference on, IEEE, 2008, pp. 220–229.

[81] Liming Zhu, Muhammad Ali Babar, Ross Jeffery, Mining patterns to support software architecture evaluation, in: Software Architecture, 2004. WICSA 2004. Proceedings. Fourth Working IEEE/IFIP Conference on, IEEE, 2004, pp. 25–34.

CHAPTER 4

Domain-driven analysis of architecture reconstruction methods

Burak Uzun, Bedir Tekinerdogan
Information Technology Group, Wageningen University, Wageningen, The Netherlands

Contents

4.1. Introduction

One of the key artifacts in the software development life cycle process is the architecture design, which represents the gross-level structure of the system. The model architecture is important for supporting the communication among stakeholders, for analysis of the design decisions, and for guiding the organizational processes [30]. Software architecture can be prescriptive in that it defines how detailed design and code should be structured. Software architecture can also be descriptive in that it reflects the structure of current design and code artifacts [1]. Very often the architecture of a system is not existing or needs to be reconstructed due to the code that has evolved. In this context SAR methods can be used to reconstruct and obtain the architecture information of the system using various sources, such as documentations, logs, and code [7].

Various different approaches have been proposed in the literature for architecture reconstruction. Since manual handling of the architecture reconstruction process is usually time consuming and costly, automatized methods and tools have been proposed by different studies in the literature. In this chapter we present a domain-driven survey of the published architecture reconstruction methods. For this we apply a domain analysis process in which we first define the generic domain model of architecture reconstruction. The domain model is represented as a set of key terms, a generic business process model, and a feature diagram that represents the common and variant features of architecture reconstruction. We also present the method for deriving concrete architecture reconstruction methods from the generic domain model. We illustrate our approach for deriving two different concrete architecture reconstruction methods.

The remainder of the chapter is organized as follows. In Section 4.2, we discuss the background for architecture reconstruction and feature-driven domain analysis by presenting the conceptual model for it. Section 4.3 presents the domain model of software architecture reconstruction methods. Section 4.4 presents the derived feature and business process models for two case studies. Section 4.5 presents the related work and Section 4.6 presents the discussion. Section 4.7 concludes the chapter.

4.2. Preliminaries

4.2.1 Software architecture reconstruction

Software architecture reconstruction is a reverse engineering process which aims to obtain the architecture information of any system using existing other source artifacts [7]. Fig. 4.1 shows a conceptual model for architecture reconstruction.

An architecture reconstruction process uses different source artifacts which are input for the SAR process such as code, detailed design, logs, documentation, and stakeholder concerns. Two basic motivations for SAR can be identified, that is, missing or incomplete architecture documentation, or an architecture drift due to evolution of the requirements. Since manual handling of architecture reconstruction is usually time consuming and costly, fully or semiautomatized methods and tools are implemented. The result of the SAR process is an architecture documentation that includes a description of a set of architecture views for addressing stakeholder concerns [27]. An architecture view conforms to an architecture viewpoint. Architecture views can be visual or textual [5].

4.2.2 Domain analysis

A well-known process for analyzing and modeling the current state of the art of a particular domain is *domain analysis*. Domain analysis is the process of identifying, capturing domain knowledge about the problem domain with the purpose of making it reusable when creating new systems [4,18,31,21]. Domain here refers to "an area of knowledge or activity characterized by a set of concepts and terminology understood by practition-

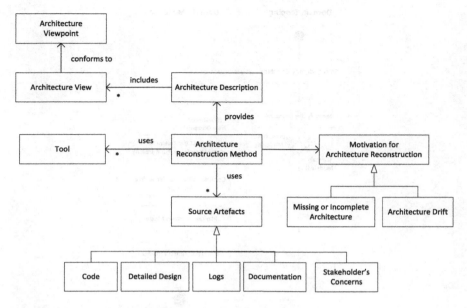

Figure 4.1 Conceptual model for architecture reconstruction.

ers in that area." Several domain analysis processes have been defined in the literature, but in general these include the steps for (1) domain scoping and (2) domain modeling. Fig. 4.2 shows the overall domain analysis process that we will adopt.

In domain scoping it is explicitly described what will be considered in the domain and likewise the scope will be determined. This will include also the selection of the primary studies that will be used to analyze the domain. In the domain modeling process, usually a commonality and variability analysis process is carried out to derive a domain model. With the commonality and variability analysis process the identified primary studies are simultaneously analyzed to identify common and variant features of a domain [32]. A *feature model* is a model that defines features and their dependencies. A feature is a system property that is relevant to some stakeholder and is used to capture commonalities or discriminate between them. Feature models are usually represented in *feature diagrams* (or tables). A feature diagram is a tree with the root representing a concept (e.g., a software system), and its descendent nodes are features. Various relationships between a parent feature and its child features (or subfeatures) are defined, including *mandatory*, *optional*, *or*, and *alternative*. A *feature configuration* is a selected set of features which describes an instance of the feature diagram. *Cross-tree constraints* further restrict the possible selections of features to define configurations. The most common cross–tree constraints are *requires* and *excludes*.

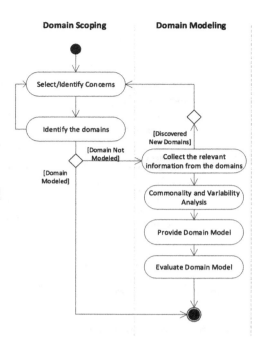

Figure 4.2 Common structure of domain analysis methods.

4.3. Domain model of architecture reconstruction methods

In this section we present the results of the domain analysis process for SAR methods. This process resulted in the domain model which consists of a feature diagram and a generic business process model. In Section 4.3.1, we present the adopted scope of the domain that we have used to derive the domain model. Section 4.3.2 presents the feature diagram for SAR methods. Finally, Section 4.3.3 presents the generic business process model.

4.3.1 Domain scoping

As discussed before, the domain scoping defines what will be considered in the domain and likewise the scope will be determined. Our key concern in this study is architecture reconstruction and in particular we focus on the process and the corresponding tools. In the scoping process we have used different reputable and most cited libraries, including IEEE Explore, ACM, Science Direct, Springer, WebOfKnowledge, and Wiley. In these platforms we have searched for papers related to software architecture reconstruction. We have created search strings including the terms "software architecture reconstruction" and "software architecture recovery" in abstract of papers. After a careful analysis of the selected papers according to exclusion criteria we have derived the list of papers

Table 4.1 Study selection criteria.

EC1	Papers in which the full text is not available
EC2	Papers that are not in English
EC3	Papers that are duplicate
EC4	Papers that were published before 1998
EC5	Papers that do not relate to software architecture reconstruction
EC6	Papers that do not have a proposed tool
EC7	Papers that are experience or survey papers

Table 4.2 Overview of search results and study selection.

Source	Number of included studies after applying search queries	Number of included studies after applying exclusion criteria EC1–EC4	Number of included studies after applying exclusion criteria EC5–EC7
IEEE Xplore	406	25	8
ACM Digital Library	25	7	0
Wiley Interscience	5	3	3
Science Direct	121	5	2
Springer	8	4	4
ISI Web of Knowledge	87	6	0
Total	652	110	17

as shown in Appendix 4.A – Primary studies. We have thoroughly read these papers and applied the commonality and analysis for deriving the common and variant concepts related to SAR. The result of the data extraction process is described using a domain which is explained in the next subsection. Table 4.1 provides our exclusion criteria for the study selection process. We further filtered out the query results according to these defined criteria. Table 4.2 presents the number of study selections after applying search queries and exclusion criteria. Seventeen studies were singled out of 652 studies after applying the exclusion criteria.

4.3.2 Feature model

Fig. 4.3 shows the feature model of SAR methods that we have derived from the domain analysis process.

A SAR method consists of the mandatory features of goal, source artifacts, adopted method, process steps, and architecture model and one optional feature of architecture model.

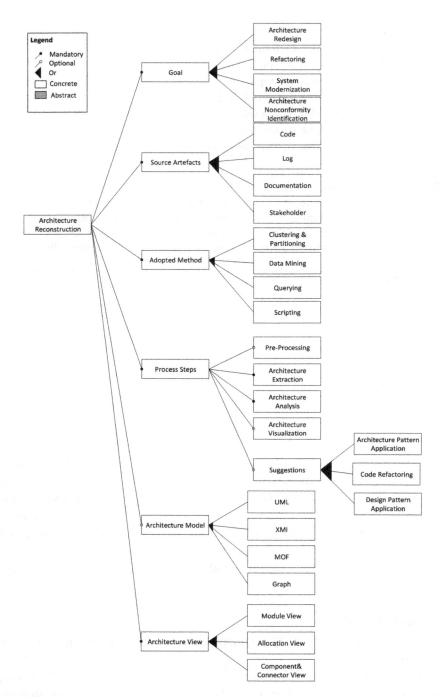

Figure 4.3 Feature model for architecture reconstruction methods.

The *goal* feature defines the aim of the SAR and why the SAR is actually applied. The goal of the SAR method can be architecture redesign, system modernization, refactoring, and architecture nonconformity identification. Architecture redesign is a process for enhancing, adapting, or revising the current architecture. Architecture redesign can be applied when underlying system architecture does not perform as desired against the current or future requirements. Refactoring is a software development process where the low-level components are rearranged without changing the functionality of the program code. Refactoring is heavily used in industry, which makes the code more readable and adaptable to further changes in the requirements. System modernization is a software process in which the current systems architecture and used technology stack are adapted to the current requirements and the current state of the system is maintained. At last, architecture nonconformity identification is a process that aims to find differences between intended and implemented architecture. These differences can be crucial and can lead to bugs, defects, or errors. One or more of the defined goals can be valid for a SAR method.

Source artifacts are artifacts that are used to (re)construct an architecture. Code, logs, documentations, and stakeholder concerns are source artifacts for SAR methods. SAR methods can use one or more of these artifacts to extract architectural elements which can be represented by architectural models. Code is an executable that is written according to a programming language. Code of systems are implemented with respect to their specifications in architecture designs. Logs are output data from systems which might give insight about system architecture. Documentation shows textual information about systems which can be architecture designs and functional and nonfunctional requirements. Stakeholders are individuals, groups, people, or organizations that have an interest in the developed system. Stakeholder concerns shape how the architecture is designed and implemented. The SAR method can have one or more of these source artifacts to reconstruct the system's architecture.

Adopted methods represent techniques applied for reconstructing the system architecture. Various methods have been implemented and proposed in the literature for SAR. These include clustering, partitioning, data mining, querying, or scripting. These methods can be used together or separately. Clustering methods consist of grouping some related data together. Partitioning is dividing data into parts that may or may not be related to each other. In SAR data mining techniques are adopted in which patterns are detected in large sets of source codes, logs, and documentation. Some of the SAR methods also utilize querying on graphs in which the architecture is represented as graph-based models. Scripting software architectures in their own domain-specific languages is another method adopted by SAR methods.

Process steps define phases used in a SAR method. SAR methods usually consist of a series of different processes including preprocessing, architecture extraction, architecture analysis, architecture visualization, and suggestions. Preprocessing is an optional step in which the source artifact might be prepared for the architecture extraction step. This can include reformatting, removing unnecessary libraries, and removing comments. Architecture extraction is performed on source artifacts; the valuable data are extracted from these artifacts and hence it is a mandatory step in all SAR methods. In the architecture analysis step architecture components are formed using the output of the extraction step. The architecture visualization step presents the reconstructed architecture in a user-friendly manner, which can be graphical or textual. Visualization of the reconstructed architecture is not always present; thus this feature is optional. Another optional feature in process steps is suggestions. Suggestions such as design pattern application, architecture pattern application, or code refactoring can be advised after the architecture is constructed.

Architecture model representation defines the representation used for describing the architecture. The adopted SAR methods in the primary studies use different representations, including Unified Modeling Language (UML), XML Metadata Interchange (XMI), MetaObject Facility (MOF), or graph representations. UML is the standardized model to represent the software architecture which contains diagrams for developers to develop, construct, and document the architectures. XMI is designed for exchanging information using extensible markup languages. Software architectures can be represented and visualized using XMI-based models. MOF is standard for model-driven engineering proposed by the Object Management Group. Graph-based models can also represent the software architecture in different forms, such as Bayesian networks, Markov chains, and state diagrams.

The *architecture view* feature represents which view is to be reconstructed using the SAR method. Architecture view is a mandatory feature in SAR methods for representing what part of the architecture will be constructed. The module view, allocation view, and component & connector view can be reconstructed in the architecture analyzing step. The module view represents how the system is structured as set of code units. The allocation view represents how the system relates the structures that are not software in the system's running environment. The component & connector view shows how the system is structured in terms of components and its relations among each other.

Using the feature diagram, we can generate many different architecture reconstruction methods by selecting the desired features. However, not all configurations might be possible in practice. Hence, cross-tree constraints are used to eliminate the invalid configurations. We have seen that if the architecture model selection feature is not chosen, then the architecture is not visualized. As architecture is not represented in any manner, it is impossible to visualize. In addition, if a model selection feature is present it is not mandatory to see the architecture visualization feature. Code source artifact must be selected if the goal of the SAR method is refactoring or architecture nonconformity identification. Hence, both of these goals are operated at code level, and code is needed. Refactoring or architecture nonconformity identification may or may not be selected

Table 4.3 Identified feature composition constraints.

Process steps. Architecture visualization **requires** *architecture model*
Goal. Refactoring **requires** *source artifact. Code source*
Goal. System modernization **requires** *source artifact. Stakeholders*
Adopted method. Querying **requires** *architecture model. Graph*
Source artifact. Log **requires** *process steps. Preprocessing*
Source artifact. Documentation **requires** *process steps. Preprocessing*
Goal. Design pattern application **requires** *source artifact. Code source*
Goal. Architecture nonconformity identification **requires** *source artifact. Code source*

if the code is selected as source artifact. Stakeholders source artifact must be chosen if system modernization goal is selected as a feature. The system modernization goal feature may or may not be selected if stakeholder source artifact is selected. The querying feature must be selected as adopted method if the architecture model is graph-based and vice versa if this constraint does not apply. The preprocessing step must be selected from process steps if log or documentation source artifacts are selected and vice versa if this constraint does not apply. The code source artifact feature must be selected if the code refactoring or design pattern application feature of process suggestions is selected and vice versa if this does not apply. Table 4.3 shows a summary of identified feature composition constraints.

4.3.3 Generic business process model

Architecture reconstruction is in essence a process activity that includes several steps linked in different ways. Based on the identified studies as discussed in the previous subsection, Fig. 4.4 provides the generic business process model for architecture reconstruction methods.

The first step in architecture reconstruction begins with a definition of a problem statement explaining what the proposed method aims to solve. According to the problem statement the required architectural information which consists of architecture views is selected. The next step is to select a method to solve the stated problem according to the needs defined in the previous step. These methods can utilize techniques from clustering, partitioning, data mining, or querying methods. Reconstruction methods might utilize a model for solving the problem or might need preprocessing of the source artifacts before the extraction of architectural elements. Models are used for representing both the abstractions in the code and reconstructed architectures. These models can be graph-, MOF-, UML-, or XMI-based. These four steps define the SAR process design describing the planning of the proposed SAR method.

Preprocessing can be done in several ways, such as reformatting and refactoring code, removing undesired libraries and dependencies, and providing intermediate models to a workflow. The successful completion of architectural element extraction is followed

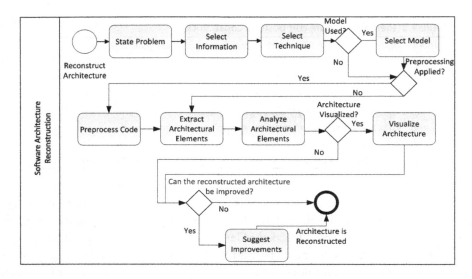

Figure 4.4 Generic business process model for architecture reconstruction.

by the analysis of the extracted elements for forming the architectural information intended in the information selection phase. The process follows by two optional steps, i.e., visualizing the reconstructed architecture and suggesting improvements on the reconstructed architecture. In the visualization step the architecture is shown using the selected architecture representation. In the suggestion step the process provides guidelines for architecture reconstruction.

4.4. Concrete architecture reconstruction method

The domain model that has been defined in the earlier section represents both the common and the variant aspects of architecture reconstruction processes. We have provided a family feature model and a generic process model that has been described after a study of multiple methods. To develop a concrete architecture reconstruction method, the family feature model and generic process model can be reused and adapted with respect to the required features. Fig. 4.5 illustrates this process.

Similar to reference models in general the approach can be used in either a prescriptive way or a descriptive way. In the prescriptive way the family feature model and the generic process can be reused to prescribe the common elements. Moreover, it can indicate the elements that might be selected as variant elements. Alternatively, in the descriptive way, the approach can be used to describe a given SAR method using the provided family feature model and the generic business process model. In the following subsections, we will show two different examples of concrete architecture reconstruction methods that are derived from the provided domain model. The first example

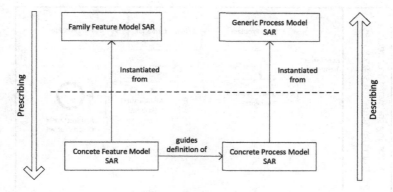

Figure 4.5 Relation between reference SAR and concrete SAR.

Table 4.4 Feature selection for recovering software architecture from the names of source files.

Feature	Selected features of SAR method
Goal	Architecture nonconformity identification
Source artifacts	Code
Adopted method	Clustering & partitioning
Process steps	Architecture extraction
	Architecture analysis
Architecture model	Not used
Architecture view	Component & connectors view

relates to the architecture reconstruction method as discussed by Anquetil et al. [2]. This is explained in Section 4.4.1. The second example describes the Art method as discussed by Fiutem et al. [11]. This is explained in Section 4.4.2

4.4.1 Recovering software architecture from the names of source files

Table 4.4 shows the features for the architecture reconstruction method of Anquetil et al. [2]. As shown in the table the goal of this method is to identify the nonconformance between the intended architecture and implemented architecture. Only the code source artifact is used in the architecture extraction step. This method adapts a clustering and partitioning method and does not generate any models. In the architecture analysis step the component & connector view of the architecture is created using the extracted information from the architecture extraction step.

Fig. 4.6 shows the concrete business process model for this SAR method that is derived from the generic business process model of Fig. 4.4. The first step is to state the problem and then select which information will be extracted from the architecture.

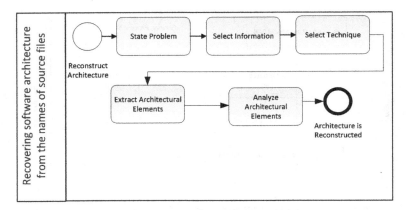

Figure 4.6 Business process model for recovering software architecture from the names of source files.

Table 4.5 Feature selection for ART: An architectural reverse engineering environment.

Feature	Selected features of the SAR method
Goal	Architecture nonconformity identification
Source artifacts	Code
Adopted method	Data mining
Process steps	Preprocessing
	Architecture extraction
	Architecture analysis
	Architecture Visualization
Architecture model	Graph
Architecture views	Module view, component & connector view

Subsequently, the reconstruction technique to be applied is selected and the steps so far actually designed the SAR method. Then the architectural elements are extracted and analyzed for reconstructing the architecture.

4.4.2 ART: an architectural reverse engineering environment

Table 4.5 shows the features for the architecture reconstruction method of Fiutem et al. [11]. The goal of this SAR method is to identify nonconformity between the architecture and code. Code of the system is used as source artifact and data mining techniques are adopted in this method. Process steps consists of preprocessing, architecture analysis, architecture extraction, and architecture visualization. The graph-based model is used for representing the architecture throughout the process. The module and component & connectors views of the architecture are reconstructed. The method proposes a nonsuggestive architectural recovery tool based on system, module, task, and data architecture

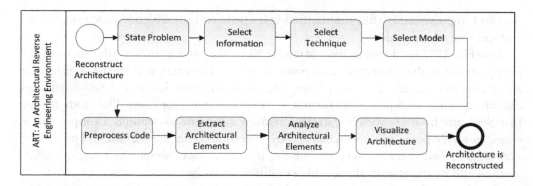

Figure 4.7 Business process model for ART: An architectural reverse engineering environment.

views for software architecture. Lexical parsers were used to model the architecture from the code into abstract syntax trees (ASTs). The user constructs a file before the architecture extraction phase. After the architecture elements are analyzed they form a graph which is visualized through a user interface.

Fig. 4.7 shows the business process model for this study. In this method, first the problem is defined. Information, technique, and model are selected according to the defined problem. Preprocessing is applied on the code and architectural elements are extracted and analyzed, different from the previous method architecture visualized in this method.

4.5. Related work

In the literature we have found similar studies but none of the studies provides a feature model and generic business process model for software architecture reconstruction. Duenas et al. [9] present experiences with architectural recovery tools. In this study, the architecture reconstruction process is modeled in an iterative manner. Also, authors compare different approaches for tackling architecture construction problems such as statistics, cluster recognition, interface discovery, and model enrichments.

Garcia et al. [13] compare six software architecture recovery techniques and the accuracy of each technique. The objective of the study was to find a suitable and reliable fully automated method for reconstruction. Moreover, Gorton et al. [14] compare the architecture reconstruction as Garcia et al. for just-in-time architecture reconstruction. They construct different architectural views using five tools and output of these tools was compared in terms of accuracy rates on the same commercial system.

Guamán et al. [15] propose a reference process framework for architecture reconstruction in which architecture reconstruction is divided into four main parts, i.e., preprocessing, extraction, analysis, and visualization. In this study the complete framework is presented with a proposed adoption of Machine Learning concepts for recommendation and improvements after the method is trained with source code, documentation, and software architecture.

Koschke [19] summarizes the current state of the art for architecture reconstruction processes with architecture viewpoints in mind. The study divides the architecture reconstruction process into two phases, i.e., reconstruction design and reconstruction execution. These two phases are further decomposed into subphases. This study reveals that SAR methods are mostly carried out from a module viewpoint. Component & connector viewpoints are also addressed by some studies in the literature. The author also suggests that there are very few studies that have documented the complete SAR method and its application along with a cost benefit analysis.

Zahid et al. [33] present a survey study for the solutions and evolutions in the SAR method. The study suggests that knowledge-based clustering methods are most commonly used for automated architecture reconstruction. The studies selected for this survey mainly concentrate on component & connector architecture views. Different clusters are formed from this viewpoint and architecture is modeled accordingly. This survey also reviews the presented solutions and discusses precision levels of the reconstructed architectures with respect to the original architecture. Due to this fact, the authors suggest that none of the proposed solutions have reconstructed architecture precisely.

4.6. Discussion

In this chapter we have adopted a domain analysis process to identify the common and variant features of SAR methods. Although we have carefully performed the domain analysis process, a number of validity threats to the study can be identified. First of all, for the development of the domain model we have not opted for a heavyweight systematic literature review process. However, we adopted a domain analysis process in which we selected a sufficient set of papers that we believed to cover the adopted features. Of course, a more in-depth systematic literature review could have resulted in more papers which could have an impact on the feature model. However, we can also state that the papers that we have selected for our study appeared to exhibit a similar set of common and variant features. This showed the convergence of the domain model and hence increased our confidence of the validity of the feature model.

The family feature model for the SAR methods defines the common and variant features that can be used to describe individual SAR methods. With the feature model we indeed derive many different alternative SAR methods, describe existing SAR methods, and support the identification of novel SAR methods. We have illustrated how we could use the feature model to describe two different SAR methods. Although possible, in this study our goal was not to provide a survey of existing SAR methods. We aim to provide this in our future work.

Similar to the representation of common and variant features we have also explicitly described the reference model for SAR methods. Here again we could identify the common and variant steps of SAR methods. The combination and usage of feature models and process models appeared to be very useful to represent the SAR methods. However, similar to the construction validity threats for the feature model, process models could be perhaps more refined if an in-depth systematic literature review was provided. From our experience the process models also appeared to converge and hence we believe that we have largely captured the currently adopted SAR process activities.

4.7. Conclusion

Software projects often need to evolve or usually do not have a well-described software architecture. For these situations it is worthwhile to extract the information from the existing artifacts to derive the architecture. In the literature several architecture reconstruction methods have been proposed albeit fragmented over different studies. The objective of this study was to provide a more comprehensive view and analysis of these methods. For this we have applied a domain-driven approach in which we applied the domain scoping and domain modeling steps. With the domain scoping process, we have identified 17 relevant primary studies that we analyzed in detail to derive the domain model. The domain model has been presented as a set of common terms of software architecture reconstruction, a generic business process model, and a feature diagram. The domain model represents a family of SAR methods and can be used to derive a concrete SAR method. We have used the domain model to derive two different SAR methods. From the domain analysis process, we conclude that the topic of SAR has been considered for a long time but is still an important area that is in development. The domain model that we have provided can be used to provide insight in the overall methods and pave the way for further research. In our future work we aim to enhance our work on architecture reconstruction and derive a comprehensive SAR method.

Appendix 4.A. Primary studies

Primary study	Reference
Anquetil, N., & Lethbridge, T. C. (1999). Recovering software architecture from the names of source files. Journal of Software Maintenance: Research and Practice, 11(3), 201–221.	[2]
El Boussaidi, G., Belle, A. B., Vaucher, S., & Mili, H. (2012, October). Reconstructing architectural views from legacy systems. In 2012 19th Working Conference on Reverse Engineering (pp. 345–354). IEEE.	[3]
Van Deursen, A., Hofmeister, C., Koschke, R., Moonen, L., & Riva, C. (2004, June). Symphony: View-driven software architecture reconstruction. In Software Architecture, 2004. WICSA 2004. Proceedings. Fourth Working IEEE/IFIP Conference on (pp. 122–132). IEEE.	[6]

(continued on next page)

(continued)

Primary study	Reference
von Detten, M. (2012, October). Archimetrix: A tool for deficiency-aware software architecture reconstruction. In Reverse Engineering (WCRE), 2012 19th Working Conference on (pp. 503–504). IEEE.	[8]
Favre, J. M. (2004, November). Cacophony: Metamodel-driven software architecture reconstruction. In Reverse Engineering, 2004. Proceedings. 11th Working Conference on (pp. 204–213). IEEE.	[10]
Fiutem, R., Antoniol, G., Tonella, P., & Merlo, E. (1999). Art: an architectural reverse engineering environment. Journal of Software Maintenance: Research and Practice, 11(5), 339–364.	[11]
Fontana, F. A., & Zanoni, M. (2011). A tool for design pattern detection and software architecture reconstruction. Information sciences, 181(7), 1306–1324.	[12]
Granchelli, G., Cardarelli, M., Di Francesco, P., Malavolta, I., Iovino, L., & Di Salle, A. (2017, April). Microart: A software architecture recovery tool for maintaining microservice-based systems. In IEEE International Conference on Software Architecture (ICSA).	[16]
Guo, G. Y., Atlee, J. M., & Kazman, R. (1999). A software architecture reconstruction method. In Software Architecture (pp. 15–33). Springer, Boston, MA.	[17]
Maqbool, O., & Babri, H. A. (2007, April). Bayesian learning for software architecture recovery. In Electrical Engineering, 2007. ICEE'07. International Conference on (pp. 1–6). IEEE.	[20]
Risi, M., Scanniello, G., & Tortora, G. (2012). Using fold-in and fold-out in the architecture recovery of software systems. Formal Aspects of Computing, 24(3), 307–330.	[22]
Riva, C. (2002). Architecture reconstruction in practice. In Software Architecture (pp. 159–173). Springer, Boston, MA.	[23]
Roy, B., & Graham, T. N. (2008, September). An iterative framework for software architecture recovery: An experience report. In European Conference on Software Architecture (pp. 210–224). Springer, Berlin, Heidelberg.	[24]
Sartipi, K., & Kontogiannis, K. (2003, September). On modeling software architecture recovery as graph matching. In International Conference on Software Maintenance, 2003. ICSM 2003. Proceedings. (pp. 224–234). IEEE. Maintenance, 2003. ICSM 2003. Proceedings. International Conference on (pp. 224–234). IEEE.	[25]
Şora, I., Glodean, G., & Gligor, M. (2010, May). Software architecture reconstruction: An approach based on combining graph clustering and partitioning. In Computational Cybernetics and Technical Informatics (ICCC-CONTI), 2010 International Joint Conference on (pp. 259–264). IEEE.	[26]
Stoermer, C., Rowe, A., O'Brien, L., & Verhoef, C. (2006). Model-centric software architecture reconstruction. Software: Practice and Experience, 36(4), 333–363.	[28]
Stringfellow, C., Amory, C. D., Potnuri, D., Andrews, A., & Georg, M. (2006). Comparison of software architecture reverse engineering methods. Information and Software Technology, 48(7), 484–497.	[29]

References

[1] S. Angelov, P. Grefen, D. Greefhorst, A classification of software reference architectures: analyzing their success and effectiveness, in: Software Architecture, 2009 & European Conference on Software Architecture. Joint Working IEEE/IFIP Conference on, WICSA/ECSA 2009, IEEE, September 2009, pp. 141–150.

[2] N. Anquetil, T.C. Lethbridge, Recovering software architecture from the names of source files, Journal of Software Maintenance: Research and Practice 11 (3) (1999) 201–221.

[3] G. El Boussaidi, A.B. Belle, S. Vaucher, H. Mili, Reconstructing architectural views from legacy systems, in: 2012 19th Working Conference on Reverse Engineering, IEEE, October 2012, pp. 345–354.

[4] K. Czarnecki, C. Hwan, P. Kim, K.T. Kalleberg, Feature models are views on ontologies, in: Software Product Line Conference, 2006 10th International, IEEE, 2006, pp. 41–51.

[5] E. Demirli, B. Tekinerdogan, Software language engineering of architectural viewpoints, in: Proc. of the 5th European Conference on Software Architecture, ECSA 2011, in: LNCS, vol. 6903, 2011, pp. 336–343.

[6] A. Van Deursen, C. Hofmeister, R. Koschke, L. Moonen, C. Riva, Symphony: view-driven software architecture reconstruction, in: Software Architecture, 2004. WICSA 2004. Proceedings. Fourth Working IEEE/IFIP Conference on, IEEE, June 2004, pp. 122–132.

[7] A. van Deursen, C. Hofmeister, R. Koschke, L. Moonen, C. Riva, Viewpoints in software architecture reconstruction, in: Proceedings. Fourth Working IEEE/IFIP Conference on Software Architecture, WICSA, 2004, pp. 122–132.

[8] M. von Detten, Archimetrix: a tool for deficiency-aware software architecture reconstruction, in: Reverse Engineering (WCRE), 2012 19th Working Conference on, IEEE, October 2012, pp. 503–504.

[9] J.C. Duenas, W.L. de Oliveira, J.A. de la Puente, Architecture recovery for software evolution, in: Software Maintenance and Reengineering, 1998. Proceedings of the Second Euromicro Conference on, IEEE, March 1998, pp. 113–119.

[10] J.M. Favre, Cacophony: metamodel-driven software architecture reconstruction, in: Reverse Engineering, 2004. Proceedings. 11th Working Conference on, IEEE, November 2004, pp. 204–213.

[11] R. Fiutem, G. Antoniol, P. Tonella, E. Merlo, Art: an architectural reverse engineering environment, Journal of Software Maintenance: Research and Practice 11 (5) (1999) 339–364.

[12] F.A. Fontana, M. Zanoni, A tool for design pattern detection and software architecture reconstruction, Information Sciences 181 (7) (2011) 1306–1324.

[13] J. Garcia, I. Ivkovic, N. Medvidovic, A comparative analysis of software architecture recovery techniques, in: Proceedings of the 28th IEEE/ACM International Conference on Automated Software Engineering, IEEE Press, November 2013, pp. 486–496.

[14] I. Gorton, L. Zhu, Tool support for just-in-time architecture reconstruction and evaluation: an experience report, in: Software Engineering, 2005. ICSE 2005. Proceedings. 27th International Conference on, IEEE, May 2005, pp. 514–523.

[15] D. Guamán, J. Pérez, J. Díaz, Towards a (semi)-automatic reference process to support the reverse engineering and reconstruction of software architectures, in: Proceedings of the 12th European Conference on Software Architecture: Companion Proceedings, ACM, September 2018, 8 pp.

[16] G. Granchelli, M. Cardarelli, P. Di Francesco, I. Malavolta, L. Iovino, A. Di Salle, Microart: a software architecture recovery tool for maintaining microservice-based systems, in: IEEE International Conference on Software Architecture (ICSA), April 2017.

[17] G.Y. Guo, J.M. Atlee, R. Kazman, A software architecture reconstruction method, in: Software Architecture, Springer, Boston, MA, 1999, pp. 15–33.

[18] K.C. Kang, S.G. Cohen, J.A. Hess, W.E. Novak, A.S. Peterson, Feature-Oriented Domain Analysis (FODA) Feasibility Study, No. CMU/SEI-90-TR-21, Carnegie-Mellon Univ. Software Engineering Inst., Pittsburgh, PA, 1990.

[19] R. Koschke, Architecture reconstruction, in: Software Engineering, Springer, Berlin, Heidelberg, 2006, pp. 140–173.

[20] O. Maqbool, H.A. Babri, Bayesian learning for software architecture recovery, in: Electrical Engineering, 2007. ICEE'07. International Conference on, IEEE, April 2007, pp. 1–6.

[21] K. Öztürk, B. Tekinerdogan, Feature modeling of software as a service domain to support application architecture design, in: Proc. of the Sixth International Conference on Software Engineering Advances, ICSEA 2011, Barcelona, Spain, October 2011.

[22] M. Risi, G. Scanniello, G. Tortora, Using fold-in and fold-out in the architecture recovery of software systems, Formal Aspects of Computing 24 (3) (2012) 307–330.

[23] C. Riva, Architecture reconstruction in practice, in: Software Architecture, Springer, Boston, MA, 2002, pp. 159–173.

[24] B. Roy, T.N. Graham, An iterative framework for software architecture recovery: an experience report, in: European Conference on Software Architecture, Springer, Berlin, Heidelberg, September 2008, pp. 210–224.

[25] K. Sartipi, K. Kontogiannis, On modeling software architecture recovery as graph matching, in: Software Maintenance, 2003. ICSM 2003. Proceedings. International Conference on, IEEE, September 2003, pp. 224–234.

[26] I. Şora, G. Glodean, M. Gligor, Software architecture reconstruction: an approach based on combining graph clustering and partitioning, in: Computational Cybernetics and Technical Informatics (ICCC-CONTI), 2010 International Joint Conference on, IEEE, May 2010, pp. 259–264.

[27] H. Sozer, B. Tekinerdogan, Introducing recovery style for modeling and analyzing system recovery, in: Proc. of Seventh Working IEEE/IFIP Conference on Software Architecture, 2008, pp. 167–176.

[28] C. Stoermer, A. Rowe, L. O'Brien, C. Verhoef, Model-centric software architecture reconstruction, Software, Practice & Experience 36 (4) (2006) 333–363.

[29] C. Stringfellow, C.D. Amory, D. Potnuri, A. Andrews, M. Georg, Comparison of software architecture reverse engineering methods, Information and Software Technology 48 (7) (2006) 484–497.

[30] B. Tekinerdogan, Software architecture, in: T. Gonzalez, J.L. Díaz-Herrera (Eds.), Computer Science Handbook, Volume I: Computer Science and Software Engineering, second edition, Taylor and Francis, 2014.

[31] B. Tekinerdogan, M. Akşit, Classifying and evaluating architecture design methods, in: M. Aksit (Ed.), Software Architectures and Component Technology: The State of the Art in Research and Practice, Kluwer Academic Publishers, Boston, 2001, pp. 3–27.

[32] B. Tekinerdogan, K. Öztürk, Feature-driven design of SaaS architectures, in: Software Engineering Frameworks for the Cloud Computing Paradigm, Springer, London, 2013, pp. 189–212.

[33] M. Zahid, Z. Mehmmod, I. Inayat, Evolution in software architecture recovery techniques a survey, in: Emerging Technologies (ICET), 2017 13th International Conference on, IEEE, December 2017, pp. 1–6.

PART 2

Methods and tools

CHAPTER 5

Monitoring model analytics over large repositories with Hawk and MEASURE

Konstantinos Barmpis[a], Antonio García-Domínguez[b], Alessandra Bagnato[c], Antonin Abherve[c]
[a]University of York, York, United Kingdom
[b]Aston University, Birmingham, United Kingdom
[c]Softeam, Paris, France[d]

Contents

[d] https://www.softeamgroup.fr/en/.

Model Management and Analytics for Large Scale Systems
https://doi.org/10.1016/B978-0-12-816649-9.00014-4

5.1. Introduction

The trend of increasing system complexity and growing global competition in the software industry is pushing companies to meet increasing demands, with small to no margin for compromising quality or time-to-market. One way to achieve this is by using model-driven engineering (MDE), which allows developers to specify systems in a domain closer to the problem and offers the automation of tasks, hence reducing the risk of defects and simplifying reuse. Sometimes, models will replace code entirely, but in other cases they will act as a first step or as an additional resource during the process. Hutchinson et al. [1] reported in 2011 three user experiences of MDE, demonstrating the benefits that this practice can have in improving communication and organizing knowledge.

Including models into the software development process presents a challenge, however. While there is ample work on tracking the growth in size, complexity, and various quality attributes of the code in scalable code metric repositories (e.g., SonarQube [2]), there has not been much work on doing the same with models. In the MDE world, models tend to be complex graphs of domain-specific entities, and therefore the metrics must be domain-specific as well, e.g., ArchiMate models will require ArchiMate metrics. Additionally, if the software development process involves working both on models and code, it may be necessary to relate the metrics of the models with those of the code.

This presents two challenges for such a system: it must be able to compute metrics for potentially very large and complex models, and it must be able to integrate and visualize together metrics from the various artifacts of the products being developed. In this chapter we will present the use of the Hawk model indexing framework [3] as part of the MEASURE[1] platform. Hawk handles the first challenge, and MEASURE handles the second one.

Hawk is a model indexer that watches over collections of model fragments and turns them into a read–only graph that can be easily and efficiently queried. Hawk provides a convenient query language, and can be used as a Java library, an Eclipse plugin, or a web service. The MEASURE platform allows for the creation and visualization of software metrics. By using Hawk as its source, MEASURE can efficiently analyze very large models, in the order of millions of model elements, allowing its use by software-intensive companies such as Bitdefender, Ericsson, Naval Group (formerly DCNS), or Montimage.

This work contributes to the field of software metrics by discussing a solution aimed at tackling both metrics integration and scalability. MEASURE offers the capability to have metrics over the entirety of a system's life cycle: from the architecture and design stages, keeping metrics over the design models used to guide the construction of a system; to system engineering and development, keeping metrics on the code as it is being developed; to system deployment, monitoring the system as it is running in the wild. Hawk, on the other hand, allows tools like MEASURE to scale to large collections of models by introducing a model indexer for offering efficient queries on such models, managed by teams of developers over widely used version control systems.

The remainder of the chapter is structured as follows. Section 5.2 further expands upon the need for software metrics for the timely creation of quality systems as well as the need for efficient model querying, when models are used as the data for such metrics. Section 5.3 introduces the technologies discussed in this chapter, i.e., the MEASURE platform and the Hawk model indexer. Section 5.4 presents the integration of these two tools by focusing on the technologies used to achieve this, and the capabilities this enables. The technical details in Sections 5.3 and 5.4 are further illustrated in Appendix 5.A, which describes an optional running example. Section 5.5 presents a thorough evaluation of the resulting system by using a real-world, evolving model from an industrial user of the MEASURE platform. Section 5.6 provides a short summary of other academic and industrial works in the field of software metrics. Section 5.7 wraps up the chapter by concluding our findings and presenting possible lines of future work for this project.

[1] http://measure.softeam-rd.eu/.

5.2. Motivation

This section will motivate the need for software measurement metrics and model indexing tools for today's software engineering projects, and define an architecture for integrating code-based and model-based metrics in a scalable and efficient manner.

5.2.1 The need for software metrics

Global competition in the software industry and the growing complexity of the systems being developed increase market pressure for new features with lower costs and shorter time-to-market. The challenge is to meet the demand while increasing or at least not sacrificing quality, and so far the solution has been to increase the use of automation in software engineering. After automating simple activities and removing manually repeated actions, it is becoming difficult to cost-effectively advance the level of automation. Automated software engineering also reduces the "gut feeling" of developers, making it more difficult to estimate and maintain the quality of the system being developed.

With this increase in use of automation and the advancement of tools and methods in software engineering, we also have to be able to measure the current performance with sufficient detail and automation, in order to improve. Currently such measurements are mainly used to support decision making, e.g., to answer questions such as "Is our product quality high enough for external release?" [4] and "Have we tested the system well enough?" [5], and the decisions are only based on a few measured aspects of the process or product. Also, these imprecise definitions of measurements have to be redefined and agreed upon again and again at the beginning of each project. As a result, the outputs of projects are often not comparable.

As such, traditional approaches for metrics and evaluation methods are not sufficient anymore; existing metrics have to be redefined and new metrics should be created based on multiple measures, to increase the reliability of the values obtained. Also, with the increasing use of model-driven engineering, a core metamodel (language) for measurement of software engineering is becoming more and more required. Due to the short cycles of iterative processes any feedback has to come quickly, so measurements have to be automated and continuous. Finally, with massive amounts of information to go through, the results have to be visualized at a suitable level of abstraction, which may be different for the various stakeholders involved.

Most of the automatically measurable software quality metrics are currently based on code-centric metrics, such as cyclomatic complexity, number of comment lines, or number of duplicated blocks [6]. Beyond code, software quality measurements should also take into account how requirements are specified and managed, and how the high-level models grow in complexity and size. Automating the measurement of the quality of these higher-level artifacts should provide useful information for assessing other quality requirements, such as maintainability, reusability, testability, or security [7].

Model metrics have not received as much attention as code metrics, and due to the domain-specific nature of these metrics, they are rarely reusable across modeling languages. A 2006 paper by Berenbach [8] showed examples of requirement model metrics by Siemens, which were used in two ways: to measure project progress (e.g., number of use cases completed) and to catch violations of design rules (e.g., circular dependencies). In 2009, Mohagheghi et al. grouped existing works into their "6C framework" of quality goals: correctness, completeness, consistency, comprehensibility, confinement, and changeability. In 2011 Monperrus et al. identified that creating new measurement software for each modeling language would be too expensive; they proposed a model-driven approach for producing these tools [9], and tested it with a model of 30,400 elements. Monperrus et al. noted in their paper the request of the OMG for proposals of a metrics metamodel (which would eventually result in the Structured Metrics Metamodel, cf. Section 5.3.1.1). In the same year, Bertolino et al. identified the need for a model-driven infrastructure for runtime monitoring of model properties, proposing a Property MetaModel (PMM) to define them [10]. A survey by Giraldo, España, and Pastor in 2014 [11] found that there was still no agreed-upon definition of model quality within MDE, but did note that some of these definitions were based on quantifiable metrics, such as structural complexity of UML diagrams.

In short, this work aims to support taking an experience like Berenbach's in Siemens (making models measurable in terms of progress and presence/absence of issues), at a reduced development cost, and at the current scales and numbers of models within companies. To obtain a general picture of the progress and quality of a software system, it will be necessary to combine metrics from all artifacts produced during its development and maintenance. One important challenge is to extend the available metrics to cover quality requirements, software design documents, modeling activities, and other activities beyond coding.

The first challenge then consists of defining metrics in a systematic way as well as creating supporting tools for measuring modern software engineering activities with respect to efficiency and quality, which can help managers during their decision making process. The second one consists of analyzing measurement results for automatically identifying what and how to improve. The last challenge consists of improving the automation in the measurement process, in order to reduce development time. A highly automated and easy-to-deploy solution that can address these three challenges is novel, as the current state of the art tools do not seem to tackle such issues.

5.2.2 The need for efficient model querying

Models are another type of artifact that can be created and developed over the life cycle of a software system. As such, they need to be tracked through metrics just like the code or the deployment of a system. In collaborative development environments, models often need to be version-controlled and shared among many developers. Over time, this may result in repositories that contain many models of considerable size. In recent years, extracting information from these large repositories has become increasingly difficult. This prompted the research community to tackle scalability in many of these facets, including persistence, querying, and the notations themselves.

One common approach for model versioning is to use file-based version control systems such as Git [12] or SVN [13]. These version control systems are robust, widely used, and orthogonal to modeling tools, the vast majority of which persist models as files. Nevertheless, since these version control systems are unaware of the structure of the models in these files, performing queries on the models requires developers to check them out locally first and read them using the appropriate modeling technology.

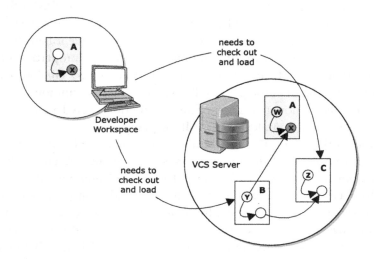

Figure 5.1 Global queries on model fragments stored in a remote VCS. The process is slowed down by the need to transfer all files over the network.

Fig. 5.1 describes a scenario where we have a very large model broken up into three fragments (*A*, *B*, and *C*), and its history is tracked through a central repository hosted on a server in our network. Assuming that the modeler is working on model *A* and needs to know who else is using *A* (which is a *global query*), the only way to answer that question would be to download over the network and load *B* and *C* into memory. This could potentially take a long time with very large models, and it would heavily impact the productivity of the modeler.

Even if all the fragments are maintained locally (for example if a version control system such as Git is used, as shown in Fig. 5.2), the issue of having to load them all into memory in order to answer the query would remain. For large enough models (or collections of model fragments) this would commonly require too many resources to be used in reasonable time, or at all.

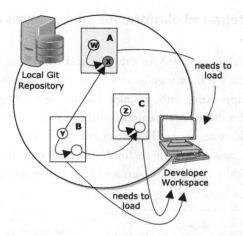

Figure 5.2 Global queries on model fragments stored in a local VCS. Network costs are avoided, but the collection of model fragments may not fit into memory.

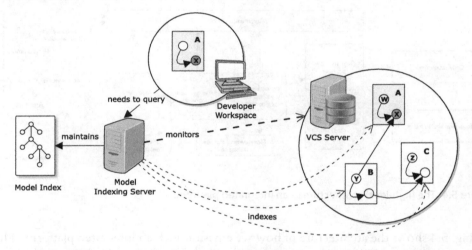

Figure 5.3 Global queries on model fragments stored in a remote VCS, using a model indexing system, adapted from [3].

In order to alleviate some of these challenges, a *model indexer* can be introduced, as shown in Fig. 5.3. This is an approach that enables efficient global model element-level queries on collections of models stored in file-based version control systems. To achieve this, a separate system is introduced (a *model indexing server*), which monitors file-based repositories and maintains a fine-grained read-only representation (graph) of models of interest, which is amenable to model element-level querying. Hence the developer queries the model indexer directly and does not have to check out or load any additional files.

5.2.3 Towards an integrated platform for joint analysis of code and model metrics

In order to efficiently tackle the need for creating and managing metrics for the entire software development life cycle, a way to cover these two needs without introducing hard constraints (like supporting only a single modeling technology) is required. To achieve this, a platform for creating, managing, and running metrics on software artifacts (whether code or models) is necessary. This platform should offer an extensible approach for adding new metrics, new modeling technologies, and new front-end visualizations, in order to support a wide range of domains and levels of expertise.

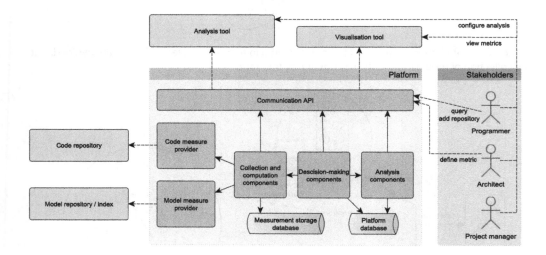

Figure 5.4 The high-level architecture of an integrated platform.

Fig. 5.4 shows the architecture of how we envision such an integrated platform. The platform will contain components for connecting to both code and model repositories, in order to obtain the latest versions of the data available there. These source data will then be used to calculate measurements that will be stored in the system. These measures are then available for both analysis and visualization components to use. Users of the system will be able to configure it for their needs by adding their repositories and defining their measures; they can then connect their visualization and analysis tools to the system and view the results/calculations on those tools.

In the next sections (Sections 5.3 and 5.4), we detail how such an architecture is realized in the MEASURE project and how it integrates with a model indexer in order to offer a highly performant approach for updating its metrics in an incremental manner.

5.3. Background

This section will introduce the main concepts governing the Hawk Model Indexer and the MEASURE platform. The interested reader is able to refer to Appendix 5.A for an in-depth look at a running example demonstrating how these components work together in practice.

5.3.1 The MEASURE platform

The main goal of the ITEA 3 Measuring Software Engineering (MEASURE) Project is to measure the quality and efficiency of software, as well as to reduce the costs and time-to-market of software engineering in Europe, by implementing a comprehensive set of tools for automated and continuous measurement. Therefore, this project aims at providing a toolset for future projects to properly measure their software quality and efficiency. More importantly, it opens a new field for innovation: the advanced analytics of the measurement data enabled by the project. To reach this ambitious goal, the project will iteratively and incrementally:

1. Define novel and improved metrics whilst developing methods and tools for automated, precise, and unbiased measurement of software engineering activities and artifacts. Ping methods and tools for automated, precise, and unbiased measurement of software engineering activities and artifacts.

2. Develop methods and tools for analyzing the (big) data produced by the continuous measurement to enable continuous improvement of performance.

3. Validate the developed metrics, approaches, and measurement tools by integrating them into software development environments and processes of the project's industrial case studies, and iteratively improve them, based on the feedback gathered from the industrial partners and other industrial actors.

4. Validate the developed tools by analyzing the data gathered from the industrial partners and measuring the impact of the improvements suggested by the analysis tools. A practical example of a measurement-based suggestion could be pointing out an area of source code not covered by automated test suite and generating new targeted test automation scripts based on manual test cases recorded during continuous measurement.

5. Support management of decision making by visualizing the results of continuous measurement at targeted level of abstraction, i.e., providing different visualization or even completely different metrics for developers and managers.

5.3.1.1 Measures and metrics

In the MEASURE platform, measures are defined in the Structured Metrics Metamodel (SMM) format, an Object Management Group (OMG) standard [14]. SMM defines "measure" as "a method assigning comparable numerical or symbolic values to entities in order to characterize an attribute of the entities," and "measurement" as "a numerical or symbolic value assigned to an entity by a measure." Unfortunately, SMM does not provide a definition for "metric." The distinction between "metric" and "measure" has been a longstanding discussion: some industrial tools consider them to be equivalent (e.g., SonarQube [2]), while the NIST Software Assurance Metrics and Tool Evaluation (SAMATE) project mentions that some authors treat "metric" as the more high-level concept (e.g., software complexity) and "measure" as the concrete and objective concept (e.g., lines of code), and some treat them the other way around [15].

Within the MEASURE platform, the Collect and Compute Components provide services to register measures defined in specific executable formats (e.g., Java JARs) derived from SMM, by using the Communication API. Registered measures are stored in the Measure Catalogue.[2] Once registered, the catalogue allows the platform to access the metadata with the unique identifier of the measure, its type (e.g., Direct Measures, Collective Measures, or Binary Measures), how to access its implementation, the format of the data produced (Unit), and a list of configuration points provided by the measure (Scope).

A measure represents a generic data collection algorithm that has to be instantiated and configured to be applied in a specific context. To be executed, the platform instantiates a measure with specific values for the configuration variables defined in its metadata file. This usually includes an identifier for the measure instance, the information related to the measurement collection schedule (i.e., when and in what interval a measurement is collected), the information required to identify the measured system and for derived measures, any information related to their required inputs and produced outputs.

The SMM standard defines two main types of measures: the direct measures which are taken from the real world, and calculated measures (Collective and Binary Measures) which are derived from direct measures. The SMM engine in the MEASURE platform allows measures to be triggered both manually and periodically over time. Once collected or calculated, the engine will store the measurements in its back-end storage. The MEASURE platform uses two databases for back-end storage: a relational database (MySQL) for its configuration (e.g., measure instances) and an Elasticsearch document index for the measurements (refer to Section 5.4.3.1 for more details).

5.3.2 Web-based interface

The SMM metrics and their measurements are exposed through a web interface. The overall goal of this interface is to provide an easy-to-use environment to assist decision making across all the stages of the life cycle of a software project.

The web interface allows users to register and manage measures, test measures (either manually or by running scheduled measures), create and manage visualizations driven

---

[2] A collection of predefined measures is available at github.com/ITEA3-Measure/Measures.

by the collected measurements, organize the visualizations by project, phase, and dashboard, and combine and relate the measurements produced by the analysis tools.

The MEASURE platform can provide graphical representations of the measurements as various plots and reports, organizing them into dashboards. Each dashboard presents a curated set of measures as graphics, tables, or specific indicators. The MEASURE project uses Kibana, an existing open source dashboard and visualization solution (refer to Section 5.4.3.2 for more details).

5.3.3 The Hawk model indexer

Hawk is a *model indexer*: it keeps an up-to-date read-only view of model collections (a *model index*) and facilitates efficient querying on them. Hawk monitors local or remote data locations (such as version control systems, local file directories, or remote URLs) and periodically synchronizes with any model files of interest in these locations.

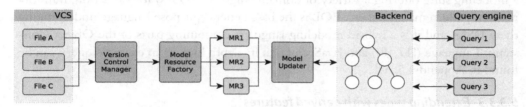

Figure 5.5 Component architecture of Hawk.

5.3.3.1 Components

Fig. 5.5 shows how the various components of Hawk interact in order to maintain such an up-to-date model index.

- **Version Control Managers**: specific for each system, these components compute the set of changed files (added, removed, or updated) with respect to the revision last indexed by Hawk (and the current latest version of the files in the system/repository).
- **Model Resource Factories**: these components provide parsers for specific model persistence formats, such as EMF models persisted in XMI or Modelio models in XML. These parsers take as input the contents of a file (provided by the version control managers) and produce as output a uniform in-memory representation ("resource") of the model.
- **Model Updater**: this component receives a resource created by the appropriate model parser and inserts/updates it into the database, through back-end-specific drivers. The structure of the store assumes that the back-end provides a mechanism for rapidly accessing specific elements using a key. This is the case for many popular

stores such as MySQL, MongoDB, or Neo4j, which include their own embedded database indices.

- **Back-end**: this component provides a labeled-property graph API on top of an existing database technology. The latest version of Hawk at the time of writing (1.2.0-rc4) supports Neo4j 2.0.5, OrientDB 2.2.30, and Greycat 11.
- **Query Engine**: this component provides a bridge between Hawk and model management tools that need to query it. Queries can be performed on both local and remote Hawk model indexers.

5.3.3.2 Model querying

Using Hawk, queries can be performed quickly and efficiently, without having to retrieve or load any of the actual model files. Such queries can be written in the Epsilon Object Language (EOL), the default query engine shipped with Hawk. Epsilon [16] is a modeling suite offering a variety of domain languages (DSLs) for validation, transformation, pattern matching, etc. EOL is the base general-purpose language underpinning these DSLs, and it is a hybrid modeling language combining parts of the Object Constraint Language ([17,18]) and JavaScript. An in-depth look at an example query can be found in Appendix 5.A.1.

5.3.3.3 Extending types with derived features

Hawk offers the capability of adding derived features [19]. They extend a type with additional features whose values are precomputed from an expression in a certain language, and then updated in an incremental manner as the models change. Derived features can noticeably speed up repeated queries with common subexpressions, as their result can be provided immediately. The granularity of such features creates a fine balance between gain in execution time and reusability of the feature; a feature using a very complex expression will likely result in a large execution time reduction, but will only be of use when that specific complex expression needs to be calculated, whereas one using a simpler expression will result in a more modest execution time reduction, but will be likely reusable in similar queries. An in-depth look at an example use of derived features can be found in Appendix 5.A.2.

5.3.3.4 Querying-as-a-service

In many scenarios, multiple people may need to query the same collection of models simultaneously, for instance when having a shared model of the requirements and design of the system and multiple stakeholders are wishing to ask things about it. It would be far more efficient to index these models once from a central location and then run queries from these indexed models. Users may also want to use queries as views on a larger model, which they may want to provide as inputs to their other tools [20].

Hawk has server versions (for Linux, Windows, and Mac) that can do this. The server exposes all functionality through an efficient web service API over HTTP, HTTPS, and/or TCP. The API is implemented with the Apache Thrift framework,[3] making it possible to choose multiple message encodings that balance bandwidth, CPU load, and language compatibility. Hawk also offers model views through the creation of pre-defined queries which return the relevant (sub)model of the original, allowing for the use of such results in further model management operations such as code generation or model validation. They can also be used for lazy loading of very large models over the network, which can greatly reduce the time needed to start browsing a model indexed by Hawk. More details on the technical aspects can be found in Appendix 5.A.3.

5.4. Monitoring model analytics over large repositories with Hawk and MEASURE

Hawk has been integrated into the MEASURE platform in order to provide it with an efficient way to retrieve the data it requires for producing its metrics. This section will focus on the changes that were necessary in order to achieve this, the challenges faced, and the functionality this integration enabled. It will also provide a basic background of any new technologies used for this purpose, for clarification purposes.

5.4.1 Integration of Hawk and MEASURE

An initial proof-of-concept for integrating Hawk and MEASURE was discussed at [21], presenting the essential steps that were performed: running Hawk as a standalone service from the MEASURE platform, allowing automatic provisioning of Hawk servers in the cloud, and adding a component to Hawk that would allow it to be deployed as a measure in the MEASURE platform, reading an SMM-based configuration and relaying the results back to the platform.

In order to run the platform, the platform administrator runs a Hawk instance to monitor the models and then installs Hawk-based metric(s) described in SMM into MEASURE (more information about SMM can be found in Section 5.3.1.1). MEASURE background jobs query Hawk periodically and insert the results into Elasticsearch (see Section 5.4.3.1). The administrator creates Kibana visualizations from the Elasticsearch data (see Section 5.4.3.2). Finally, the administrator adds the visualizations to the MEASURE dashboard.

This preliminary work, while enabling a basic integration, revealed that Hawk was not suited for model monitoring as it only held the latest version of the model in its index. As such, investigation into using a new back-end for Hawk was performed, concluding that the use of a temporal store would be beneficial in this regard.

---

[3] https://thrift.apache.org/.

Table 5.1 Implemented time-aware primitives in the Hawk EOL dialect, for a model element or type x.

Operation	Syntax
All versions, from newest to oldest	`x.versions`
Versions within a range	`x.getVersionsBetween(from, to)`
Versions from a timepoint (included)	`x.getVersionsFrom(from)`
Versions up to a timepoint (included)	`x.getVersionsUpTo(from)`
Earliest/latest version	`x.earliest, x.latest`
Next/previous version	`x.next, x.prev/x.previous`
Version timepoint	`x.time`

5.4.2 Temporal databases

The MEASURE platform produces metrics based on the history of the project it is monitoring, for example for each commit made in a version control system. In order for Hawk to have the capability of indexing the entire evolution of a model and answering historical queries, a new type of back-end was necessary: a temporal graph database, where nodes and edges evolve over time and disk space is saved by only storing the changes from one version to the next. Recent versions of Hawk include such a back-end, based on Greycat[4] graph databases. Greycat implements the temporal graph concept presented in [22], which is a versioned graph with copy-on-write storage.

Two new query engines were created to expose the time awareness of the new back-end, building on top of the existing EOL query engine. The first one, the "time-aware" query engine, extended model instances and types with new properties and operations to navigate their versions, as shown in Table 5.1. Model instances are considered to have a life span from their creation to their destruction, and a new version is created every time a reference or property is changed. Types are "immortal," but they do produce new versions every time an instance is created or deleted.

An initial version of the time-aware query engine was first presented in [23], which was dedicated to queries that required comparing adjacent versions. However, the MEASURE platform is more interested in evaluating a metric on every version of a model. To simplify this task, a new "timeline" query engine was implemented on top of the time-aware query engine for this paper. The query engine takes the EOL query and reruns it across the history of the model, producing a sequence of (timestamp, value) pairs. This allows users to write queries in a time-agnostic way and still reap the benefits of the time-aware back-end.

With these two new query engines, it is possible to perform temporal queries in Hawk. An in-depth look at the Hawk temporal database back-end can be found in Appendix 5.A.4.

[4] https://greycat.ai/.

5.4.3 Software metric visualization

With the ability to store all revisions of a model, Hawk can now provide the necessary information to MEASURE for producing metrics. This section will introduce the technologies used by the MEASURE platform for obtaining, calculating, and visualizing metrics as well as present an example of how this is done with Hawk.

5.4.3.1 Elasticsearch

MEASURE will gather the necessary data from its measure instances (in this case Hawk) and calculate any relevant metrics. These metrics will be stored in an Elasticsearch instance in order to be amenable to visualization.

Elasticsearch[5] is a web service built on top of Lucene,[6] one of, if not the, most popular open source enterprise search engines to date. Alongside Logstash,[7] a data collection and log parsing engine, and Kibana, an analytics and visualization platform (discussed below, in Section 5.4.3.2), it forms an integrated solution referred to as the "Elastic Stack" (formerly the ELK stack).

Lucene was designed to add clear value to businesses in the most practical of ways, such as parsing and visualizing old log files that had suddenly become of interest. Since then, Elasticsearch has grown to provide scalable, near real-time search and to be distributed, in the sense that search indices are divided onto shards hosted on separate nodes with a coordinator used to delegate operations to the correct shard(s). Besides all Lucene features made available through the Java API, Elasticsearch also offers gateways for long-term storage of indices in case of server failures and real-time GET requests, ensuring compatibility with NoSQL datastores.

5.4.3.2 Kibana

Since MEASURE metrics are stored in Elasticsearch, they can be easily visualized in Kibana,[8] an open source tool that allows conducting analysis and building visualizations of Elasticsearch data through a graphical user interface. Kibana does not require delving into technical details of queries (although it is still possible in order to obtain more fine-grained control over the output), and hence data visualizations can become more accessible to regular users. The additional business value of Kibana shines with several visualizations combined on a dashboard, enabling the explorations of multiple views (coordinated with one another) at once. An example of how Hawk data are stored in Elasticsearch can be found in Appendix 5.A.5.

---

[5] https://www.elastic.co/products/elasticsearch.
[6] http://lucene.apache.org/.
[7] https://www.elastic.co/products/logstash.
[8] https://www.elastic.co/products/kibana.

5.5. Case study: the DataBio models

This section presents a use-case of running the MEASURE platform using Hawk as its source of data. Since the majority of the industrial partners using Hawk work with sensitive data, a compromise had to be made between using a smaller model of real-world data available to the MEASURE partners from the ITEA3 consortium, or using a larger synthetic model. A decision was made to use the real-world model as it will provide similar insights to a larger synthetic model with regards to this evaluation.

Specifically, a set of real-world ArchiMate models that had been developed through the SOFTEAM Constellation [24] collaborative platform was selected. ArchiMate is an Open Group specification [25] which provides an independent modeling language for Enterprise Architecture that is supported by different tool vendors and consulting firms. The ArchiMate Specification provides instruments to enable Enterprise Architects to describe, analyze, and visualize the relationships among business domains in an unambiguous way.

The selected set of ArchiMate models represent the enterprise architecture of the DataBio technology platform.[9] DataBio is an H2020 European research project focused on the creation of a Data-Driven Bioeconomy, focusing on empowering the production of natural raw materials while taking into account concerns about responsible production and consumption, by leveraging information technologies and Big Data approaches.

The ArchiMate models of the chosen case study were developed using the BA Enterprise Architect edition of the Modelio modeling tool [26]. Modelio is an open source tool which supports the main standards: UML, BPMN, SysML, and so on. The BA Enterprise Architect edition focuses on enterprise architecture modeling and supports the TOGAF, ArchiMate, BPMN, and UML standards.

SOFTEAM Constellation servers expose Modelio model repositories as Subversion version control systems. The models themselves are divided into projects and then into fragments. The fragments are further broken up into objects, which are stored as plain XML files. Objects refer to each other through their globally unique identifiers, which are long strings of randomly allocated letters and numbers. This approach is a good fit for Hawk, which is intended to provide efficient global querying capabilities for models that have been fragmented into many files that are tracked through a remote version control system. Hawk was also used within the MEASURE project case studies to index the models of Naval Group and SOFTEAM, with millions of elements, using Modelio Constellation servers.

5.5.1 Research questions

This case study aimed to provide some evidence on two key research questions:

[9] http://archive.is/MVaYO.

- **RQ1**: Can Hawk produce model analytics from the full history of "in the wild" repositories?

  So far, Hawk had indexed the latest version of various synthetic and real-world models, as shown in prior work on integration with commercial modeling tools [27] and on stress testing of remote model querying APIs [28]. A recent paper had indexed a small repository with over 800 versions of a single synthetic model file [23] with the new Greycat back-end, but the DataBio models were much larger and had been collaboratively developed in an organic manner outside the control of the Hawk developers. It was necessary to check if Hawk would handle this case gracefully, or if it would not scale to the longer history and more complex models present in the repository.

- **RQ2**: Is the solution flexible enough to provide metrics richer than simpler class counts?

  The size of the model or the number of instances of a particular class are simple metrics which can quickly tell us how the model is growing in size, but they are not very nuanced. In this research question, we will consider the ease of creating more advanced metrics which can evaluate the evolving complexity of the model, or may highlight the number of design rule violations that may occur at certain points.

5.5.2 Method

The study was performed on a Lenovo Thinkpad X1 Carbon laptop with an i7-6600U CPU, using an SSD and running Oracle Java 8u102 over an Ubuntu Linux 18.04 OS, using the standard 4.15.0-38-generic kernel. Java was given a starting and maximum amount of 8 GB of heap (-Xms8g -Xmx8g) to speed up the process.

The study followed the following steps:

1. SOFTEAM provided the authors with read-only access to the two DataBio projects in their Constellation server. The first DataBio project went through 936 revisions, while the second DataBio project went through 214 revisions. A checkout of the latest version of the first project is 28MB, containing 6636 model elements,[10] while a checkout of the second project is 16MB, with 3789 model elements. These are reasonable projects for an enterprise architecture model, but they are not "huge" models as one would see from other domains (e.g., reverse engineering from code).

2. The two DataBio projects were indexed by Hawk into a Greycat temporal graph, using the time-aware indexer and updater components. Greycat was used through its own RocksDB back-end, with Snappy compression. The Greycat temporal graph grew to 4 GB on disk: interestingly, an XZ-compressed backup of this temporal graph only took up 229 MB. It appears that the internal compression of

---

[10] As measured with Hawk v1.2.0-rc3-53-g618b2341, with query return Model.allContents.size;.

the RocksDB back-end focuses on speed rather than compression, but the specific tradeoff may need to be reexamined: this is configurable to some degree.

3. A set of queries was selected for the ArchiMate models. At the time of writing, the official MEASURE measure repository lacked specific queries for ArchiMate models, but it did include two queries that could be reused: one counting the number of requirements created within the Modelio Analyst metamodel, and one counting the number of Package instances in the model.

It was easy to define queries to count the total number of elements in the model and to count the total number of ArchiMate requirements. However, these would not constitute answers for RQ2. To prove that Hawk could implement richer metrics, some of the lightweight "element criticality" ArchiMate metric proposals of Singh and van Sinderen [29] were implemented in EOL. The original metrics were intended to highlight the most critical elements, but MEASURE was expecting a single number evaluating the entire model, so the average criticality was used as a proxy for the complexity of the model.

Two of the metrics were adapted: relationship score (which gives points to each element depending on the types of their outgoing relationships to other elements) and importance score (which multiplies the relationship score with a modifier based on the type of element being evaluated). Appendix 5.B shows a snippet of the implementation of the average relationship score metric and explains it in detail.

4. The queries were bundled into an SMM-based measure: a standard Java library built with Maven, bundling the various EOL query files as resources. The measure connects to Hawk through the Thrift API and invokes the query selected in the configuration, returning the desired results. The measure itself is rather simple, requiring only 218 lines of code for its implementation.

5. The Hawk SMM measure was deployed into the Measure Catalog of a locally running MEASURE platform instance (as seen in Fig. 5.6), and six measure instances were created to expose each of the EOL queries (see Fig. 5.7).

6. The Hawk-based SMM measures were executed, providing the data points at the time of writing. However, this would produce visualizations with only one data point: at the time of writing, the MEASURE platform did not have direct support for requesting metrics about older versions.

7. To solve the lack of historical data, it was decided to use the temporal capabilities of Hawk. A new "timeline" query engine was added to Hawk, which would run each EOL query repeatedly over all the identified versions of the repository and produce a list of (timestamp, query result) pairs. All queries were run through this timeline query engine, and the results were saved into text files.

8. Once the full timeline queries had been run, a separate Java program was used to read the (timestamp, query result) pairs and feed those back into the Elasticsearch

Figure 5.6 MEASURE platform: registration of Hawk SMM measure into catalog.

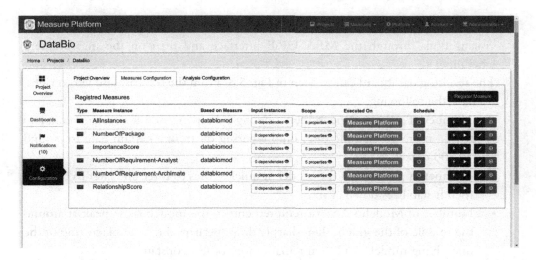

Figure 5.7 MEASURE platform: creation of Hawk SMM measure instance.

database used by MEASURE through the Bulk API.[11] This was a fast process: inserting the results for each of the queries never took longer than 1 second in the test environment.

9. At this point, the measures were deployed in the MEASURE platform, and both current and historic data were available. The last step was to create the Kibana-based

---

[11] https://www.elastic.co/guide/en/elasticsearch/client/java-api/5.4/java-docs-bulk.html.

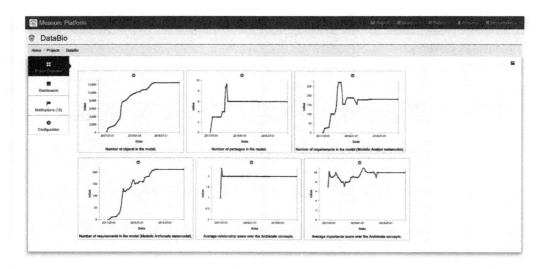

Figure 5.8 MEASURE platform with DataBio metrics.

visualizations through the MEASURE interface and turn on the measurement scheduling.

The resulting dashboard can be seen in Fig. 5.8. From the top row and left to right, we have the following:

- Number of objects in the model: showing a sharp increase at a certain point and then a more gradual increase, until tapering off from a certain date to around 12,500 objects.
- Number of packages in the model: it quickly rises to three and later to ten, and then it stabilizes down at six.
- Number of Modelio Analyst requirements in the model: these peak at around the middle of the graph, then sharply drop (perhaps due to a refactoring of the underlying models) and then remain more or less constant.
- Number of ArchiMate requirements: unlike the Modelio ones, these increase at roughly the same rate as the number of objects in the model.
- Average relationship score: surprisingly, this had some changes at the beginning, and later stabilized very consistently at 2.3. This may point to a reasonably consistent mix of relationship types throughout the project.
- Average importance score: this did see a considerable level of change through-out the project. When compared against the relationship score, this may suggest that while the relationship mix is stable, the mix of ArchiMate types may change slightly throughout the history of the model.

5.5.3 Discussion

In this section, we will answer RQ1 and RQ2 and mention some of the threats for validity of the presented work. The positive results for RQ1 and RQ2 will be discussed first, followed by a commentary of some of the limitations in those positive results.

5.5.3.1 RQ1: scalability with history

Regarding RQ1, Hawk was indeed able to index the full history of the DataBio project and answer queries about the various versions of the models. Hawk is able to reevaluate the same query easily over the history of the project through a new "timeline" query engine, and it can also use an extended dialect of EOL to allow the same query to compare various versions.

However, at the time of writing, the current implementation of Hawk could be improved in terms of space and time. With the current Greycat RocksDB back-end, the full temporal graph took 4 GB of disk space and required several hours to be built. Based on the observations made during the study, there are several simple improvements that could be made to improve this: stronger compression algorithms are available in RocksDB, and proxy resolution (which heavily dominated the indexing times, over fragment insertion and remote file downloading) could be optimized to only consider proxy references from nodes that are alive at the relevant point in time.

5.5.3.2 RQ2: flexibility for complex domain-specific metrics

Regarding RQ2, it was possible to express the ArchiMate metrics of Singh and van Sinderen in EOL. These metrics are more complex and nuanced than the usual "number of instances of Type" examples seen in the literature. After all, EOL is a full-fledged object query and management language, and has been used to implement complex model-to-model transformations long before Hawk was created. The MEASURE repository has a collection of other "rich" EOL-based metrics, such as an "overall class complexity" index, a "class dependency" ratio, or a "model abstractness" index.

Regardless, there are some limitations in the study. It was not possible to verify if the metrics matched the experience of the modelers during the DataBio project: they may have been able to inform us why the model size spiked at a certain point, or why the average relationship score may have remained stable after a certain point in time.

Another limitation in the study is the apparent lack of publicly available quantitative metrics of model quality for the ArchiMate models. There have been other studies about quantitative analysis of ArchiMate enterprise architecture models (such as [30]), but these seem to be mostly focused on process simulation rather than model quality.

5.5.3.3 *Threats to validity*

Due to time and data availability limitations, it was not possible to test the integration of the model-based Hawk metrics with those from other MEASURE data sources, such as SonarQube, which is focused on the code.

This could have been done with another MEASURE component, the Metrics INtelligence Tool (MINT) [31], which can apply Machine Learning algorithms to detect correlations between metrics and make suggestions for an improved metrics collection plan.

It would also have been interesting to check its integration with the QualityGuard metrics monitoring component of the MEASURE platform,[12] which allows for defining thresholds that may trigger notifications.

5.6. Related projects

This section will discuss other projects that enable the extraction and/or collection of metrics from large software projects. Unlike Section 5.2.1, which refers to academic foundational work on software metrics, this section focuses on industrial-strength tools at the level of MEASURE, and their industrial and/or academic applications.

5.6.1 SonarQube

SonarQube [2], [32] is a code quality platform. It offers tools for continuous monitoring and improvement of code as well as software metrics that can be visualized in its dashboards. SonarQube's metrics are obtained using its source-code analyzer component that feeds the relevant data to the application server responsible for visualizing it. It supports various programming languages like Java, C++, and Python, it offers both open source and commercial licenses, with varying capabilities,[13] and it is used by over 100 companies, including Starbucks, Ericsson, and Ocado Technology.[14]

Works like [33] have not only provided insight into the usefulness of software metrics and SonarQube in particular, but also demonstrated how integrating it with other tools (such as the Understand[15] tool) can provide extra capabilities to this tool, in this case adding the ability for SonarCube to obtain analysis results from code rules analyzed by external tools, and visualize them alongside its own. SonarQube has also already been integrated with MEASURE, providing MEASURE with the capability to perform the code-level quality metrics such as the ones summarized above.

---

[12] https://github.com/ITEA3-Measure/QualityGuardAnalysis.
[13] https://www.codacy.com/pricing.
[14] https://www.ocadotechnology.com/.
[15] https://scitools.com/.

5.6.2 OpenHub

OpenHub[16] (formerly Ohloh) is an online platform offering metrics for open source projects. It contains metrics such as popularity and activity of projects, users, and organizations, as well as tools for comparing multiple projects, programming languages, and repositories. OpenHub currently monitors almost half a million projects and over 350 organizations. It has a REST-based API for obtaining its data that can be freely used by third-party tools to create or combine their own metrics. OpenHub covers a variety of system life cycle areas such as development and usage, but does not cover the analysis or deployment stages.

5.6.3 CROSSMINER

CROSSMINER[17] [34] is a European Union research project under the H2020 initiative. It is a collaboration of various academic and industrial partners providing a platform for metrics on open source projects. The main component in CROSSMINER is the knowledge base. This stores a collection of metrics on monitored open source projects that periodically update, keeping their evolution over time. The aim is to offer both the ability to monitor existing projects as well as obtain exploratory information regarding the use of new open source components.

There are three main research areas with respect to metrics that CROSSMINER tackles:

1. *source code analysis tools* extract knowledge from the source code of a collection of open source projects;
2. *natural language analysis tools* extract metrics related to the communication channels and bug tracking systems of open source projects;
3. *system configuration analysis tools* analyze system configurations providing DevOps-level metrics for considered open source projects.

Finally, work on advanced integrated development environments allows developers to benefit from the CROSSMINER knowledge base and analysis tools from their development environment, including alerts and recommendations, to help them improve their productivity. CROSSMINER covers a variety of system life cycle areas, including development, analysis, and usage, with the ability to cover further areas like deployment with the addition of relevant plugins to read and analyze the appropriate data.

Similarly to MEASURE, it offers the ability to add new repositories to be monitored and metrics to be calculated. It also has the ability to monitor noncode-related sources such as bug tracking systems and communication channels (such as forums or mailing lists) and creates a homogeneous store for accessing and analyzing these data.

---

[16] https://www.openhub.net.
[17] https://www.crossminer.org/.

CROSSMINER aims at providing out-of-the-box metrics for various stages of the development life cycle through its source code analysis, natural language processing, and configuration analysis partners. This can allow for its easier use by nonexperts who will use the available drivers and metrics for analyzing their projects. On the other hand, it does not currently have a way to monitor any model-based life cycle artifacts, as new custom drivers for such technologies would have to be implemented and relevant metrics that fit this type of data would have to be defined. Finally, it uses its own Java API for defining metrics (i.e., metric providers), that the users would have to learn and implement, in order to add any new metrics that the platform does not currently support.

5.6.4 Elasticsearch

MEASURE is not the only metrics platform using internally Elasticsearch (see Section 5.4.3.1) for storage and visualization. There is a number of other industrial and academic projects that also use it.

5.6.4.1 Beats

Beats[18] offers a platform for managing a central repository of metrics hosted on Elasticsearch. Metrics related to various tasks like logging, network usage, data audits, or system-level metrics like resource usage or uptime can be provided and stored in Elasticsearch for visualization.

The Beats project implements a standard API and a set of premade, officially supported "data shippers" that send data from the monitored entities back into Elasticsearch. These may be metrics, if using the Metricbeat shipper: these could be levels of resource usage (CPU/RAM), or workload levels on the server. The Heartbeat shipper provides uptime metrics on servers as well. Other shippers are not limited to metrics. For instance, Filebeat ships logs to a centralized location, and Packetbeat ships network packets.

Beats has an important user community, and the project has a process to allow users to share their data shippers [35]. Custom Beats are created in the Go programming language and need to implement the "Beater" Go interface provided by the Beats project. Some examples include Amazon product information, Docker container statistics, and GitHub repository activity.

5.6.4.2 Elastic Machine Learning

Elastic Machine Learning (EML [36]) is an optional extension for the Kibana component of the Elastic technology stack. EML can be told to create and maintain a baseline of the typical behavior of certain metrics, using Machine Learning technologies. This

---

[18] https://www.elastic.co/products/beats.

baseline can be used to detect abnormalities (e.g., sudden spikes in the requests per second for a website, or intrusions in a network) or to forecast future values. This is similar to the *analysis* components in MEASURE, which use the data recorded elsewhere to produce further insights.

5.6.4.3 GrimoireLab

Another open source metrics platform based on the ELK stack is GrimoireLab [37], developed by the Spanish Bitergia software analytics company. GrimoireLab includes components for measuring code size and quality, issue tracking, and community activity, among other aspects. However, it does not have any components for measuring metrics of models, much like CROSSMINER.

GrimoireLab is divided into a number of components. Among others, Perceval retrieves data from various data sources, Arthur coordinates Perceval instances, Sorting Hat correlates heterogeneous user account systems into unified identities, and Kibiter is a custom fork of Kibana "to work on new ideas for metric and data visualization." Bitergia displays on the GrimoireLab website dashboards created for the Samsung Open Source Group and the Cauldron.io project, among others.

5.6.4.4 Academic uses of elasticsearch

There is a plethora of academic work using Elasticsearch and the Elastic stack (formerly ELK stack)[19] for various types of software metrics and monitoring.

CERN seems to be a very heavy user of the Elastic Stack. Hamilton, Gonzalez Berges, Schofield, and Tournier [38] discussed a use case of the Elastic Stack used to run statistics on data from 200 applications in CERN.[20] These applications included a variety of systems like electrical networks, cryogenic system parameters, and magnetic outputs. Filebeat data shippers sent logs to Logstash queues, which would extract the relevant information from the logs and index it into Elasticsearch. Kibana was used to produce visualizations and dashboards from the Elasticsearch indices.

Lassnig, Toler, Vamosi, et al. [39] presented the ATLAS analytics system from CERN, which uses Hadoop, Elasticsearch, and Jupyter as back-ends. This work applied Machine Learning techniques to create models for network statistics that are used to enhance job scheduling and data brokering for the Worldwide Large Hadron Collider Computing Grid.[21] In the current version of EML, some of this work may not have been necessary.

CERN introduced the ELK stack into their software development processes as well, using it to collect code quality information produced by the Lizard tool on the code for their ATLAS project [40]. Their work mentioned several types of metrics, such as lines of code, cyclomatic complexity, or function decision depth.

---

[19] https://www.elastic.co/products.
[20] https://home.cern/.
[21] http://wlcg.web.cern.ch/.

Beyond CERN, Voit, Stankus, Magomedov, and Ivanova [41] looked at full-text search and visualization platforms that could support the deanonymization of unauthenticated users in web applications, while scaling up to "Big Data" levels. After benchmarking several other options (such as Sphinx, Apache Solr, or Xapian) and comparing their features, the authors concluded that Elasticsearch was the best performing tool that had all the features they required.

As a large and active open source project, the metrics of Elasticsearch itself have been studied as well. The authors of [42] analyzed the Experience until Bug Introduction Change (EuBIC), Time to Notify (TTN), and Time to Fix (TTF) maintenance metrics. Unlike prior work that had estimated TTN through an algorithm, this work used manually annotated bug reports and manually located BICs (commits that introduced the defect). TTN could be considered to be an indicator of the health of the maintenance of a software project. The Elasticsearch and Nova projects were studied, finding a mean TTN of 312 days for Elasticsearch and 431 days for Nova. Further study showed that the real values of TTN were worse than the optimistic estimates produced by the most common algorithm used to find the BIC and that there was a negative correlation in Nova between EuBIC and TTN: more experienced developers took less time to notify bugs.

5.7. Conclusions

This work presented the integration of Hawk as part of the MEASURE platform for obtaining software metrics from large collections of evolving (versioned) models developed by teams of developers. It discussed the various technologies used, focusing on the innovations that were necessary for achieving the full functionality of producing metrics for the entire lifespan of such models through their entire version history.

Through the use of this integration we were able to index the full history of a set of medium-sized real-world models stored in a Subversion version control system and to provide relevant metrics throughout their development process.

The study triggered some extensions and internal improvements in the Hawk model indexer, such as the new timeline query engine. It was observed that indexing time was largely dominated by the proxy reference resolution process, over the insertion of the Modelio model files themselves, and that the compression of the Greycat database could be improved. Our first line of work from these results is to optimize both aspects to reduce time and storage requirements.

It was possible to use MEASURE as is, without any specific changes. However, it was necessary to manually feed the underlying Elasticsearch database through the bulk API. At the time of writing, the MEASURE platform did not support computing metrics of old versions through the web interface. It would be useful to extend the MEASURE platform with the ability to tag metrics as "time-aware," providing a starting date for their computation.

During the DataBio case study, it was noted that not many quality or complexity metrics were available for ArchiMate enterprise architecture models. In addition to several typical "how many instances of type X" and model size metrics, two metrics by [29] that highlight critical elements were adapted to EOL. MEASURE partners have shown interest in the creation of new metrics for ArchiMate, which is another avenue of future work. The MEASURE platform could also be extended to cover "top X elements" visualizations, listing the most critical elements in a model, or the models with the most warnings or design rule violations.

Acknowledgments

The research leading to these results was partially funded by the ITEA3 project 14009 (MEASURE) and H2020 project DataBio. DataBio has received funding from the European Union Horizon 2020 Research and Innovation Programme under grant agreement No. 732064.

Appendix 5.A. Running example

In order to better understand the internal structure of Hawk, we will describe it through a running example. This example shows how a simple UML class diagram can be indexed in Hawk: Fig. 5.9 shows the source class diagram, which contains two UML packages: one with the class "DrawingProgram," and one with the class "Shape."

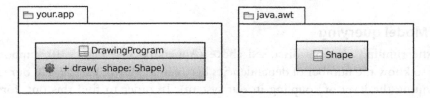

Figure 5.9 Running example – indexing a UML model in Hawk: UML class diagram.

Fig. 5.10 shows the resulting model index, which is made up of four types of elements.

Metamodel elements (top two rows of nodes on the right side) represent the various concepts of UML-like packages, classes, properties, and data types.

Model elements (next two rows of nodes) contain the necessary model data from the source files, such as their attributes, their references, as well as links to their type (UML metaclass).

File nodes (last row of nodes) contain information about the source files of the various model nodes, such as the filename, the revision currently stored in Hawk, and the modeling technology this file conforms to.

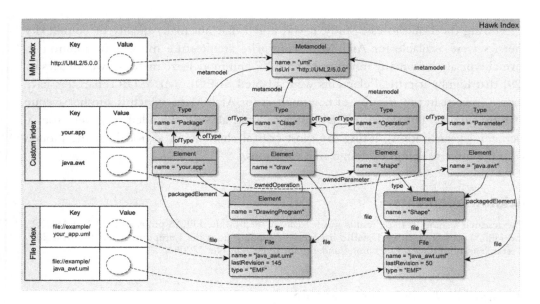

Figure 5.10 Running example – indexing a UML model in Hawk.

Node indices (left-hand side) are used to provide starting points for queries on Hawk, such as using one or more files or one or more metamodels to scope the query, or even custom indices using element attributes as keys.

5.A.1 Model querying

Using the running example discussed above (Appendix 5.A, Fig. 5.10), suppose we wanted to know the number of dependencies between the UML packages in our model (representing the level of coupling in our system). In order to find this out, for each package, we would have to go through all of the classes in that package and all of the operations these classes define, and see if any of their parameters refers to a class found in another package.

Listing 5.1 shows an implementation of this query in EOL. In this listing we can see that we are counting the number of dependencies, stored in an integer variable *total* (line 1). For each *Package* in the model (line 2), we keep a note of the package name (line 4) and we navigate to all the *Classes* in that package (line 5) and all the *Operations* in those classes (line 6). From those operations, we look at any *Parameters* they have (line 7) and for each such parameter we take note of the name of the package it is found in (line 8). In line 8, we use composite reference navigation to go from the parameter to its type, then from this type to its package (the eContainer reference is a special keyword used to navigate from a contained model element to its container, in this case from a *Class* to its *Package*). Finally, we see if the resulting package is different from our original

Listing 5.1: EOL query for computing the number of interpackage dependencies.

```
1   var total = 0;
2   for(p in Package.all) {
3       var targetDeps = new Set;
4       var pName = p.name;
5       for(c in p.packagedElement) {
6           for(o in c.ownedOperation) {
7               for(op in o.ownedParameter) {
8                   var dependantPackageName = op.type.eContainer.
                        name;
9                   if(dependantPackageName <> name) targetDeps.
                        add(dependantPackageName);
10      } } }
11      total += targetDeps.size;
12  }
13  return total;
```

package, and if it is we add it to the list of dependencies of our current package (line 9), using this unique set of dependencies to calculate the total amongst all our packages (line 11).

Since Hawk stores the model as a graph, performing this query simply requires navigating the appropriate edges, such as the *packagedElement* references to find the classes belonging to a package. As Hawk contains a global view of the entire model, it can solve such queries without the need for fetching or loading any parts of the model that may not be found locally. For example, if a developer only maintains local copies of the UML packages they are interested in, they can query any part of the overall UML model without the need to retrieve or load any model files locally.

5.A.2 Derived features

Using the running example discussed above (Appendix 5.A, Fig. 5.10), taking the same query as previously (finding the number of dependencies between the UML packages in our model), we can precompute the number of dependencies of a package to others, and store this result in each package. Hence our query would only require, for each package, to retrieve this number and add it to a running total. Since the derivation logic is computed at update time, this results in a much shorter query execution time due to the fact that fewer elements and references need to be navigated through during the query execution.

The relevant EOL query would look like this:

```
1   Package.all.deps.sum;
```

In this listing we use the feature *deps* which is a derived attribute of the type *Package* (the operation *.sum* takes a collection of elements and produces a single number resulting from the addition of all the numbers in that collection). In order to calculate the value of this derived feature (at update time) we use the following EOL expression:

```
1   return self.packagedElement.ownedOperation. flatten
2       .ownedParameter. flatten
3       .type.eContainer.name
4       . reject (n | self.name == n)
5       . asSet . size ;
```

This listing has a similar logic to the original listing found in Appendix 5.A.1, using composite reference navigation instead of imperative loops[22] (*flatten* is a keyword used in EOL which converts a collection of collections into a flattened collection of elements). Since this derived attribute is calculated for each *Package*, the keyword *self* is used to refer to the package itself. Finally, the operation *.reject(…)* removes elements in a collection (in this case in the collection of package names) based on the Boolean result of the expression in brackets (in this case removing all elements that are equal to the name of the originating package), and the operation *.asSet* converts a collection with possible duplicate elements into a set, hence containing no duplicates.

5.A.3 Querying-as-a-service

Taking the running example, performing the interpackage dependency query using Hawk's remote API, we will receive a reply in JSON format[23] such as the one shown in Listing 5.2. The information in this listing is as follows:

- Lines 1 and 18 denote the start and the end of the Json message.
- Line 2 contains metadata information:
    - the number *1* denotes the Thrift protocol version,
    - "fetchAsyncQueryResults" is the name of the invoked operation,
    - the number *2* denotes the message type (1: request, 2: response),
    - and the number *3* denotes the sequence number of the current request–response pair: this would denote that this is the response for the third query performed on this Hawk instance.
- Line 3 contains the response type: 0 is "normal query results," 1 is "invalid query," and 2 is "failed query."
- Line 4 denotes that the contents of this response are of type record.

---

[22] It is worth noting that the original listing in Appendix 5.A.1 could have been written in this manner as well, but a more imperative style was chosen for the sake of clarity. The Epsilon execution engine would run both styles in the same manner.

[23] This serialization format is not supposed to be human-readable: users would use a JavaScript library generated by Thrift for this specific API instead.

Listing 5.2: Reply from the Hawk Query API, in Thrift JSON format.

```
1   [
2     1,"fetchAsyncQueryResults ",2,3,
3     {"0":
4       {"rec ":
5         {
6           "1":{
7             "rec ":{
8               "4":{
9                 "i32 ":1
10              }
11            }
12          },
13          "2":{" i64 ":144},
14          "3":{" tf ":0}
15        }
16      }
17    }
18  ]
```

- Line 5 denotes the start of the actual result (record), and line 15 its end.
- Line 6 denotes the start of the first field of the result, containing the actual data, and line 12 its end.
- Line 7 denotes that the contents of the result are of type record.
- Line 8 denotes that the result is of type int (1 is boolean, 2 is byte, 3 is short, 4 is int, 5 is long).
- Line 9 contains the result value, of type i32 (32-bit integer) and value 1.
- Line 13 contains the second field of the result, a 64-bit integer denoting the execution time of this query (in this case 144 milliseconds).
- Line 14 contains the third field of the result, a "tf" (Boolean) value denoting whether this query was canceled (or ran to completion), in this case 0 meaning false, i.e., the query ran to completion.

5.A.4 Temporal databases

Using the running example, Fig. 5.11 shows an example of how the model would look like using a Greycat back-end. Note that this figure includes two future versions of this model, each of which has a single change for demonstration purposes. Various unchanging elements have been omitted to reduce clutter, such as the indices and the metamodel and type nodes.

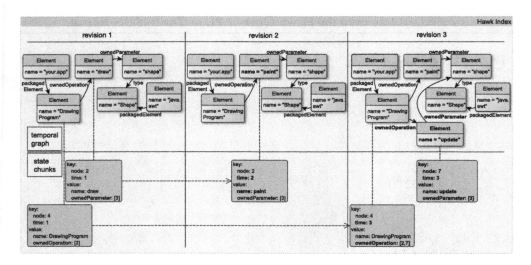

Figure 5.11 Running example – indexing a UML model in Hawk Greycat.

In Greycat, the graph is stored as a collection of nodes: conceptual identifiers mapped to specific *state chunks* depending on the *world*[24] and *timepoint* chosen to visit it. Nodes have a *lifespan* between two specific *timepoints*, and within that lifespan they may take on a sequence of state chunks. Each state chunk appears at a specific timepoint and overrides any previous state chunk. Hence, instead of storing three full graphs (one for each revision), we only need to create new state chunks for the relevant changes made to the model at each revision. As we can see in Fig. 5.11, a new state chunk is made for the "draw" node when it is renamed to "paint" and similarly state chunks are made for the "Drawing Program" and "update" nodes when the "update" element is created, as the "Drawing Program" node now has a reference to it. It is worth noting that state chunks for unchanged elements are not shown in the image (but will exist in the database), in order to reduce clutter.

With the two new query engines described in Section 5.4, it is possible to perform temporal queries in Hawk. For example, it is possible to run the interpackage dependency query on all three revisions of the model, and hence monitor how it changes as the model evolves over time.

5.A.5 Storage of Hawk data into elasticsearch

Using the running example presented here, Fig. 5.12 shows what an Elasticsearch instance would contain for the three revisions of our model shown in Fig. 5.11. This

---

[24] Greycat uses the concept of a *world*, allowing forking of the current database in order to simulate what-if scenarios.

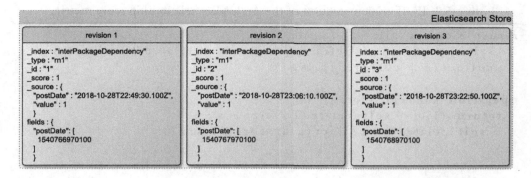

Figure 5.12 Running example – documents stored in Elasticsearch.

instance is holding the three values for the metric *interPackageDependency* which is the same as our query used for the running example. Each measure in Elasticsearch is stored as a document containing its current value as well as other metadata:

- *_index* contains the name of the metric.
- *_type* contains the name of the measure instance.
- *_id* contains the identifier of the measure instance.
- *_source* contains a document with the time this value was taken as well as the value itself.
- *fields* contains an alternative format of the timestamp.

For this simple example we can see that the value remains at 1 throughout the revisions as we never add or remove a new Package dependency (revision 3 adds a dependency to an already dependent package so does not affect the total). For a more detailed example and its integration with Elasticsearch and Kibana, refer to Section 5.5, where a real-world model is used as a use case for this integration. Section 5.5.2 in particular includes a screenshot of a visualization that would be built on Kibana using these stored documents (Fig. 5.8).

Appendix 5.B. EOL-based ArchiMate metric implementation

Listing 5.3 shows our Epsilon Object Language-based implementation of the ArchiMate relationship score metric by Singh and van Sinderen, averaged over all concepts [29]:

- Lines 1–3 go through every concept in the current version of the model, leaving only those with outgoing relationships and a clear name, and compute the average score.
- Lines 5–8 implement the original formula from Singh and van Sinderen [29], $RS(E) = (1/n) \sum_{k=1}^{n} W_k$, where E is the element, n is the number of outgoing relationships, and W_k is the weight given to the kth relationship based on its type.

Listing 5.3: EOL implementation of the ArchiMate relationship score metric.

```
1   return Concept.all
2     .select(c|c.Name.isDefined() and not c.relatedTo.isEmpty)
3     .collect(c | c.score()).avg();
4
5   operation Concept score() {
6     return (1.0 / self.relatedTo.size) *
7       self.relatedTo.collect(a|a.score()).sum();
8   }
9
10  operation Sequence avg() {
11    return self.sum() * 1.0 / self.size();
12  }
13
14  operation Composition score()   { return 8; }
15  operation Aggregation score()   { return 7; }
16  -- ... other scoring relationships ...
17  operation Flow score()          { return 1; }
18  -- ... other nonscoring relationships ...
19  operation Relationship score() { return 0; }
```

- Lines 10–12 implement the concept of "average" of a sequence of numbers in EOL.
- Lines 14 and onwards provide the weights suggested by Singh and van Sinderen for each of the ArchiMate relationship types.

Overall, the relationship score metric was implemented in 26 lines of EOL (excluding whitespace and comments), and the importance score metric required 63 lines of EOL.

References

[1] J. Hutchinson, M. Rouncefield, J. Whittle, Model-driven engineering practices in industry, in: Software Engineering (ICSE), 2011 33rd International Conference on, IEEE, 2011, pp. 633–642 [Online]. Available: http://ieeexplore.ieee.org/xpls/abs_all.jsp?arnumber=6032504. (Accessed 7 May 2016).

[2] SonarSource, SonarQube, https://www.sonarqube.org/, Mar. 2019. (Accessed 26 March 2016).

[3] K. Barmpis, D.S. Kolovos, Hawk: towards a scalable model indexing architecture, in: Proceedings of the Workshop on Scalability in Model Driven Engineering, BigMDE '13, ACM, Budapest, Hungary, ISBN 978-1-4503-2165-5, Jun. 2013, 6.

[4] T.J. Vijay, M.G. Chand, H. Done, Software quality metrics in quality assurance to study the impact of external factors related to time, International Journal of Advanced Research in Computer Science and Software Engineering 7 (1) (Jan. 2017) 221–224, https://doi.org/10.23956/ijarcsse/V6I11/0114, ISSN 2277-6451, 2277-128X [Online]. Available: http://ijarcsse.com/docs/papers/Volume_7/1_January2017/V6I11-0114.pdf. (Accessed 13 March 2019).

[5] D. Graham, I. Evans, E.V. Veenendaal, R. Black, Foundations of Software Testing: ISTQB Certification, 2nd revised edition, Cengage, ISBN 978-1-84480-989-9, Jan. 2008 (in English).

[6] S.A. SonarSource, Metric definitions, archived at http://archive.is/qyEW1, Jan. 2018. (Accessed 13 March 2019).

[7] M.-C. Lee, Software quality factors and software quality metrics to enhance software quality assurance, British Journal of Applied Science & Technology (ISSN 2231-0843) 4 (21) (Jan. 2014) 3069–3095, https://doi.org/10.9734/BJAST/2014/10548 [Online]. Available: http://www.sciencedomain.org/abstract.php?iid=541&id=5&aid=4777. (Accessed 19 March 2019).

[8] B. Berenbach, G. Borotto, Metrics for model driven requirements development, in: Proceedings of the 28th International Conference on Software Engineering, ICSE '06, Shanghai, China, ACM, New York, NY, USA, ISBN 978-1-59593-375-1, 2006, pp. 445–451 [Online]. Available: http://doi.acm.org/10.1145/1134285.1134348. (Accessed 19 March 2019).

[9] M. Monperrus, J.-M. Jézéquel, B. Baudry, et al., Model-driven generative development of measurement software, Software & Systems Modeling 10 (4) (Oct. 2011) 537–552, https://doi.org/10.1007/s10270-010-0165-9, ISSN 1619-1366, 1619-1374 [Online]. Available: http://link.springer.com/10.1007/s10270-010-0165-9. (Accessed 13 March 2019).

[10] A. Bertolino, A. Calabro, F. Lonetti, et al., Towards a model-driven infrastructure for runtime monitoring, in: E.A. Troubitsyna (Ed.), Software Engineering for Resilient Systems, in: Lecture Notes in Computer Science, Springer, Berlin, Heidelberg, ISBN 978-3-64224124-6, 2011, pp. 130–144.

[11] F.D. Giraldo, S. España, O. Pastor, Analysing the concept of quality in model-driven engineering literature: a systematic review, in: 2014 IEEE Eighth International Conference on Research Challenges in Information Science, RCIS, May 2014, pp. 1–12.

[12] Software Freedom Conservancy, Git project page, https://git-scm.com/, Feb. 2019. (Accessed 20 March 2019).

[13] Apache Software Foundation, Subversion, https://subversion.apache.org/, Mar. 2019. (Accessed 20 March 2019).

[14] Object Management Group, The software metrics meta-model specification 1.2, https://www.omg.org/spec/SMM/1.2/, Mar. 2018. (Accessed 26 March 2019).

[15] NIST SAMATE project, Metrics and measures, https://samate.nist.gov/index.php/Metrics_and_Measures.html. (Accessed 13 March 2019). Archived at http://archive.is/980Dm, Feb. 2019.

[16] D.S. Kolovos, L. Rose, A. Garcia, R. Paige, The Epsilon Book, 2008 [Online]. Available: http://www.eclipse.org/epsilon/doc/book/.

[17] The object constraint language specification [Online]. Available at: http://www.omg.org/spec/OCL/2.3.1/, 2012. (Accessed 1 June 2012).

[18] E. Willink, Aligning OCL with UML, in: Proceedings of the Workshop on OCL and Textual Modelling, Electronic Communications of the EASST, 2011.

[19] K. Barmpis, D.S. Kolovos, Evaluation of contemporary graph databases for efficient persistence of large-scale models, Journal of Object Technology 13 (3) (Jul. 2014) 1–26, https://doi.org/10.5381/jot.2014.13.3.a3 (in English).

[20] A. Garcia-Dominguez, K. Barmpis, D.S. Kolovos, et al., Stress-testing remote model querying APIs for relational and graph-based stores, Software & Systems Modeling (ISSN 1619-1374) (Jun. 2017), https://doi.org/10.1007/s10270-017-0606-9.

[21] O. Al-Wadeai, A. Garcia-Dominguez, A. Bagnato, et al., Integration of hawk for model metrics in the measure platform, in: MODELSWARD 2018 - Proceedings of the 6th International Conference on Model-Driven Engineering and Software Development, SciTePress, Jan. 2018, pp. 719–730 (in English).

[22] T. Hartmann, F. Fouquet, G. Nain, et al., A native versioning concept to support historized models at runtime, in: J. Dingel, W. Schulte, I. Ramos, et al. (Eds.), Model-Driven Engineering Languages and Systems, Springer International Publishing, Cham, ISBN 978-3319-11653-2, 2014, pp. 252–268.

[23] A. García-Domínguez, N. Bencomo, L.H.G. Paucar, Reflecting on the past and the present with temporal graph-based models, in: Proceedings of the 13th International Workshop on Models@run.time, Copenhagen, Denmark, in: CEUR Workshop Proceedings, vol. 2245, 2018, pp. 46–55.

[24] A. Abhervé, M. Almeida, Modelio project management server Constellation, in: E. Di Nitto, P. Matthews, D. Petcu, A. Solberg (Eds.), Model-Driven Development and Operation of Multi-Cloud Applications: The MODAClouds Approach, Springer International Publishing, Cham, ISBN 978-3-319-46031-4, 2017, pp. 113–122.

[25] The Open Group, Archimate™ 3.0.1 specification, http://pubs.opengroup.org/architecture/archimate3-doc/, 2017.

[26] Modeliosoft, Modelio: the open source modeling environment, https://www.modelio.org/, Feb. 2019.

[27] A. Garcia-Dominguez, K. Barmpis, D.S. Kolovos, et al., Integration of a graph-based model indexer in commercial modelling tools, in: Proceedings of the ACM/IEEE 19th International Conference on Model Driven Engineering Languages and Systems, ACM Press, Saint Malo, France, ISBN 978-1-4503-4321-3, 2016, pp. 340–350. (Accessed 17 October 2016).

[28] A. Garcia-Dominguez, K. Barmpis, D.S. Kolovos, et al., Stress-testing remote model querying APIs for relational and graph-based stores, Software & Systems Modeling (Jun. 2017) 1–29, https://doi.org/10.1007/s10270-017-0606-9, ISSN 1619-1366, 1619-1374 [Online]. Available: https://link.springer.com/article/10.1007/s10270-017-0606-9. (Accessed 3 July 2017).

[29] P.M. Singh, M.J. van Sinderen, Lightweight metrics for enterprise architecture analysis, in: W. Abramowicz (Ed.), Business Information Systems Workshops, vol. 228, Springer International Publishing, Cham, ISBN 978-3-319-26761-6, 2015, pp. 113–125.

[30] M.-E. Iacob, H. Jonkers, Quantitative analysis of enterprise architectures, in: D. Konstantas, J.-P. Bourrières, M. Léonard, N. Boudjlida (Eds.), Interoperability of Enterprise Software and Applications, Springer-Verlag, London, ISBN 978-1-84628-151-8, 2006, pp. 239–252.

[31] S. Dahab, E. Silva, S. Maag, et al., Enhancing software development process quality based on metrics correlation and suggestion, in: Proceedings of the 13th International Conference on Software Technologies, SCITEPRESS - Science and Technology Publications, Porto, Portugal, ISBN 978-989-758-320-9, 2018, pp. 120–131.

[32] G. Campbell, P. Papapetrou, Sonarqube in Action, Manning Publications, ISBN 9781617290954, 2013.

[33] J. García-Munoz, M. García-Valls, J. Escribano-Barreno, Improved metrics handling in sonarqube for software quality monitoring, in: S. Omatu, A. Semalat, G. Bocewicz, et al. (Eds.), Distributed Computing and Artificial Intelligence, 13th International Conference, Springer International Publishing, Cham, ISBN 978-3-319-40162-1, 2016, pp. 463–470.

[34] A. Bagnato, K. Barmpis, N. Bessis, et al., Developer-centric knowledge mining from large open-source software repositories (CROSSMINER), in: M. Seidl, S. Zschaler (Eds.), Software Technologies: Applications and Foundations, Springer International Publishing, Cham, ISBN 978-3-319-74730-9, 2018, pp. 375–384.

[35] Elasticsearch BV, Community beats, https://www.elastic.co/guide/en/beats/libbeat/current/community-beats.html. (Accessed 3 April 2019). Archived at http://archive.is/Stpen, Apr. 2019.

[36] Elasticsearch BV, Elasticsearch machine learning, https://www.elastic.co/es/products/stack/machine-learning. (Accessed 9 March 2019). Archived at http://archive.is/oDDh9, Mar. 2019.

[37] Bitergia, GrimoireLab project page, https://chaoss.github.io/grimoirelab/, 2017.

[38] J. Hamilton, M. Gonzalez Berges, B. Schofield, J.-C. Tournier, SCADA statistics monitoring using the Elastic stack (Elasticsearch, Logstash, Kibana), in: TUPHA034, 2018, 5 pp. [Online]. Available: http://cds.cern.ch/record/2305659.

[39] M. Lassnig, W. Toler, R. Vamosi, et al., Machine learning of network metrics in ATLAS distributed data management, Journal of Physics. Conference Series 898 (6) (2017) 062009 [Online]. Available: http://stacks.iop.org/1742-6596/898/i=6/a=062009.

[40] A. Washbrook, Continuous software quality analysis for the ATLAS experiment, Journal of Physics. Conference Series (ISSN 1742-6596) 1085 (Sep. 2018) 032047, https://doi.org/10.1088/1742-6596/1085/3/032047. (Accessed 3 April 2019).

[41] A. Voit, A. Stankus, S. Magomedov, I. Ivanova, Big data processing for full-text search and visualization with elasticsearch, International Journal of Advanced Computer Science and Applications 8 (12) (2017) 76–83.

[42] G.R. Perez, G. Robles, J.M.G. Barahona, How much time did it take to notify a bug? Two case studies: ElasticSearch and Nova, in: 2017 IEEE/ACM 8th Workshop on Emerging Trends in Software Metrics, WETSoM, May 2017, pp. 29–35.

CHAPTER 6

Model analytics for defect prediction based on design-level metrics and sampling techniques

Aydin Kaya[a], **Ali Seydi Keceli**[a], **Cagatay Catal**[b], **Bedir Tekinerdogan**[b]
[a]Department of Computer Engineering, Hacettepe University, Ankara, Turkey
[b]Information Technology Group, Wageningen University, Wageningen, The Netherlands

Contents

6.1. Introduction

Software defect prediction is a quality assurance activity which applies historical defect data in conjunction with software metrics [1–3]. It identifies which software modules are defect-prone before the testing phase, and then quality assurance groups allocate more testing resources into these modules because the other group of modules, called nondefect-prone, will not likely cause defects based on the output of the prediction approach [4]. Most of the studies in literature use classification algorithms in Machine Learning to classify software modules into two categories (defect-prone and nondefect-prone) [1].

Prior defects and software metrics data are mostly stored in different repositories, namely, bug tracking systems and source code repositories, and the processing of these data depends on the underlying platforms. Sometimes the mapping of each defect into the appropriate source code introduces errors and it is not a trivial task. While this defect

Model Management and Analytics for Large Scale Systems
https://doi.org/10.1016/B978-0-12-816649-9.00015-6
125

prediction activity provides several benefits to organizations, only a limited number of companies have yet adopted this approach as part of their daily development practices. One strategy to increase the adoption of these approaches is to develop new models to increase the performance of existing defect prediction models.

In this study, we focus on data sampling techniques for the improvement because defect prediction datasets are always unbalanced, which means that approximately 10–20% of modules belong to the defect-prone category. While there are several attempts to apply sampling techniques in this domain, there is not an in-depth study which analyzes several classification algorithms in conjunction with design-level metrics and sampling approaches. Hence, we performed several experiments on ten public datasets by using six classification algorithms and design-level metrics.

The object-oriented Chidamber–Kemerer (CK) metrics suite, which includes six class-level metrics, is used in this study. Since datasets include additional features such as lines of code, we performed another case study to compare prediction models having the total set of features. We adopted the following classification algorithms because they have been widely used in defect prediction studies: AdaBoostM1, Linear Discriminant, Linear Support Vector Machine (SVM), Random Forest, Subspace Discriminant, and Weighted-kNN (W-kNN). The following six performance evaluation parameters are used to evaluate the performance of models: Area under ROC Curve (AUC), Recall, Precision, F-score, Sensitivity, and Specificity [5].

We analyzed the effects of the following sampling methods on the performance of defect prediction models using design-level metrics: ADASYN, Borderline SMOTE, and SMOTE. We compared their performance against the ones built based on the unbalanced data.

Our research questions are given as follows:

- RQ1: Which sampling techniques are more effective to improve the performance of defect prediction models?
- RQ2: Which classifiers are more effective in predicting software defects when sampling techniques are applied?
- RQ3: Are design-level metrics (CK metrics suite) suitable to build defect prediction models when sampling techniques are applied?

The contribution of this chapter is two-fold:

- The effects of sampling techniques are investigated for software defect prediction problem in detail.
- The effects of six classification algorithms and design-level software metrics are evaluated on ten public datasets.

This paper is organized as follows: Section 6.2 explains the background and related work, Section 6.3 shows the methodology regarding our experimental results, Section 6.4 explains the experimental results, Section 6.5 provides the discussion, and Section 6.6 shows the conclusion and future work.

6.2. Background and related work

Software defect prediction, a quality assurance activity, identifies defect-prone modules before the testing phase and therefore, more testing resources are allocated to these modules to detect defects before the software deployment. It is a very active research field which has attracted many researchers in the software engineering community since the middle of the 2000s [4,1].

Machine Learning algorithms are widely used in these approaches and the prediction models are built using software metrics and defect labels. Software metrics can be collected at several levels, i.e., the requirements level, design level, implementation level, and process level. In the first case study of this paper, we applied the CK metrics suite, which is a set of design-level metrics. CK metrics are explained as follows:

- The Weighted Methods per Class (WMC) metric shows the number of methods in a class.
- The Depth of Inheritance Tree (DIT) metric indicates the distance of the longest path from a class to the root element in the inheritance hierarchy.
- The Response For a Class (RFC) metric provides the number of available methods to respond a message.
- The Number Of Children (NOC) indicates the number of direct descendants of a class.
- The Coupling Between Object classes (CBO) metric indicates the number of noninheritance-related classes to which a class is coupled.
- The Lack of COhesion in Methods (LCOM) metric indicates the access ratio of the class' attributes.

Software metrics are features of the model and the defect labels are class labels. From a Machine Learning perspective, these prediction models can be considered as classification models since we have two groups of data instances, namely, defect-prone and nondefect-prone. Defect prediction datasets are imbalanced [6], which means that most of the modules (i.e., 85–90%) in these datasets belong to the nondefect-prone class. Therefore, classification algorithms in Machine Learning might not detect the minority of data points due to the imbalanced characteristics of datasets.

The Machine Learning community has done a lot of research on the imbalanced learning so far [7,8], but the empirical software engineering community has not evaluated the impact of these algorithms in detail yet compared to the Machine Learning researchers. Algorithms in imbalanced learning can be shown in the following four categories [9]:

1. Subsampling [10]: With subsampling, data distribution is balanced before the classification algorithm is applied. The following four main approaches exist in this category [9]:
   (a) Under-sampling: A subset of the majority class is selected to balance the distribution of the data points, but this might cause loss of some useful data.

(b) Over-sampling: A random replication of the minority class is created, but this might cause an over-fitting problem [11].

(c) SMOTE [10]: This is a very popular over-sampling method and it successfully avoids over-fitting by generating new minority data points with the help of interpolation between near neighbors.

(d) Hybrid methods [12]: These methods combine several subsampling methods to balance the dataset.

2. Cost-sensitive learning [13]: A false negative prediction is more costly than a false positive prediction in software defect prediction. This approach uses a cost matrix which specifies the costs of misclassification for each class, and then this matrix is applied for the optimization of the training data [9]. Since misclassification costs are not publicly available, their applications require the knowledge of domain experts.

3. Ensemble learning [14]: The generalization capability of different classifiers are combined, and then the new classifier, which is called ensemble of classifiers or multiple classifiers system (MCS), provides a better performance than the individual classifiers used to design it. Bagging [15] and Boosting [16] are some of the most popular algorithms used in this category. AdaBoost is shown in one of the top ten algorithms in data mining [17,9].

4. Imbalanced ensemble learning [11]: These algorithms combine the subsampling methods with the ensemble learning algorithms. If the over-sampling method is used in the Bagging approach instead of random sampling in Bagging, this is known as OverBagging [18]. There are several approaches which use the same idea [19–21].

In this study, we used algorithms from subsampling (ADASYN, Borderline SMOTE, and SMOTE algorithms) and ensemble learning (AdaBoostM1 algorithm) categories. We did not experiment with algorithms in the cost–sensitive learning category because there is not an easy way to identify the costs for the analysis. In addition to these algorithms, we used additional classification algorithms in conjunction with imbalanced learning algorithms. Since AdaboostM1 is an ensemble learning algorithm and we used it in conjunction with subsampling techniques, we can say that we also utilized from the category called imbalanced ensemble learning.

We applied two additional subsampling approaches, called ADASYN and Borderline SMOTE, in our study compared to the study of Song et al. [9]. Also, we applied two additional classification algorithms, called Linear Discriminant and Subspace Discriminant, which were not analyzed in the study of Song et al. [9].

6.3. Methodology

6.3.1 Classification methods

Six classification methods have been investigated in this study. These are Random Forest, Adaboost, SVM, Linear Discriminant Analysis (LDA), Subspace Discriminant, and

W-kNN. Random Forest, Adaboost, and Subspace Discriminant are ensemble classification methods which combine single classifiers to obtain better predictive performance. Single classifiers during our experiments are SVM, LDA, and W-kNN. These classifiers are widely used base methods for ensemble classification. A brief description of these ensemble and single Machine Learning algorithms are provided in the following subsections.

6.3.2 Ensemble classifiers

The first ensemble method used in our experiments is Random Forest. Random Forest is a combination of multiple decision trees [22]. Bootstrapping is applied for sample selection for each tree in the forest. Two-thirds of the selected data is used to train a tree, and the classification is performed with the remaining data. Majority voting is applied to obtain the final prediction result. The Random Forest algorithm counts the votes from all the trees and the majority of the votes is used for the classification output. Random Forest is easy to use, it prevents over-fitting, and it stores the generated decision tree cluster for the other datasets.

The second ensemble method is Subspace Discriminant, which uses linear discriminant classifiers. Subspace Discriminant utilizes a feature-based bagging. In this method, feature bagging is applied to reduce the correlation between estimators; however, the difference compared to the bagging is that the features are randomly subsampled, having a replacement for each learner. Learners that are specialized on different feature sets are obtained.

The final ensemble method is Adaboost. The AdaBoost method, proposed by Freund and Schapire [23], utilizes boosting to obtain strong classifiers by combining the weak classifiers. In boosting, training data are split into parts. A predictive model is trained with the parts of the model and this model is tested with one of the remaining parts. Then, a second model is trained with the false predicted samples of the first model. This process is repeated. New models are trained with the samples falsely predicted by previous models.

6.3.3 Single classifiers

The first and most well-known single classifier used in our experiments is SVM. SVM is a common method among Machine Learning tasks [24]. In this method, the classification is performed by using linear and nonlinear kernels. The SVM method aims to find the hyperplane that separates the data points in the training set with the farthest distance.

The second single classifier is LDA. LDA projects a dataset into a lower-dimensional feature space to increase the separability between classes [25]. The first step of LDA is the computation of the variances between class features to measure the separability. The second step is to calculate the distance between the mean and samples of each class.

The final step is to construct the lower-dimensional space that maximizes the interclass variance and minimizes the intraclass variance.

Our last single classifier method is W-kNN. W-kNN is an extension of the k-Nearest Neighbor algorithm. In the standard kNN, influences of all neighbors are the same although they have different individual similarity. In W-kNN, training samples which are close to the new observation have more weights than the distant ones [26].

6.4. Experimental results

During the experiments, we applied three data sampling methods and six classification methods on ten publicly available datasets. Three-fold cross-validation was applied for the evaluation of approaches. Each experiment was repeated fifty times. The experimental results are compared based on five evaluation metrics (AUC, Recall, Precision, F-score, and Specificity). The features are grouped as CK features, and ALL features include all features presented in datasets. The selected datasets are ant, arc, ivy, log4j, poi, prop, redactor, synapse, xalan, and xerces.

6.4.1 Experiments based on CK metrics

During these experiments, CK features were used and results were evaluated according to this feature set. The average scores of the classification methods on all datasets are presented in Fig. 6.1 and Table 6.1. Random Forest and Adaboost are the most successful classifiers based on the AUC metric (0.7396, 0.7269). The Adaboost classifier provides the best performance to identify the positive samples (i.e., minority class) (recall: 0.6083), however its performance is the worst for the identification of negative samples (i.e., majority class) (specificity: 0.6903). Linear Discriminant, Subspace Dis-

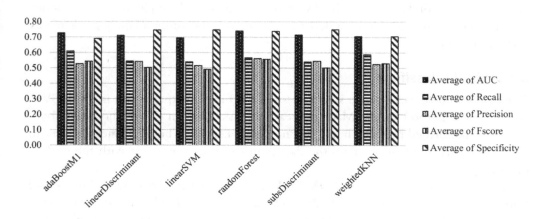

Figure 6.1 Average scores of the classifiers trained and tested with CK features.

Table 6.1 Average scores of the classifiers trained and tested with CK features.

Classifiers	Average of AUC	Average of recall	Average of precision	Average of F-score	Average of specificity
AdaBoostM1	0.7269	0.6083	0.5265	0.5429	0.6903
Linear Discriminant	0.7098	0.5435	0.5408	0.5019	0.7446
Linear SVM	0.6943	0.5397	0.5148	0.4908	0.7465
Random Forest	0.7396	0.5658	0.5614	0.5576	0.7381
Subs Discriminant	0.7157	0.5414	0.5458	0.5026	0.7495
Weighted-kNN	0.7057	0.5888	0.5252	0.5301	0.7062

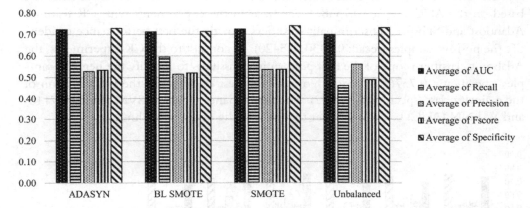

Figure 6.2 Average scores of the sampling methods with CK features.

Table 6.2 Average scores of the sampling methods with CK features.

Sampling method	Average of AUC	Average of recall	Average of precision	Average of F-score	Average of specificity
ADASYN	0.7232	0.6065	0.5282	0.5342	0.7286
BL SMOTE	0.7121	0.595	0.5146	0.5208	0.7138
SMOTE	0.7247	0.5933	0.5374	0.5375	0.7412
Unbalanced	0.7013	0.4634	0.5628	0.4914	0.7332

criminant, and Linear SVM exhibited similar performance and their performance is better to identify the negative samples.

The average scores of the classification methods grouped under sampling methods on all datasets are presented in Fig. 6.2 and Table 6.2. AUC and Specificity performances of the classifiers on balanced and unbalanced data are similar with improvement/decline of 1–2%. Besides, sampling methods improve the Recall metric by 13%. The effect of sampling methods is obvious on the identification of the positive samples, which is the minority class. Based on the AUC metric, SMOTE is the most successful sampling method on the datasets including CK features.

6.4.2 Experiments based on ALL metrics

In these experiments, all of the features were used, and the results are presented according to this feature set [27]. Features and datasets can be accessed from the SEACRAFT repository http://tiny.cc/seacraft. Jureczko and Madeyski [27] used the CKJM tool to collect the 19 metrics, i.e., CK metrics suite, QMOOD metrics suite, Tang, Kao and Chen's metrics, and cyclomatic complexity, LCOM3, Ca, Ce, and LOC metrics. They collected defects using the BugInfo tool.

The average scores of the classification methods on all datasets are presented in Fig. 6.3 and Table 6.3. Random Forest and Adaboost are the most successful classifiers based on the AUC metric (0.7558, 0.7363), as in the experiments with CK features. Adaboost and Subspace Discriminant classifiers provide the best performance to identify the positive samples (recall: 0.5430, 0.5420). In contrast to the CK experiments, the Adaboost classifier is one of the best performing classifiers to identify the negative samples (specificity: 0.7576). Random Forest is the best classifier for the identification of the majority class (specificity: 0.7692). The performance of Linear SVM (AUC: 0.7116) and W-kNN (AUC: 0.6776) is lower than that of the other algorithms.

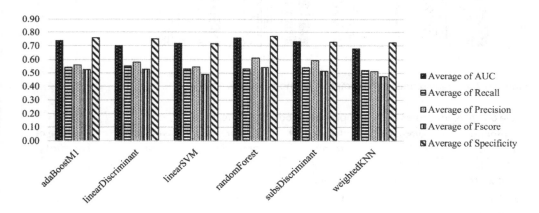

Figure 6.3 Average scores of the classifiers trained and tested with CK features.

Table 6.3 Average scores of the classifiers trained and tested with ALL features.

Classifier	Average of AUC	Average of recall	Average of precision	Average of F-score	Average of specificity
AdaBoostM1	0.7363	0.5430	0.5608	0.5256	0.7576
Linear Discriminant	0.6994	0.5530	0.5809	0.5276	0.7497
Linear SVM	0.7161	0.5309	0.5469	0.4892	0.7150
Random Forest	0.7558	0.5325	0.6116	0.5431	0.7692
Subs Discriminant	0.7298	0.5420	0.5946	0.5154	0.7262
Weighted-kNN	0.6776	0.5208	0.5138	0.4757	0.7223

The average scores of the classification methods grouped under sampling methods on all datasets are presented in Fig. 6.4 and Table 6.4. The performance of classifiers with respect to AUC and Specificity parameters on balanced (ADASYN and BL SMOTE methods) and unbalanced data are similar, with improvement/decline of 1–3%. Besides, ADASYN and BL SMOTE sampling methods improve the recall metric by 14%; however, this effect is reversed when the SMOTE method is considered. The average AUC value drops from 0.7445 to 0.6357 in the case of the SMOTE data sampling method. Precision score is the highest for the classifiers on unbalanced data. The effect of sampling methods is obvious on the identification of the positive samples which belong to the minority class. In contrast to CK experiments, SMOTE is the worst performing sampling method on the datasets including ALL features, and the recall value drops from 0.5025 to 0.3613 when SMOTE is applied. Therefore, SMOTE does not help to identify the instances belonging to the minority class when all the features are used.

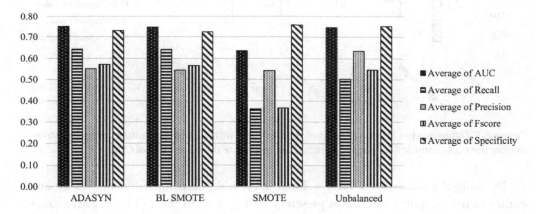

Figure 6.4 Average scores of the sampling methods with ALL features.

Table 6.4 Average scores of the sampling methods with ALL features.

Sampling method	Average of AUC	Average of recall	Average of precision	Average of F-score	Average of specificity
ADASYN	0.7499	0.643	0.5514	0.5728	0.7296
BL SMOTE	0.7465	0.6414	0.5443	0.5667	0.7233
SMOTE	0.6357	0.3613	0.5418	0.3663	0.7572
Unbalanced	0.7445	0.5025	0.6349	0.5452	0.7499

6.4.3 Comparison of the features

In this section, the experiments with CK and ALL features are compared. By using the values from Table 6.1 and Table 6.3, the comparison of average outcomes of the classification methods is presented in Fig. 6.5. In this figure, negative values denote the superiority of ALL features, while positive values denote the superiority of CK features. While the CK features set is more successful on the identification of the positive samples, the ALL features set is better on the negative samples. It seems that the difference with respect to AUC and F-score values is low. Besides, classifiers trained with ALL features provide better precision scores.

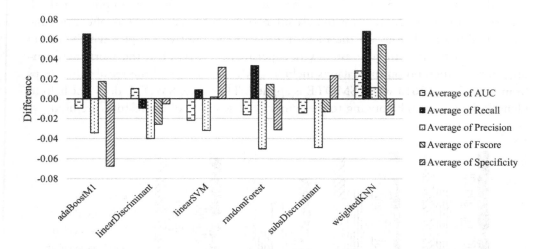

Figure 6.5 Difference between outcomes of the classifiers with CK and ALL features. Negative values denote the superiority of ALL features, positive values denote the superiority of CK features.

By using the values from Table 6.2 and Table 6.4, the comparison of average outcomes of the sampling methods is presented in Fig. 6.6. In this figure, negative values denote the superiority of ALL features, while positive values denote the superiority of CK features. While the classifiers trained with balanced data using the SMOTE method provided better AUC, F-score, and recall outcomes on CK features, the other sampling methods are more beneficial on ALL features and provided better outcomes for all scores.

In Fig. 6.7, the difference between outcomes of the classifiers with CK and ALL features is presented on the unbalanced datasets. In this figure, negative values denote the superiority of ALL features, positive values denote the superiority of CK features. The outcomes of the classifiers with ALL features are better except the W-kNN classifier. ALL features are more preferable than CK features when it comes to dealing with unbalanced datasets.

In Table 6.5, statistical significance results obtained by McNemar's test are presented. All the classifier/sampling method combinations are compared as CK vs. ALL features for all datasets. In the table, classification methods are grouped for each data sampling

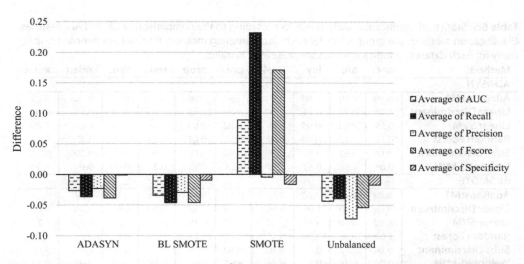

Figure 6.6 Difference between outcomes of the sampling methods with CK and ALL features. While negative values denote the superiority of ALL features, positive values denote the superiority of CK features.

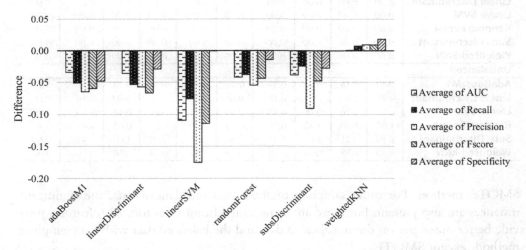

Figure 6.7 The difference between outcomes of the classifiers with CK and ALL features on the unbalanced datasets. Negative values denote the superiority of ALL features, positive values denote the superiority of CK features.

method. P-values are shown for each dataset. The significant values ($P < 0.05$) are indicated as italic and boldfaced.

Based on Table 6.5, Fig. 6.5, and Fig. 6.6, we conclude that the SMOTE sampling method provides 100% significant values for our experiments, and this significance shows the superiority of the data with CK features when it is balanced using the

Table 6.5 Statistical significance analysis results according to the comparison of CK and ALL features. Classification methods are grouped under each data sampling method. *P*-values are denoted respectively for each dataset. Significant values are presented in italics.

Methods	ant	arc	ivy	log4j	poi	prop	red.	syn.	xalan	xerces
ADASYN										
AdaBoostM1	*0.00*	*0.01*	*0.00*	*0.00*	0.30	*0.00*	*0.00*	*0.00*	*0.00*	*0.02*
Linear Discriminant	*0.00*	0.10	0.79	*0.00*	*0.00*	0.52	*0.00*	0.12	*0.00*	*0.00*
Linear SVM	0.55	*0.01*	*0.05*	*0.00*	0.02	*0.04*	*0.00*	*0.01*	*0.00*	*0.04*
Random Forest	0.17	0.12	*0.00*	0.01	0.52	*0.01*	0.19	0.66	0.32	0.87
Subs Discriminant	*0.00*	0.07	0.47	*0.00*	*0.00*	0.14	*0.00*	0.63	*0.00*	0.20
Weighted-kNN	0.08	1.00	0.52	0.37	0.68	0.92	0.26	*0.00*	*0.00*	0.39
BL SMOTE										
AdaBoostM1	*0.00*	*0.01*	*0.00*	*0.00*	0.23	*0.00*	*0.00*	0.23	*0.01*	0.64
Linear Discriminant	*0.00*	*0.03*	1.00	*0.00*	*0.00*	0.88	*0.00*	0.90	*0.00*	*0.00*
Linear SVM	*0.02*	*0.02*	0.49	*0.02*	*0.00*	0.46	*0.00*	0.20	*0.00*	0.71
Random Forest	1.00	*0.03*	0.26	0.17	0.72	0.73	*0.00*	*0.05*	*0.03*	0.40
Subs Discriminant	*0.00*	0.58	0.65	*0.00*	*0.00*	0.49	*0.00*	0.15	*0.00*	0.87
Weighted-kNN	0.70	0.30	0.28	0.44	0.68	0.06	0.88	0.82	*0.00*	0.48
SMOTE										
AdaBoostM1	*0.00*	*0.00*	*0.00*	*0.00*	*0.00*	*0.00*	*0.00*	*0.00*	*0.00*	*0.00*
Linear Discriminant	*0.00*	*0.00*	*0.00*	*0.00*	*0.00*	*0.00*	*0.00*	*0.00*	0.23	*0.00*
Linear SVM	*0.00*	*0.00*	*0.00*	*0.00*	*0.00*	*0.00*	*0.00*	*0.00*	*0.00*	*0.00*
Random Forest	*0.00*	*0.00*	*0.00*	*0.00*	*0.00*	*0.00*	*0.00*	*0.00*	*0.00*	*0.00*
Subs Discriminant	*0.00*	*0.00*	*0.00*	*0.00*	*0.00*	*0.00*	*0.00*	*0.00*	*0.00*	*0.00*
Weighted-kNN	*0.00*	*0.00*	*0.00*	*0.00*	*0.00*	*0.00*	*0.00*	*0.00*	*0.00*	*0.00*
Unbalanced										
AdaBoostM1	*0.01*	0.18	0.43	0.37	0.39	0.62	*0.06*	0.37	0.41	0.32
Linear Discriminant	0.85	1.00	0.82	0.18	*0.00*	*0.03*	*0.00*	*0.03*	0.16	*0.00*
Linear SVM	0.17	0.16	*0.00*	0.45	*0.00*	0.55	*0.00*	0.49	0.60	0.35
Random Forest	1.00	*0.05*	0.32	0.26	0.90	*0.00*	0.80	0.63	0.45	0.47
Subs Discriminant	*0.03*	0.21	0.10	0.32	*0.00*	0.53	*0.01*	1.00	0.44	*0.00*
Weighted-kNN	0.92	0.78	0.68	0.10	*0.01*	0.42	0.78	0.58	0.56	0.06

SMOTE method. For other sampling methods and unbalanced data, the significant instances are also present, but there are some insignificant ones too. ALL features provide better outcomes on the unbalanced data and the balanced data with data sampling methods except SMOTE.

6.5. Discussion

In this study, we performed several experiments to evaluate the effect of design-level metrics and data sampling methods on the performance of defect prediction models. Research questions of this study were introduced in Section 6.1, and in this section we present our responses to each of them.

- RQ1: Which sampling techniques are more effective to improve the performance of defect prediction models?

The answer to RQ1: Regarding the CK metrics suite, the SMOTE data sampling method is the best one; for the ALL metrics suite, the other data sampling methods are preferable.

- RQ2: Which classifiers are more effective in predicting software defects when sampling techniques are applied?

  The answer to RQ2: Regarding the CK metrics, the Random Forest algorithm is the most effective one with respect to the AUC value. Adaboost is the second best performing algorithm and it is also the best algorithm to identify the minority class instances.

  Regarding the ALL metrics set, again the Random Forest algorithm provides the best performance and Adaboost is the second best performing one. Adaboost and Linear Discriminant are the best performing algorithms for the identification of minority class instances. For the detection of majority class instances, Adaboost is the best classifier.

- RQ3: Are design-level metrics (CK metrics suite) suitable to build defect prediction models when sampling techniques are applied?

  The answer to RQ3: We showed that the CK features set is more successful for the identification of minority class instances and the ALL features set is better for the identification of majority class instances. When CK metrics are used with the Random Forest algorithm, the best performance was achieved using the CK metrics suite. When additional features are added to the CK metrics suite, the performance of a Random Forest-based model with respect to AUC value improved from 0.7396 to 0.7558.

Potential threats to validity for this study are evaluated from four dimensions, namely, conclusion validity, construct validity, internal validity, and external validity [28]. Regarding the conclusion validity, we applied a three-fold cross-validation evaluation technique 50 times (N $\star$ M cross-validation, N = 3, M = 50) to get statistically significant results and avoid randomness during the experiments. In addition to this validation technique, we performed McNemar's test to evaluate the statistical significance of our results. Five evaluation parameters were used to report the experimental results and ten public datasets were investigated. For the construct validity dimension, we can state that we applied the widely used ten public datasets for the software defect prediction problem. Regarding the internal validity, we investigated three single classifiers, three ensemble techniques, and three data sampling techniques in this study to evaluate the impact of classifiers, metrics suite, and data sampling techniques on software defect prediction. While we selected the best performing algorithms, other researchers might analyze the other techniques with different parameters and therefore, they can obtain different results. For the external validity dimension, we must stress that our conclusions are valid on datasets explained in the paper and results might be different on a different set of datasets.

6.6. Conclusion

Software defect prediction is an active research problem [29,9,30–33]. Hundreds of papers on this problem have been published so far [32]. In this study, we focused on the impact of data balancing methods, ensemble methods, and single classification algorithms for software defect prediction problems. We empirically showed that design-level metrics can be used to build defect prediction models. In addition, when design-level metrics (CK metrics suite) are applied, we demonstrated that the SMOTE data sampling approach can improve the performance of prediction models. Among other ensemble methods, we observed that the Adaboost ensemble method is the best one to identify the minority class (defect-prone) samples when the CK metrics suite is adopted. When all metrics in the dataset are applied, the Adaboost algorithm provides the best performance to identify the majority class instances. In the near future, we will perform more experiments when we access more data balancing methods and more datasets on this issue.

References

[1] C. Catal, B. Diri, A systematic review of software fault prediction studies, Expert Systems with Applications 36 (4) (2009) 7346–7354.
[2] C. Catal, U. Sevim, B. Diri, Metrics-driven software quality prediction without prior fault data, in: Electronic Engineering and Computing Technology, Springer, 2010, pp. 189–199.
[3] C. Catal, A comparison of semi-supervised classification approaches for software defect prediction, Journal of Intelligent Systems 23 (1) (2014) 75–82.
[4] T. Hall, S. Beecham, D. Bowes, D. Gray, S. Counsell, A systematic literature review on fault prediction performance in software engineering, IEEE Transactions on Software Engineering 38 (6) (2012) 1276–1304.
[5] M. Sokolova, N. Japkowicz, S. Szpakowicz, Beyond accuracy, F-score and ROC: a family of discriminant measures for performance evaluation, in: Australian Conference on Artificial Intelligence, 2006.
[6] S. Wang, X. Yao, Using class imbalance learning for software defect prediction, IEEE Transactions on Reliability 62 (2) (2013) 434–443, https://doi.org/10.1109/TR.2013.2259203.
[7] H. He, E.A. Garcia, Learning from imbalanced data, IEEE Transactions on Knowledge and Data Engineering 9 (2008) 1263–1284.
[8] H. He, Y. Ma, Imbalanced Learning: Foundations, Algorithms, and Applications, John Wiley & Sons, 2013.
[9] Q. Song, Y. Guo, M. Shepperd, A comprehensive investigation of the role of imbalanced learning for software defect prediction, IEEE Transactions on Software Engineering (2018), https://doi.org/10.1109/TSE.2018.2836442, in press.
[10] N.V. Chawla, K.W. Bowyer, L.O. Hall, W.P. Kegelmeyer, Smote: synthetic minority over-sampling technique, Journal of Artificial Intelligence Research 16 (2002) 321–357.
[11] M. Galar, A. Fernandez, E. Barrenechea, H. Bustince, F. Herrera, A review on ensembles for the class imbalance problem: bagging-, boosting-, and hybrid-based approaches, IEEE Transactions on Systems, Man and Cybernetics. Part C, Applications and Reviews 42 (4) (2012) 463–484.
[12] E. Ramentol, Y. Caballero, R. Bello, F. Herrera, SMOTE-RSB*: a hybrid preprocessing approach based on oversampling and undersampling for high imbalanced data-sets using SMOTE and rough sets theory, Knowledge and Information Systems 33 (2) (2012) 245–265.

[13] C. Elkan, The foundations of cost-sensitive learning, in: International Joint Conference on Artificial Intelligence, vol. 17, Lawrence Erlbaum Associates Ltd, 2001, pp. 973–978.

[14] C. Zhang, Y. Ma, Ensemble Machine Learning: Methods and Applications, Springer, 2012.

[15] L. Breiman, Bagging predictors, Machine Learning 24 (2) (1996) 123–140.

[16] Y. Freund, R.E. Schapire, A decision-theoretic generalization of on-line learning and an application to boosting, Journal of Computer and System Sciences 55 (1) (1997) 119–139.

[17] X. Wu, V. Kumar, J.R. Quinlan, J. Ghosh, Q. Yang, H. Motoda, G.J. McLachlan, A. Ng, B. Liu, S.Y. Philip, et al., Top 10 algorithms in data mining, Knowledge and Information Systems 14 (1) (2008) 1–37.

[18] R. Barandela, R.M. Valdovinos, J.S. Sánchez, New applications of ensembles of classifiers, Pattern Analysis & Applications 6 (3) (2003) 245–256.

[19] S. Wang, X. Yao, Diversity analysis on imbalanced data sets by using ensemble models, in: Computational Intelligence and Data Mining, 2009. CIDM'09. IEEE Symposium on, IEEE, 2009, pp. 324–331.

[20] C. Seiffert, T.M. Khoshgoftaar, J. Van Hulse, A. Napolitano, Rusboost: a hybrid approach to alleviating class imbalance, IEEE Transactions on Systems, Man and Cybernetics. Part A. Systems and Humans 40 (1) (2010) 185–197.

[21] N.V. Chawla, A. Lazarevic, L.O. Hall, K.W. Bowyer, Smoteboost: improving prediction of the minority class in boosting, in: European Conference on Principles of Data Mining and Knowledge Discovery, Springer, 2003, pp. 107–119.

[22] I. Barandiaran, The random subspace method for constructing decision forests, IEEE Transactions on Pattern Analysis and Machine Intelligence 20 (8) (1998) 1–22.

[23] Y. Freund, R. Schapire, N. Abe, A short introduction to boosting, Journal of Japanese Society for Artificial Intelligence 14 (771–780) (1999) 1612.

[24] J. Friedman, T. Hastie, R. Tibshirani, The Elements of Statistical Learning, vol. 1, Springer Series in Statistics, Springer New York Inc., New York, NY, USA, 2001.

[25] R.A. Fisher, The use of multiple measurements in taxonomic problems, Annals of Eugenics 7 (2) (1936) 179–188.

[26] K. Hechenbichler, K. Schliep, Weighted k-Nearest-Neighbor Techniques and Ordinal Classification, Discussion Paper 399, SFB 386, Ludwig-Maximilians University Munich, 2004, URL http://nbn-resolving.de/urn/resolver.pl?urn=nbn:de:bvb:19-epub-1769-9.

[27] M. Jureczko, L. Madeyski, Towards identifying software project clusters with regard to defect prediction, in: Proceedings of the 6th International Conference on Predictive Models in Software Engineering, PROMISE '10, ACM, New York, NY, USA, 2010, 9, URL http://doi.acm.org/10.1145/1868328.1868342.

[28] W. Claes, R. Per, H. Martin, C. Magnus, R. Björn, A. Wesslén, Experimentation in software engineering: an introduction, Available online: http://books.google.com/books.

[29] J. Nam, W. Fu, S. Kim, T. Menzies, L. Tan, Heterogeneous defect prediction, IEEE Transactions on Software Engineering.

[30] K.E. Bennin, J. Keung, P. Phannachitta, A. Monden, S. Mensah, Mahakil: diversity based oversampling approach to alleviate the class imbalance issue in software defect prediction, IEEE Transactions on Software Engineering 44 (6) (2018) 534–550.

[31] A. Agrawal, T. Menzies, Is better data better than better data miners?: on the benefits of tuning smote for defect prediction, in: Proceedings of the 40th International Conference on Software Engineering, ACM, 2018, pp. 1050–1061.

[32] D. Bowes, T. Hall, J. Petrić, Software defect prediction: do different classifiers find the same defects?, Software Quality Journal 26 (2) (2018) 525–552.

[33] X. Yu, M. Wu, Y. Jian, K.E. Bennin, M. Fu, C. Ma, Cross-company defect prediction via semi-supervised clustering-based data filtering and MSTrA-based transfer learning, Soft Computing 22 (10) (2018) 3461–3472.

CHAPTER 7

Structuring large models with MONO: Notations, templates, and case studies

Harald Störrle

QAware GmbH, München, Germany

Contents

7.1. Introduction

Conceptual models can be considered knowledge bases [25,26]. As such, they represent financial and intellectual investment, sometimes at a considerable scale. Such investment can be worthwhile through two mechanisms. Firstly, building up the knowledge base is valuable in and by itself already, because making knowledge explicit forces the modelers to clarify tenuous assumptions: building a model amounts to learning about the domain being modeled. Secondly, the artifacts holding a conceptual model allow persisting and distributing the knowledge they represent more easily. Of course, being

Model Management and Analytics for Large Scale Systems
https://doi.org/10.1016/B978-0-12-816649-9.00016-8

able to retrieve knowledge from a model effectively is essential. For small and medium models, full-text search and manually browsing a model is sufficient: it is fast, simple, and effective, and it needs no specific tool support. As models grow, however, this approach becomes inadequate. Rising up to this challenge, many approaches for advanced model querying have been proposed over the last decades (e.g., [3,33,34]; see [21] for a recent comprehensive account of the field). These approaches are often limited with respect to expressive power, general applicability, tool availability, and not least quality of user experience [1]. None of them has been widely adopted. A more pragmatic solution is to create a tree-like model outline with defined structures and names for the diagrams, model elements, and submodels. This approach to structuring knowledge is already widely used for directory structures on any conventional hierarchical file system.

There are indications that applying guidelines improves model quality [18], so it is no wonder there are several guidelines for modeling for different modeling languages like UML [15], UML-RT [8], BPMN and similar languages [6], and Matlab/Simulink [13]. There are also guidelines on various other aspects of modeling, including diagram size and layout [12,23,30], naming model elements [10], and model syntax and style [4]. Finally, the spectrum of guidelines ranges from very low-level detailed guides like the guidelines from dSpace or the MAAB up to very high-level, almost philosophical accounts such as [22]. However, apart from [31], no guidelines are readily available for structuring models. There are guidelines for setting up directory structures for *coding* projects, though, such as the Maven guidelines[1] or the various Common Java Practices,[2] underlining the need for this type of guideline, and highlighting the gap that is there for modeling – the gap we aspire to fill with this chapter.

To this end, we report two case studies, one from industry and one from academia, about outline structures for large models. In order to be able to describe the model outlines precisely and succinctly, we introduce the Model Outline NOtation (MONO). First, however, we elaborate what we mean by model size and modeling in the large, and we analyze the requirements for effective structures for large models.

7.2. Modeling in the large

Handling "Big Data" is very different from handling smaller amounts of data, and large models are very different from small models along several dimensions:

- **Volume:** First of all, large models are made up of a large number of model elements and their relationships. A larger model means more work, more modelers, and more time spent modeling.

---

[1] maven.apache.org/guides/introduction/introduction-to-the-standard-directory-layout.html.

[2] www.javapractices.com, specifically www.javapractices.com/topic/TopicAction.do?Id=269.

- **Views:** Second, large models often have many diagrams. Note that this depends on the context: in a BPM or Matlab setting, "model" and "diagram" are synonymous terms, and every model element belongs to exactly one diagram. In a UML setting, however, there may be any number of diagrams portraying any set of model elements. Thus, model elements may belong to any number of diagrams. Again, more diagrams means more work.
- **Visibility:** Third, large models have a large audience. Modelers create it, domain experts review it, developers consult it to inform their implementation, and so on. These people differ in their intents and backgrounds, and not catering for all of them reduces the value of the model, endangering the investment a model represents. Thus, a large size of the audience implies a higher demand for consistency and finish than for smaller models. Also, large audiences often go hand in hand with geographical distribution, which increases effort all by itself.
- **Variation:** Fourth, the amount of change applied to a model reflects the amount of learning and evolution that took place during the model's lifetime. Larger amounts of change corresponds to greater intellectual and financial investment.

As a corollary, the greater an investment is, the longer it takes to recoup that investment. Longer amortization durations invite external changes to become effective, e.g., changes of the tooling and modeling language, evolution of the original and the intent underlying the model, and the overall project context. This drift adds to the natural erosion of the orderliness of models over time ("model decay"), representing a potential threat to the value of a model.

All of these characteristics represent important aspects of model size that are significantly different for large models than they are for small models. Table 7.1 documents what amounts to small or large models in these dimensions. These dimensions are relatively easy to measure, if access to the projects and its artifacts is available. In 2006 we conducted a survey on large models [29]. The results are summarized in Fig. 7.1, which shows the views and the volume as x- and y-positions, respectively. Where data are available, we have also charted the visibility as the circle size, and the variation as a flag inspired by wind strength representations on marine weather charts. Two of these are the case studies presented in Sections 7.5 and 7.6.

Table 7.1 Dimensions of model size: The magnitude of the majority of the metrics determines the overall size of a model.

Dimension of size	S	M	L	XL	XXL
VOLUME	$\ldots 10^2$	$\ldots 10^4$	$\ldots 10^6$	$\ldots 10^8$	$> 10^8$
VIEWS	1	$\ldots 10^2$	$\ldots 10^4$	$\ldots 10^6$	$> 10^6$
VISIBILITY	1	$\ldots 10$	$\ldots 50$	$\ldots 100$	> 100
VARIATION	$< 10\%$	$\ldots 100\%$	$\ldots 500\%$	$\ldots * 10$	$> * 10$

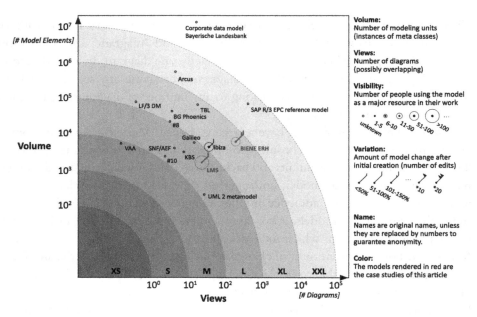

Figure 7.1 Results from a 2006 survey on model sizes: *x*- and *y*-position show the number of diagrams (views) and model elements (volume), respectively. The size of the circles represents the number of people that worked on or with the model (visibility), the flags represent the amount of change a model has endured (variation). Data for the latter two dimensions are incomplete.

7.3. Structuring big models

7.3.1 The need for structuring big models

Given the characteristics of large models discussed in the previous section, it is clear that many model creation and management tasks of large models are difficult in comparison to small models. This extends to the most basic activities associated to modeling: adding and retrieving knowledge to/from the knowledge base that is a model. For a small model, it is easy to see, literally, whether a certain fact is contained in the model, or not. It is just as easy to add a new nugget of knowledge in a way that it can be found again. For large models, that does not work.

Large *volume* means that a simple full–text search often yields too many false positives to be effective, and many *views* means that there are too many diagrams to be able to browse the model manually with reasonable effort. There have been many approaches to provide powerful search facilities on models that are suitable for large audiences [21,33], though their usability is still a problem largely unaddressed [1]. Therefore, high *visibility* means that there are many model users for whom more advanced search facilities are not accessible (cf. "End–user Modelers", [2]). Employing external documentation to guide knowledge discovery is not an economically viable option in the face of frequent *vari-*

Table 7.2 Requirements for solutions to structuring large-scale models.

Effectiveness	PERSISTENCE	conducive to storing and retrieving knowledge
	COLLABORATION	supports diverse and distributed contributors
Practicality	USABILITY	easy to understand and use by diverse users
	UNIVERSALITY	cover common scenarios and environments
Durability	SCALABILITY	works for models that grow larger than expected
	STABILITY	remain fully functioning despite drift and decay

ation, and keeping discovery documentation in sync with the model implies substantial effort.

Similar problems arise for other, more advanced model management activities. So, clearly, there is a need to organize large models in a way to accommodate the effect of large size. We identify the following requirements for an approach to address these issues (see Table 7.2 for a summary):

- **Persistence:** Clearly, the solution must provide a way to easily store and retrieve knowledge in large models. As a consequence, it should reduce the number of misplaced model elements, and thereby reduce the density of model clones [32,24].
- **Collaboration:** Next, it must support the collaboration of diverse and distributed teams, with differing intents and capabilities.
- **Usability:** Modeling is difficult in itself. Any solution to structuring a model should make modeling easier, not more difficult. As most modelers are experts in the respective domain that is being modeled rather than in conceptual modeling, the solution must be very simple to understand and apply. Complex features are not widely accepted.
- **Universality:** It should be applicable in a wide range of contexts in terms of application scenarios, modeling languages, methodologies, and modeling tools. Tool-specific features are of limited use.
- **Scalability:** The solution must scale up to large models, that is, hundreds and thousands of diagrams and model elements. As a rule of thumb, if a model persists, it grows an order of magnitude beyond what is initially expected.
- **Stability:** Finally, large models can have a long enough lifetime for the environment to evolve beyond what could be imagined at the time of kick-off. A successful structuring solution must remain fully functioning despite evolution, drift, model decay, and growth.

7.3.2 Means for structuring big models

The problem of structuring large models is similar to the need of large organizations to organize large corporate file shares. There, standardized directory structures are commonly found. This is a simple, universally applicable, and globally used solution.

We observe that all modeling tools and languages have some kind of grouping mechanism similar to file directories. Such model grouping mechanisms can be used to create an aggregation hierarchy similar to directory trees. We argue that such a solution addresses the requirements listed above in a particularly cost-effective way.

✔ **Effectiveness:** Obviously, the solution works for storing and retrieving electronic files (*Persistence*), and it works for thousands of people collaborating in large corporations (*Collaboration*). We see no reason why it should not work for model elements and diagrams.

✔ **Practicality:** Clearly, directory trees are readily understood and applied by very many people (*Usability*). It does not use any complex features or advanced tooling, and is thus generally applicable (*Universality*).

✔ **Durability:** Aggregation structures apparently scale up to very large structures (*Scalability*) and seem to be stable under decades of drift, decay, and growth (*Stability*).

In modeling languages and tools, different kinds of aggregation hierarchies have been proposed. First, there are static structures, like the aggregation tree readily found in modeling tools for UML, EPCs, or BPMN (see Fig. 7.3 for an example). Some modeling languages even have specific model elements for structuring models, e.g., `Package`[3] in UML, or the model groups of ADONIS. Second, many tools offer dynamic grouping, e.g., views or collections that are computed based on filters or queries. Changing the underlying model can be reflected automatically in updated views. Also, views can overlap, which is clearly an advantage over static structures. However, dynamic views are strongly dependent on tools and require extra facilities for sharing and persistence. This approach is thus vulnerable to drift and evolution. Third, naming conventions can be used to establish groups of model elements or submodels. This structuring mechanism has the disadvantage that it is difficult and error-prone to change or extend naming conventions once in place. However, they can be used in addition to static (or dynamic) model structures to establish additional relationships or structures.

In summary, we believe that static aggregation structures are the best solution to organize big models. In the next section, we define the basic terminology for model structuring, provide examples, and show how model structures can be described visually.

7.4. Describing and specifying model structures

From our point of view, models are trees with three kinds of nodes: diagrams, model elements, and submodels. A model, then, is an individual root unit into which other model elements, diagrams, or submodels are nested. There are two major paradigms how such trees can be organized.

---

[3] In the remainder, we often use the UML metamodel as an example, highlighting its metaclasses typographically by initial capital letters, camel-caps, and monospace font, e.g., as in `ControlNode`.

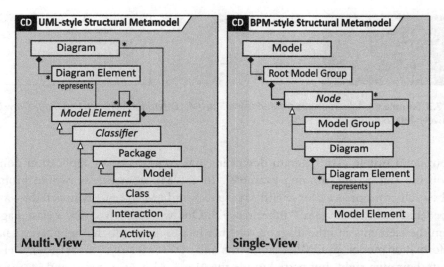

Figure 7.2 There are two modeling paradigms with different conceptual understandings of model structures: in a multiview setting (e.g., UML), `ModelElements` are nested recursively and diagrams are attached (left). In a single-view paradigm (e.g., BPM or Matlab/Simulink), diagrams are leaves of the model structure tree, and model elements are parts of diagrams (right).

On the one hand, there is the *multiview paradigm* typically found in software engineering, and exemplified by UML with model elements that act as containers and may be nested recursively to form a containment tree. Diagrams and their elements appear only as the leaves of this tree, and there is an n–to–1 element between model elements and diagram elements. That is to say, model elements may occur in no diagrams, or in several diagrams. For instance, in UML, the top–level root unit is a `Package` into which more `Classifiers` like `Package`, `Class`, `Activity`, or `Interaction` are nested recursively.

On the other hand, there is the *single-view paradigm*, which is commonly found in BPM or Matlab/Simulink contexts. For instance, ADONIS offers `Root model groups` and `Model groups` as tool facilities outside the language proper to define the containment structure with diagrams at the leaves of this tree. The model elements proper are contained in the diagram, not in the structural containment tree as such. In particular, diagrams and model elements are not nested recursively. Fig. 7.2 summarizes and juxtaposes the multi- and single-view model structure paradigms.

If a model does not have any nested submodels we call it a *simple model*, and its type is solely determined by the kinds and numbers of model elements and diagrams contained in the model, along with any constraints and restrictions applied to them. A model with nested model elements is called a *compound model*. Its type is determined not just by the model elements and diagrams it contains, but also by the submodels, and possibly by their multiplicity. As a consequence, the structure of a model is a tree where each inner node carries an (implicit) type.

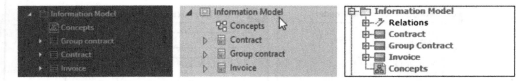

Figure 7.3 Sample model structures created using *StarUML, Enterprise Architect,* and *MagicDraw UML,* respectively (left to right).

Consider a simple class diagram describing the concepts of an application domain (we will use this as our running example). In a UML context, we would probably describe each concept as a `Class` with `Attributes` and `Operations`, connected by `Associations` and `Generalizations` ("inheritance"). One would likely create a class diagram and join the elements and the diagram together in a `Package` which might carry a name like **Information model**. In UML, `Operations` and `Attributes` of `Classes` are model elements in their own right, but nested inside the `Class`. Also, `Operations` and `Attributes` may contain other model elements like `Parameters` and `Visibilities`, and so on. In fact, whole `Classes` may be nested inside other `Classes`. In a modeling guideline, we would want to specify this constellation exactly, along with any naming conventions, multiplicity constraints, and other side conditions. Clearly, using prose is not a very convenient tool even for the simplest, most traditional flat type of model.

Alternatively, one might actually model the desired structure in a modeling tool and capture the resulting model structure. Fig. 7.3 shows the aggregation trees from three popular UML tools, i.e., MagicDraw UML, StarUML, and Enterprise Architect.[4] While this is a straightforward way to create templates in a given setting, it comes with several disadvantages. First, actually creating model structures first means considerable effort and creates tool-specific artifacts. For instance, consider the different handling of `Associations` by the three tools in Fig. 7.3: MagicDraw groups them in a virtual node "Relations," while StarUML and Enterprise Architect hide them altogether. Enterprise Architect also hides the `Generalization`. Secondly, model outlines created this way are only examples, not generic specifications. Any variation points and constraints would still have to be described in prose, adding more effort.

In order to allow for tool-independent, generic descriptions of model structures with less overhead, we propose MONO (Model Outline NOtation), a visual notation to describe model outlines. The notation is straightforward and can easily be emulated with any drawing tool. Fig. 7.4 shows our running example model outline **Information model**. On the left of Fig. 7.4, we show a one-to-one translation of the model outlines shown in Fig. 7.3. On the right, we present a more abstract specification of this model structure. Fig. 7.6 defines the MONO metamodel.

---

[4] See www.nomagic.com/products/magicdraw, staruml.io, and sparxsystems.com, respectively.

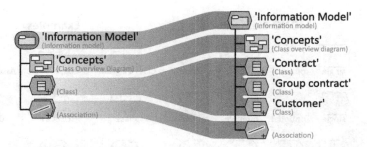

Figure 7.4 Refinement from the generic Information Model outline specification (left) to a concrete outline instance (right), both expressed visually with MONO. The model outline specified here would be realized in concrete modeling tools as shown in Fig. 7.3.

MONO distinguishes three kinds of entities, i.e., models, diagrams, and model elements, represented by graphemes with different color and shape, where gray rectangles represent diagrams, amber hexagons represent model elements, and blue roundtangles represent (sub)models. Entities may be further differentiated into subtypes by nesting an icon. Entities may be equipped with a name and/or type annotation, e.g., a specific name or type, but more complex expressions are also possible (see below). Stacking two entity icons expresses multiplicity; refinement by entities of the same kind is indicated by the plus sign in the bottom right corner. Lists of entities that are understood or defined elsewhere are represented by a named entity icon with three dots. The names of elements can be specified using a combination of constant strings in single quotes, character classes as all caps, named variables in brackets, and some regular expression-like multiplicities. Additionally, white spaces and round brackets can be used for legibility. Fig. 7.5 summarizes the elements of our notation.

Elements may be further differentiated into subtypes by nesting an icon. For instance, different types of diagrams may be distinguished by the icon in the gray box. If the icon is filled white it is just a visual identifier that serves to distinguish one element from another. If the icon is yellow (light gray in print version), however, it refers to a particular model element. When used with a model entity, it indicates the type of model element used as the container. For instance, the **Process model** in Fig. 7.5 is nested inside a `Package` element. Conversely, the **System structure** model type does not indicate what kind of model element it is nested into. In this case, the container is also a `Package`, though, as this is the default. For added clarity, we suggest to also provide an explicit type annotation below the naming constraint. This may refer to an existing type of model element or diagram, or a user-defined one, including names of submodels. Instead of providing a single type, comma-separated lists may be used.

With this notation, we can now express more complex model outlines effectively. Fig. 7.7 demonstrates how MONO can be used to define recursive model structures, allowing for reuse and succinct declarations. Outline (A) defines the most generic model

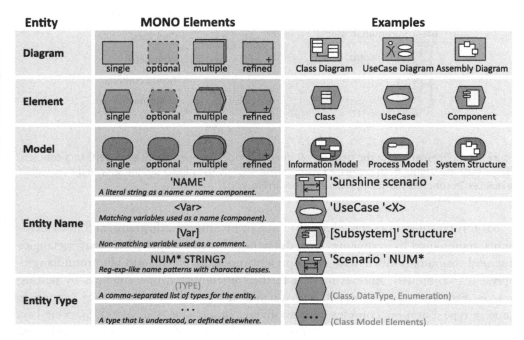

Figure 7.5 Elements of MONO: the graphemes for the three kinds of entities are distinguished by color and shape. Stacking two icons expresses multiplicity, homogenous nesting is indicated by the plus sign in the bottom right corner.

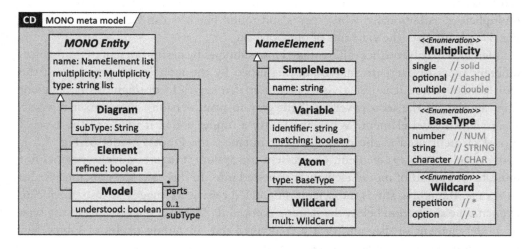

Figure 7.6 The metamodel of MONO as a UML class diagram: there are three types of MONO entities, which have names, multiplicities, and types. The names may be compounds of name element-like strings and variables. The types are (lists of) type names and may be complemented by different kinds of subtypes.

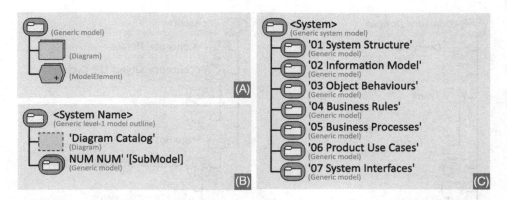

Figure 7.7 Examples of reusing model outlines in a modular way: (A) a Generic model outline with any number of diagrams and elements, possibly refined; (B) Generic Level-1 Model Outline with name constraints on submodels; and (C) System Model outline with a set of submodels with defined names, but undefined content.

structure: a package with any number and any kind of diagram and model element, with no naming restrictions. Outline (B) defines a top–level structure where the root is a `Package` containing an optional diagram named **Diagram catalog**, and any number of generic submodels as defined by (A). The names of the submodels are made up of two numeric characters, a space, and an arbitrary, meaningful name for a submodel. Outline (C) defines a first–level model outline with concrete names. Observe that outline (C) satisfies the constraints imposed by outline (B). So, instances of outline (C) are also instances of outline (B), but not the other way around.

A further model is shown in Fig. 7.8. It defines domain model structures which contain both structure and behavior. By structuring, it allows for larger models while maintaining a manageable structure. The domain model defines an overview diagram and detail diagrams, a `Package` for data type declarations with a diagram as index, the usual model elements for class models, and a set of submodels to elaborate on particularly complex domain concepts. Those special domain concepts come with an overview diagram, a `StateMachine` for the object life cycle, and, optionally, a `Package` with `Interactions`. Note that the details for the `Interactions` and the `StateMachine` are not made explicit here, as they are understood. In fact, all of the detail structure of the `Interactions` and the `StateMachine` could have been omitted.

7.5. Case study 1: Library Management System (LMS)

The Library Management System (LMS) is a large UML model used mainly for academic teaching. Table 7.3 provides an overview of the most important structural and size metrics. This case study exemplifies the multiview model structuring paradigm.

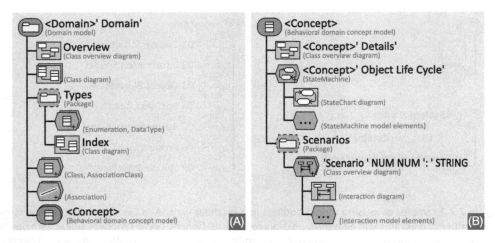

Figure 7.8 Outline specifications for Domain models: (A) a generic model for complex domains, packaging type declarations, and complex individual concepts; (B) submodel for complex concepts with life cycles.

7.5.1 Model creation and evolution

The LMS was created as a teaching example. It originated in 2009, when four students created it as their course work assignment in an MSc-level 13-week Requirements Engineering course taught by the author at the Technical University of Denmark (DTU, www.dtu.dk). After that, the model was sanitized and extended, to serve as a template solution for students in subsequent instances of the course. The model evolution lasted until 2015, when the author ceased to teach that course. During that time, the model was continuously adapted, improved, and extended by a sequence of teaching assistants, a PhD-student, and the author. The model was also used as a template and study example in other modeling-related courses taught by the author, and as sample data in several research projects.

The model is expressed in UML 2.4 using a full-blown MagicDraw modeling environment, including the TeamWork Server and several actively used plugins. In line with the mainly educational intent of the model, a wide set of notations and concepts of UML were used, which is not necessarily representative of industrial usage. Also, the focus was on syntactical correctness and methodological effectiveness rather than domain fidelity.

7.5.2 Model usage

The LMS model has mainly been used as a teaching example in several courses at the MSc- and BSc-level where modeling played an important role. Students in those courses were asked to create a large model, or parts of a model, for some application

Table 7.3 A summary description of the Library Management System (LMS) model.

LANGUAGE	UML 2.4
TOOL	MagicDraw 16.6-17.0.3
FILE	MDXML (XMI 2.1), 2.59MB
SIZE	3,850 model elements, 82 diagrams, 2.53MB, ca. eight authors/500 users, approx. 80% change
PURPOSE	The models were originally created as course work, and then upgraded to serve as a template for students in subsequent courses. The model was also used as sample data in several research projects.
HISTORY	Started in 2009 as part of a student project, the model was enhanced and extended in several iterations until 2015.
MODELERS	Four MSc students (first version); subsequent versions by several teaching assistants, a PhD student, and the author.
INFORMATION	Five class diagrams
	268 Properties, 67 Associations, 58 Operations, 33 Classes, 21 Generalizations, 10 Enumerations, 5 DataTypes
SNAPSHOTS	Three object diagrams
	171 Slots, 82 Values, 52 InstanceSpecifications
FUNCTIONALITY	32 use case diagrams
	44 Use Cases, 7 Actors, 5 Includes-relationships
PROCESSES	30 activity diagrams
	236 Actions, 472 ActivityNodes, 460 Flows, 138 ControlNodes, 98 ObjectNodes, 74 Parameters, 62 Swimlanes, 38 Events
SYSTEM	Three assembly diagrams
	54 Ports, 46 Connectors, 19 Components
BEHAVIOR	Five state machine diagrams
	49 Transitions, 19 States, 11 PseudoStates, 7 Regions
SCENARIOS	One interaction diagram
	14 Messages, 5 Lifelines
STRUCTURING	410 packagedElement, 43 Packages
	Nesting depth: avg. 2.38/max. 5
AUXILIARY	832 Literals, 21 Constraints, 3 Comments

domain. The resulting model was a major part of the course assignment, so it was fairly important to the students and they demanded detailed guidance on what to create. In order to provide this guidance the students were given the LMS model together with a detailed report explaining its content in prose. This case study also served as a very detailed example for many aspects of modeling, and was used as the running example in all course materials.

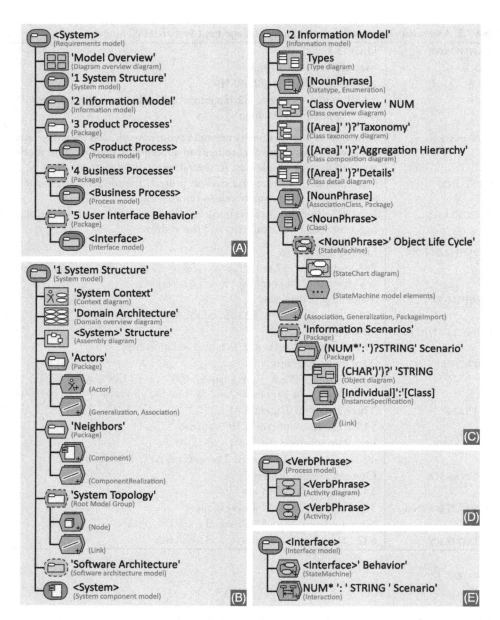

Figure 7.9 Part 1 of the LMS model outline (see Fig. 7.10 for pt. 2): the top-level outline (A) is refined by four distinct model kinds that can (and do) exist individually (B)–(E).

7.5.3 Model structure

The outline of the LMS model is shown in Fig. 7.9. The guiding metaphor for this model outline is that of a course syllabus, which is of course a consequence of the main

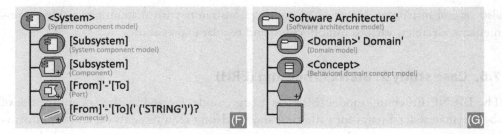

Figure 7.10 Part 2 of the LMS model outline (see Fig. 7.9 for pt. 1): the construction-level parts of the overall model (F), (G).

usage scenario of this model as a pedagogic tool. This impacts the model outline in several ways.

First, the top-level Packages reflect different kinds of models that can be used independently in diverse contexts: courses with different focus and scope can use different constellations of submodels. Second, the top-level Packages (see Fig. 7.9A) have fixed names and start with a number. That way, it is easier to identify parts of the model without leaving room for misunderstandings. The initial numbers ensure that these packages always appear in a specific order, reflecting the course syllabus. Third, the names and scopes of submodels and entities are chosen to be as generic as possible so that they can indeed be used independently. In a more restricted environment with specific constraints on the size, purpose, and domain of the model, the outline could be more streamlined to better match the environment. Conversely one might argue that a generic, standardized structure is better suited to accommodate project drift and other characteristics of large models (see our discussion in Section 7.2).

The first two submodels – 1 System Structure and 2 Information Model – are "true" submodels in the sense that they are elaborate substructures that could exist independently (see Fig. 7.9B and C). The submodels 3–5, on the other hand, are just containers for sets of simpler structures (see Fig. 7.9D and E; note that one of the submodel types is used twice).

Submodel 1 System Structure exhibits a recursive structure that allows to nest models arbitrarily; see Fig. 7.9B and Fig. 7.10F. Note that in Fig. 7.10F, subsystems can be either flat (i.e., simple components) or refined, in which case a new "System component model" may be instantiated. Of course, only one of the two should be done for any system, but MONO is not sufficient to express this constraint. It has to be expressed externally, e.g., by a prose description. Fig. 7.10F allows construction-level submodels without constraining them.

Submodel 2 Information Model shows a particular kind of rich class model with optional object life cycles for classes, and scenarios consisting of ensembles of Instance-Specifications and object diagrams to model specific states. In this submodel, there are also several instructive instances of naming constraints with matching variables, commentary variables, various multiplicities, and regular expressions.

7.6. Case study 2: BIENE Erhebung (ERH)

The BIENE Erhebung model (ERH) is a large model created by the tax authorities of the German federal states for gathering and validating requirements of, and for informing construction and deployment of, administrative systems concerned with levying taxes.[5] Table 7.4 provides an overview of the most important structural and size metrics. This case study exemplifies the single-view model structuring paradigm.

7.6.1 Model creation and evolution

The BIENE project aspires to replace the software for handling all tax-related payments and the associated processes in the federal states of Germany. Those systems have been in operation since the 1950s and 1960s, and have been under heavy maintenance ever since to accommodate multiple layers of jurisdiction and regulation, from the European Union level via national and state levels to organizational rules of the tax authorities of the 16 federal states of Germany. Most of these systems were written in COBOL and Assembler for mainframe computers. They still run today and will likely continue to do so for years to come. So, it is not unreasonable to expect a BIENE lifetime in the same range. For political reasons, BIENE is a federated project with teams working in several cities. BIENE is a very large, mission-critical, and high-risk project with a long and painful history of failed previous attempts.[6]

In the first six months of the BIENE project, a *Domain-Specific Modeling Language* (DSML) was created together with the domain experts. It offers a syntax inspired by UML but with a much smaller and more concise metamodel, plus several specialties to address domain needs. The DSML is implemented using the ADONIS BPM metamodeling tool (then a commercial tool, now open source [16]; see www.adonis-community.com/en). In addition to the language proper and the modeling tool to support it, a development process was tailored for BIENE, following the VM'97 standard that was mandatory in Germany at the time for large public projects like BIENE [11]. Also, several more detailed guidelines on modeling, tool usage, and other aspects were created when the need emerged. Altogether, the package comprising the language, tool, method, and guidelines will henceforth be called the *environment*. After

---

[5] The model is entirely created in German. In the interest of accessibility, all names have been translated to English for this paper. "Erhebung" translates to collection or levying.

[6] Originally started in 1991, it was restarted in 2001 and again in 2005, at which point the setup reported here was implemented. From 1991 to 2005, the project has delivered *"50,000 pages of documentation and 1.6 million lines of mostly useless code"* [5, p. 23] (our translation). Depending on the source, project cost estimates range in the hundreds of millions of euros, plus 4,500 million euros in unclaimed taxes [5,14,19].

Table 7.4 A summary description of the BIENE ERH model.

LANGUAGE	Custom-created UML-like DSML
TOOL	Adonis 3.5NT
FILE TYPE	ADL
SIZE	11,661 model elements, 523 diagrams, 21.4MB, ca. 35 active and 300 passive users, approx. 200% change
PURPOSE	The models were created with three intentions: (1) to specify system requirements; (2) to validate requirements across organizations and groups; and (3) to inform developers.
HISTORY	In 2004, the project was initiated by creating the modeling language in the meta-CASE environment ADONIS. Modeling proper started mid-2004 and lasted until at least 2014.
MODELERS	Authors: approx. 30 civil servants plus five modeling coaches
	Users: approx. 220 civil servants providing domain expertise and approx. 80 developers and architects
INFORMATION	48 class diagrams
	527 Associations, 352 Classes, 40 AssociationClasses, 30 Packages, 15 Aggregations, 1 Generalization
FUNCTIONALITY	109 use case diagrams
	341 Functions, 170 UseCases, 76 Actors, 64 Includes-relationships
PROCESSES	231 activity diagrams, 46 function pools
	2,931 Flows, 1,500 Actions, 1,149 ControlNodes, 16 Swimlanes
SYSTEM	37 assembly diagrams, 46 system structure diagrams
	263 Ports, 180 Systems, 175 Connectors, 8 RelaisPorts
ORGANIZATION	Five organization charts
	19 Clerks, 13 Roles, 6 ReportsTo-relationships
STRUCTURING	371 Model Groups, 1,270 PartOf-, 64 Contains-relationships
	Nesting depth: avg. 6.53/max. 8
AUXILIARY	482 Plan Heads, 349 Notes

the initial six-month development period, small adjustments were continuously made to the environment for four years.

7.6.2 Model usage

After the initial development of the BIENE modeling environment, it was deployed to *domain teams* who would each work independently on a particular application subdomain. Every domain team consisted of a lead modeler, one or two additional modelers, an architect, and several developers. Every domain team would co-opt domain and technology experts as they were needed, particularly a *modeling coach*. That way, the very diverse background, tasks, and levels of expertise of the teams could be addressed.

Every team would kick off a *domain project* with a three-day workshop in which a coach trained one to four modeling teams simultaneously. There, domain teams initiated their submodels and started working with it, documenting requirements and constructing a solution. When a domain project had reached a degree of completeness, extensive and distributed reviews were conducted to ensure that the needs of all BIENE clients were met. After satisfying any additional requirements the development would ensue, followed by integration testing. In the meantime, the modeler had started the next modeling assignment, which in practice meant that modeling teams were involved in up to four different domains at any given time, all in different stages of completion.

Apart from requirements gathering and software development, the models were also used for the engineering test cases, creating exploratory prototypes, project management and controlling, and informing the migration planning. In some cases, where substantial domain knowledge was truly missing, models would also be used to document the insights generated by reverse engineering old code. In a nutshell: the BIENE project was truly model-based, and models were first-class citizens.

For the core modeling team, it was essential that topic teams could work independently on their respective parts of the model and achieve completeness and correctness. For the construction crew, it was important that the model would be detailed and stable enough to effectively inform the software construction and testing.

Fig. 7.11 shows the domain architecture of the tax levying system under construction in 2008, ca. four years into the project. Every line of text represents a business process or function, every rectangle represents an application domain or subsystem (the darker the shading the deeper the nesting). It defines the complete scope of the functionality associated with the tax levying domain. This domain architecture is the result of a comprehensive survey of required functionality. While complete in scope, it is not final in terms of the degree of detail: only the two parts with dashed heavy outline are complete in that sense, too; the other parts were not completed at the time. A typical domain project would comprise realizing around five of these processes and functions in one iteration.

7.6.3 Model structure

The outline of the ERH model is shown in Fig. 7.13. The top-level model outline reflects the project setup: the first four `RootModelGroups` implement a staged pipeline for developing independent subsystems, where submodels of unrelated application domains are created in group 0 and moved on to groups 1–3 as they mature. Therefore, `RootModelGroups` 0 through 3 have the same second-level structure; see Fig. 7.13B.

0 Workspace   is the area for models under construction.

1 Under Review   stores models under review for alignment to stakeholder needs.

2 Accepted   contains models after review and during software construction.

3 Constructed   acts as documentation of completed submodels.

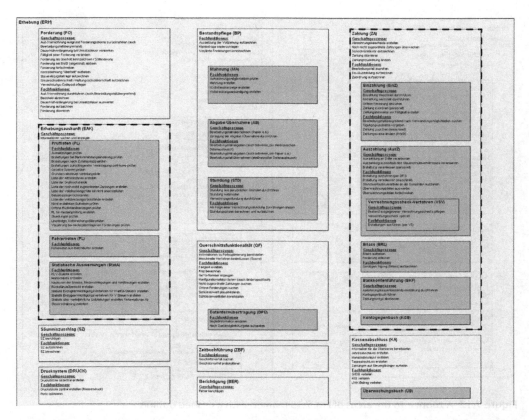

Figure 7.11 The ERH domain architecture as of 2008, presented as the Domain architecture diagram in Fig. 7.13C. The original was an A0-size poster, an enlarged part is shown in Fig. 7.12.

At the second level of the model outline the decisive influence of the project purpose comes into view: the single most important goal of BIENE was to achieve consensus across all stakeholders and ensure ownership and acceptance from all the authorities involved. Therefore it was vital that the project was run in several organizations, which led to a strict decoupling of subprojects for subdomains so that impediments in one of them would not delay others. In terms of the model outline, this was achieved by creating independent submodels for different domain teams and factoring out common concerns, such as the data structures that all domain projects shared. So, Fig. 7.13B defines ModelGroups for the interface of one domain project to others (0 Context), and for models common to all subdomains (1 Shared models and 2 Pools).

The third outline level of ERH uses the same structural metaphor as the second level of the LMS model, namely, that submodels correspond to chapters of a book. Thus, there are strong structural similarities: ERH Context model and System model (Fig. 7.13C and E) combined are similar to the LMS System model (Fig. 7.9, B),

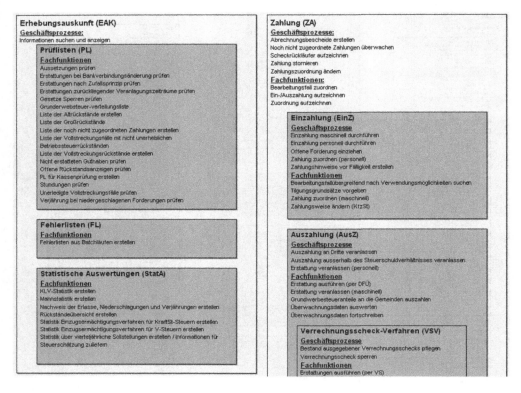

Figure 7.12 A zoom into the areas of Fig. 7.13 marked with a dashed outline.

and the ERH Information model (Fig. 7.13H) is similar to the LMS Information model (Fig. 7.9C). They differ in details, though, e.g., the scenarios in ERH are less expressive than those in LMS.

`RootModelGroups` 4 and 5 support the involvement of a large number of people only loosely attached to the project. These provisions have been made necessary by the strongly distributed nature of the project.

4 Project Info   informs participants about the project organization, process, and progress as well as dependencies between work units and project plans. It also holds tutorials and guidelines on the modeling language and methodology.

5 Playground   allows for experiments by inexperienced users and temporary modeling sketches without harming the production models.

The model outline also reflects the tooling environment. The BIENE project selected ADONIS as the modeling tool at an early stage. Since ADONIS had been created with Business Process Modeling in mind, which means that the primary entity are diagrams, not model elements – model elements are only parts of diagrams. The concepts for model structuring (`ModelGroups` and `RootModelGroups`) are not part of the modeling

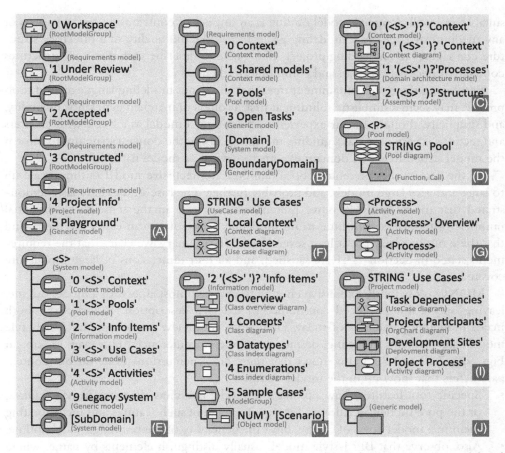

Figure 7.13 Outline of the ERH model: the top-level outline (A), (B) reflects the process model and uses recursion to cope with system size (C). Due to the modeling paradigm embedded in the ADONIS tool, this outline is structured by model groups and diagrams only, while model elements play no role.

language as such. Also, there is no multiview modeling as is commonly found in UML modeling tools, so all model elements are contained in *exactly* one diagram. Thus, it is redundant to specify model element types, and specifying a diagram type is completely sufficient. Many of the constraints and details we had to specify explicitly in the LMS model outline in Fig. 7.9 are implicit in the language definition in the ERH outline.

7.7. Discussion

Standardizing model structures makes it easier to locate knowledge in a model, not least because it makes it easier to identify the right place for adding and changing a model, which in turn reduces the deleterious effect of drift. Therefore, establishing a suitable and well-designed model outline is an important contribution to the success of any medium or large-scale modeling project, just like the quality of a directory structure contributes to any coding project. Being able to describe such directory structures concisely is necessary to plan and communicate them.

Employing the basic containment trees that all serious modeling languages and tools provide may seem a simplistic solution at first. It offers unmatched generality, flexibility, and simplicity, though. In our experience, describing a model as a book with chapters and sections is very useful as a guiding metaphor. Using the domain structure to inform the model structure allows domain experts to navigate a model naturally.

As the case studies presented above demonstrate, recursive model outlines scale up to large and very large models, and probably further. The case studies have also demonstrated that model outlines as we presented them scale in the sense that substantial numbers of modelers with little to no modeling background can easily use them, and that these outlines may last at least several years. In summary, we claim that the requirements stated in Section 7.3.1 are satisfied, and all four dimensions of large models are covered.

MONO is obviously expressive enough to describe most aspects of a model outline, leaving only a few advanced constraints for external description. It also covers both major modeling paradigms of today, UML-style multiview modeling, and BPM-style models-are-diagrams modeling, despite their inherent differences. However, looking at Figs. 7.9, 7.10 and 7.13 also shows that there are repercussions from various context parameters that show up in the model outlines.

- Specific tool features show up, like the model overview diagrams in MagicDraw, or the pool models in Adonis. Taking advantage of such features and incorporating them in a model does, of course, limit the generality of the resulting model outline.
- Also, observe that BPM-style models usually distinguish elements by name, where UML-style models identify elements by name and type, or internal identifier. So, some sets of name constraints may not be consistent in all environments (e.g., Fig. 7.9F).
- The organizational structure and constraints of the BIENE project directly impact a great deal of the ERH outline. Or, conversely, the simple model outlines we propose can actually accommodate and support project goals.
- The modeling paradigm shows up at first sight in the outlines: where the ERH outline shows mainly blue roundtangles (submodels) and some gray rectangles (diagrams), the LMS outline shows a lot more amber hexagons (model elements). In other words: the ERH outline is defined mostly in terms of diagrams and (recursive) submodels while the LMS is defined mainly by model elements and their relationships. This is a byproduct of using diagrams as containers and embedding a great amount of the structural constraints in their definitions. It is also a pattern that is typical when contrasting BPM-style model outlines with UML-style model outlines.

Irrespective of the modeling paradigm, naming conventions are very important when facing long lifetime or large audiences. Failing to put them into place at the very beginning turns into a burden later.

7.8. Conclusions

7.8.1 Summary

Modeling projects benefit from structuring their models just like all projects benefit from structuring their artifacts. As models grow, the need becomes more and more pressing. Also, model structure templates are useful for guiding junior modelers and domain experts. Finally, modeling is inherently difficult. Providing a standardized model outline takes one burden away from modelers and lets them focus on the task at hand: collecting and creating new knowledge in order to clarify murky questions and resolve controversy.

Despite these benefits, experience and advice on how to best structure models is not publicly available. As a first step towards true best practices, we report on two large models and explain their outline structures. We describe these outlines using a dedicated precise visual notation that we define and introduce for this occasion. The notation is a necessary prerequisite to comparing, communicating, and consolidating model outline knowledge, which we hope to encourage by this chapter.

7.8.2 Related work

In the Business Process Modeling arena, structuring large models has been an issue for a long time. For instance, ARIS – probably the most widely used modeling language in the IT industry – comes with several notations supporting high-level views of process models. Practical guidebooks complement them by practical guidance (see, e.g., "segmenting models" in [9, Section 11.3.3]).

Similarly, UML provides notations for hierarchical decompositions of large diagrams such as Activity Overview Diagrams. Other notations like StateChart Diagrams had hierarchical decomposition built into them from the start. Some tools also add specific features for providing more structure in large models. For instance, MagicDraw comes with "Overview Diagrams" where modelers can create diagrams whose graphemes represent other diagrams. A different path is pursued by VisualParadigm, which offers diagram layers [27,28] to tackle complexity within one diagram.

In the academic research literature, there is very little information available on means of structuring models, i.e., concepts, formalisms, and tools. Notable exceptions are [17] and [36] for formal method-inspired approaches and [7] for a more hands-on engineering approach.

Another angle on model structuring comes from recent work model slicing [20,35]. A model slice is a submodel that is closed under certain metaassociations and contains a seed model element (or set of model elements), the so-called slicing criterion.

However, we perceive a dearth of scientific work on how models ought to be structured, that is, studies, best practices, or guidelines, or even just examples or experience reports, let alone standards or systematic studies on how to best structure models. Apart from our own previous work [31], we are not aware of any such research. Although we have anecdotal reports of some relevant internal material in organizations that do large-scale modeling, no such material seems to be published, e.g., as an internal report or an industry white paper.

7.8.3 Contributions

This paper has four major contributions. First, we elaborate what "size" means for models, propose a conceptual framework and some metrics to measure the size of models, and also provide interpretations for the figures yielded by these metrics. Second, we propose MONO as a way to visually define the structure of a model, either for documenting the outline of an existing model, or for designing and planning new model structures. Third, we present two case studies of large and very large models and show how MONO can be applied to describe their model outlines. Allowing for recursive definitions, MONO offers a substantial amount of scalability. Fourth, between the introductory examples and the case studies, we provide many small and large examples of how models can actually be structured. These examples can not yet be considered best practices, as they lack independent confirmation. Yet, they are a starting point that people can refer to and elaborate on. We hope that this chapter will help to raise the visibility of this issue and spawn new ideas for model structures.

7.8.4 Limitations and future work

While we are confident that the notation and model outlines presented here will hold up under scrutiny, they obviously need independent validation before they can be considered reliable. Also, we have only presented case studies for specific coordinates in the space of models and modeling languages. It is currently unclear, just how far the outlines and structuring means we have proposed will carry. In particular, scaling *down* there will be a point at which the effort of defining a dedicated model outline will exceed the benefit derived from it. Where this point is, exactly, is an open question. Also, we have not validated MONO on Matlab/Simulink models, which is a highly relevant modeling environment in industrial practice.

We see three major areas for future work. First, there is a need to formalize MONO such that model outline checkers can be created. Among other things, it would check the consistency of a given outline and whether a given model conforms to a predefined model outline. Second, putting MONO into practice requires tool support, both for editing the visual notation and for checking it against model instances. Third, it would be desirable to extend the notion of a submodel and the conformance checking in such a way as to define a proper notion of modularity of models. This would be a necessary prerequisite for sharing models based on an interface while hiding details.

References

[1] S. Abrahão, et al., User experience for model-driven engineering: challenges and future directions, in: ACM/IEEE 20th Intl. Conf. Model Driven Engineering Languages and Systems, MODELS, IEEE, 2017, pp. 229–236.

[2] Vlad Acreţoaie, Model Manipulation for End-User Modelers, PhD thesis, Tech. Univ. Denmark, Dept. Appl. Math. and Comp. Sci., 2016.

[3] D.H. Akehurst, B. Bordbar, On querying UML data models with OCL, in: Martin Gogolla, Chris Kobryn (Eds.), Proc. 4th Intl. Conf. Unified Modeling Language, UML, in: LNCS, vol. 2185, Springer, 2001, pp. 91–103.

[4] Scott Ambler, The Elements of UML Style, Cambridge University Press, 2003.

[5] Dirk Aspendorf, Absturz von Amts wegen, Die Zeit 30 (July 15, 2004).

[6] Jörg Becker, Michael Rosemann, Christoph von Uthmann, Guidelines of business process modeling, in: Wil van der Aalst, et al. (Eds.), Proc. Intl. Conf. Business Process Management, BPM, in: LNCS, vol. 1806, Springer, 2000, pp. 30–49.

[7] Jean Bézivin, et al., A canonical scheme for model composition, in: Proc. 2nd Eur. Conf. Model Driven Architecture-Foundations and Applications, ECMDA-FA '06, in: LNCS, vol. 4066, Springer, 2006, pp. 346–360.

[8] Tuhin Kanti Das, Jürgen Dingel, Model Development Guidelines for UML-RT, Tech. rep. 2016-628, School of Computing, Queen's University, Kingston, Canada, 2016.

[9] Rob Davis, Business Process Modelling with ARIS - A Practical Guide, Springer, 2001.

[10] Florian Deißenböck, Markus Pizka, Concise and consistent naming, in: Proc. 13th Intl. Ws. Program Comprehension, IWPC'05, IEEE, 2005.

[11] Wolfgang Dröschel, Manuela Wiemers, Das V-Modell 97, Oldenbourg, 1999.

[12] H. Eichelberger, K. Schmid, Guidelines on the aesthetic quality of UML class diagrams, Information and Software Technology 51 (12) (2009) 1686–1698.

[13] Tibor Farkas, Christian Hein, Tom Ritter, Automatic evaluation of modelling rules and design, in: 2nd Ws. Prom Code Centric to Model Centric Software Engineering: Practices, Implications and ROI, 2006.

[14] Das Fiscus-Projekt steht vor dem Scheitern (The Fiscus project is about to fail), in: Computerwoche, January 2, 2002.

[15] Frank Hilken, et al., Towards a catalog of structural and behavioral verification tasks for UML/OCL models, in: Proc. Nat. Conf. Modellierung, Gesellschaft für Informatik eV, 2016.

[16] Stefan Junginger, et al., Ein Geschäftsprozessmanagement- Werkzeug der nächsten Generation – ADONIS: Konzeption und Anwendungen, Wirtschaftsinformatik 42 (5) (2000) 392–401.

[17] Pierre Kelsen, Qin Ma, A modular model composition technique, in: D.S. Rosenblum, Gabriele Taentzer (Eds.), Proc. ISth Intl. Conf. Fundamental Approaches to Software Engineering, FASE'10, Springer, 2010, pp. 173–187.

[18] Christian F.J. Lange, et al., An experimental investigation of UML modeling conventions, in: 9th Intl. Conf. Model Driven Engineering Languages and Systems, MoDELS'09, in: LNCS, vol. 4199, Springer, 2006, pp. 27–41.

[19] Peter Mertens, Schwierigkeiten mit IT-Projekten der öffentlichen Verwaltung, Informatik-Spektrum 32 (1) (Feb. 2009).

[20] Christopher Pietsch, et al., Incrementally slicing editable submodels, in: Proc. 32nd Intl. Conf. Automated Software Engineering, ASE, IEEE Press, 2017, pp. 913–918.

[21] Artem Polyvyanyy (Ed.), Process Querying Methods, Springer, 2018, http://link.springer.com/article/10.1007/tbd.

[22] Reinhard Schuette, Thomas Rotthowe, The guidelines of modeling-an approach to enhance the quality in information models, in: Conceptual Modeling-ER'98, Springer, 1998, pp. 240–254.

[23] Harald Störrle, Diagram size vs. layout flaws: understanding quality factors of UML diagrams, in: M. Genero, A. Jedlitschka, M. Jørgensen (Eds.), Proc. 10th Intl. Conf. Empir. Softw. Eng. and Measurement, ESEM, ACM, 2016.

[24] Harald Störrle, Effective and efficient model clone detection. Essays dedicated to Martin Wirsing on the occasion of his emeritation, in: Rocco De Nicola, Rolf Hennicker (Eds.), Software, Services and Systems, in: LNCS, Springer, 2014.

[25] Harald Störrle, How are conceptual models used in industrial software development? A descriptive survey, in: E. Mendes, K. Petersen, S. Counsell (Eds.), Proc. 21st Intl. Conf. on Evaluation and Assessment in SE, EASE, ACM, 2017.

[26] Harald Störrle, Implementing knowledge management in agile projects by pragmatic modeling, in: Ina Schaefer, et al. (Eds.), Proc. Fachtagung Modellierung, Gesellschaft für Informatik, 2018, pp. 233–244.

[27] Harald Störrle, Improving model usability and utility by layered diagrams, in: Ina Schaefer, et al. (Eds.), Proc. Ws. Modeling in SE (MiSE), ICSE Companion, ACM, 2018.

[28] Harald Störrle, Improving modeling with layered UML diagrams, in: Joaquim Filipe, et al. (Eds.), Proc. 1st Intl. Conf. Model-Driven Engineering and Software Development, SCITEPRESS, 2013, pp. 206–209.

[29] Harald Störrle, Large scale modeling efforts: a survey on challenges and best practices, in: Wilhelm Hasselbring (Ed.), Proc. IASTED Intl. Conf. Software Engineering, Acta Press, ISBN 978-0-88986-641-6, 2007, pp. 382–389.

[30] Harald Störrle, On the impact of layout quality to the understanding of UML diagrams, Software & Systems Modeling 17 (1) (2018) 115–134, https://doi.org/10.1007/s10270-016-0529-x, accepted 2016-04-23.

[31] Harald Störrle, Structuring very large domain models: experiences from industrial MDSD projects, in: Ian Gorton, Carlos Cuesta, Muhammad Ali Babar (Eds.), Proc. 4th Eur. Conf. Sw. Architecture (ECSA): Companion Volume, ACM, 2010, pp. 49–54.

[32] Harald Störrle, Towards clone detection in UML domain models, Software & Systems Modeling 12 (2) (2013) 307–329, http://link.springer.com/article/10.1007/s10270-011-0217-9.

[33] Harald Störrle, VMQL: a generic visual model query language, in: Martin Erwig, Robert DeLine, Mark Minas (Eds.), Proc. IEEE Symposium on Visual Languages and Human-Centric Computing, VL/HCC, IEEE CS, 2009, pp. 199–206.

[34] Harald Störrle, Vlad Acreţoaie, VM*: a language for end-user model transformation, in: Artem Polyvyanyy (Ed.), Process Querying Methods, Springer, 2018, pp. 61–90, Chap. 3, http://link.springer.com/article/10.1007/tbd.

[35] Gabi Taentzer, et al., A formal framework for incremental model slicing, in: Alessandra Russo, Andy Schürr (Eds.), Intl. Conf. Fundamental Approaches to Software Enqineerinq, FASE, Springer, 2018, pp. 3–20.

[36] Ingo Weisemöller, Andy Schürr, Formal definition of MOF 2.0 metamodel components and composition, in: K. Czarnecki, et al. (Eds.), Proc. 11th Intl. Conf. Model Driven Engineering Languages and Systems, Springer, 2008, pp. 386–400.

CHAPTER 8

Delta-oriented development of model-based software product lines with DeltaEcore and SiPL: A comparison

Christopher Pietscha, Christoph Seidlb, Michael Niekeb, Timo Kehrerc

aDepartment of Electrical Engineering and Computer Science, University of Siegen, Siegen, Germany
bInstitute of Software Engineering and Automotive Informatics, Technische Universität Braunschweig, Braunschweig, Germany
cDepartment of Computer Science, Humboldt-Universität zu Berlin, Berlin, Germany

Contents

8.1. Introduction

Model–based software development (MBSD) has become a widespread approach to implement software, especially for embedded systems. In MBSD, models replace source

167

code as primary development artifacts. Thus, implementing a system means creating one or several models specifying the behavior of the system using modeling languages such as Simulink, ASCET, a subset of the UML, or domain-specific modeling languages. Complex model-based systems must often be delivered in a large number of variants with slightly different functionality to satisfy customer demands, e.g., in the automotive domain. *Software product line* (*SPL*) engineering is a methodology to represent such a family of software systems in terms of the functionality shared by all variants (commonalities) and the functionality which is specific for an individual variant or for a subset of variants (variabilities) [24]. On the conceptual level, configurable functionality is usually represented in terms of *features* and the configuration logic to determine configurations (valid combinations of features) is specified in a *feature model* [15,4]. On the realization level, features must be implemented within reusable artifacts, e.g., as changes to models in accordance with a feature being (de)selected. On this basis, an SPL can generate a variant for any valid feature combination. A model-based SPL (MBSPL) combines MBSD and SPL engineering by specifying variability in models and generating models as variants. Techniques for implementing and maintaining MBSPLs are key to support the model management for large-scale model-based systems evolving into many versions and variants.

Delta modeling is a transformational approach to implement MBSPLs [32]. In essence, an MBSPL is implemented by a core model, representing one concrete variant, and a set of interrelated delta modules defining transformations to realize individual features. To specify a delta module, a dedicated delta language is used that aligns with the core concepts of the targeted language; e.g., DeltaJava [29,47] uses transformation operations targeting especially the language concepts of Java. A variant is derived by applying one or several delta modules onto the core model in accordance with a selected configuration.

While the foundations of delta-oriented MBSPLs are well understood, realizations within tools are faced with the following challenges: (1) dedicated delta languages are required for each modeling language, which are tedious to create; (2) an SPL must be defined, which includes the creation of delta modules; (3) an SPL needs to be maintained and developed further as part of software evolution; and (4) variants need to be derived by applying the delta modules.

These are serious obstacles for applying delta modeling in practice. In this chapter, we present a set of essential capabilities that should at least be offered by an integrated development environment for delta-oriented MBSPLs in order to address these challenges. In particular, we analyze how these capabilities are provided by *DeltaEcore* [33, 39] and *SiPL* [25], two tool suites for the delta-oriented development MBSPLs which are implemented based on the *Eclipse Modeling Framework* (*EMF*) [9] and which are designed to be adaptable to arbitrary EMF-based modeling languages. We highlight the individual strengths of both tool suites and compare them regarding their capabilities. With these contributions, for academics, we illustrate the state of practice, e.g., as basis

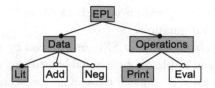

Figure 8.1 Feature model for Expression Product Line (EPL) [30].

for implementing results of their ongoing research and, for practitioners, we provide means to select and apply the most adequate tool in a given scenario by delta-oriented development of MBSPLs.

The rest of the chapter is organized as follows: Section 8.2 introduces a small yet illustrative MBSPL taken from the literature used as running example throughout the chapter. Section 8.3 briefly recalls the basic ideas of delta-oriented implementation of MBSPLs from a conceptual point of view, focusing on the development tasks of creating an MBSPL from scratch and managing its evolution over time. Experienced readers familiar with the basic concepts of delta-oriented MBSPLs may skip individual aspects of this section. Subsequently, Section 8.4 describes how these development tasks are supported by the tool suites DeltaEcore and SiPL. Section 8.5 discusses and contrasts essential capabilities of both tool suites regarding their underlying concepts, technologies, and workflows, with the ultimate goal of assessing their suitability for different application scenarios. Section 8.6 demarcates our tool suites from related work and Section 8.7 concludes the chapter along with an outlook on future work.

8.2. Running example

In *SPL engineering*, a feature constitutes a user-visible functionality of a variant. *Feature modeling* is a widespread approach to define all valid feature combinations, also referred to as *problem space*. Each valid feature combination (i.e., a *configuration*) specifies exactly one *variant* of the SPL. A feature model can be expressed as a *propositional formula* over all features or using a graphical notation [4]. Each feature is either *mandatory* or *optional* and may contain sets of subfeatures grouped by *and*, *or*, and *alternative groups*.

Fig. 8.1 shows a feature model for an *Expression Product Line* (*EPL*; cf. [30]), which serves as running example throughout this chapter. It consists of two and-groups *Data* and *Operations*. The feature *Data* declares three kinds of expressions, i.e., literal (*Lit*), addition (*Add*), and negation (*Neg*). The feature *Operations* defines two kinds of operations, which can be performed on expressions, represented by the features *Print* and *Eval*. The print operation writes expressions to the console, while the eval operation evaluates the expression and returns the result. The mandatory features with a gray background represent the *core features* of the SPL, i.e., the features that are common to each variant of the SPL. The optional features with a white background represent the variability of the SPL leading to eight configurations.

Commonalities and variabilities of an SPL are realized by developing a generic architecture for all variants and implementing reusable artifacts, commonly referred to as *solution space*. Moreover, a suitable *variation mechanism* (see Section 8.3) must be chosen, which specifies how the problem space will be mapped onto the solution space and how to derive a specific variant by tailoring realization artifacts to functionality of the features in the selected configuration.

Variation mechanisms for MBSPLs can be classified into *annotative, compositional*, and *transformational* approaches [10]. Annotative approaches form a so-called 150% model, which contains all possible variations of a realization artifact within an SPL and, during variant creation, removes those parts that are not needed by the features selected in a configuration. Compositional approaches focus on a base model common to all variants of an SPL and create composition modules with specific additions to the base module so that a variant can be created by copying the base model and superimposing the respective composition modules for the features selected in a configuration. Finally, transformational approaches use a core model of an SPL and specify transformation modules containing instructions to add, modify, and remove elements so that, during variant creation, the changes associated with the features selected in a configuration can be performed to realize the desired functionality.

8.3. Delta modeling for MBSPLs

Delta modeling is a transformational approach, in which the solution space is described by a *core model* and a set of *delta modules* specifying changes to the core model in terms of adding, modifying, and removing elements. Unlike annotative and compositional approaches, which use specialized SPL representations of realization artifacts, delta modeling usually uses a regular variant of a realization artifact as core model (e.g., the most commonly requested variant), which permits the use of regular tools for specification and analysis. However, the flexibility of delta modeling also permits that the core model is a partial variant and that it does not comprise all core features of the SPL.

In the following, we assume that model-based technologies are used for implementing the realization artifacts of the EPL introduced in Section 8.2. More precisely, each variant of the EPL is implemented as a structured data model using the EMF [9]. EMF allows the specification of *structured data models*, including the specification of operations, for building applications by generating respective *Java code* from the model.

8.3.1 Delta language creation

A *delta module* consists of calls of predefined *delta operations*, referred to as *delta actions*, an *application condition* (*when*-clause), and may reference *required* delta modules (*after*-clause).

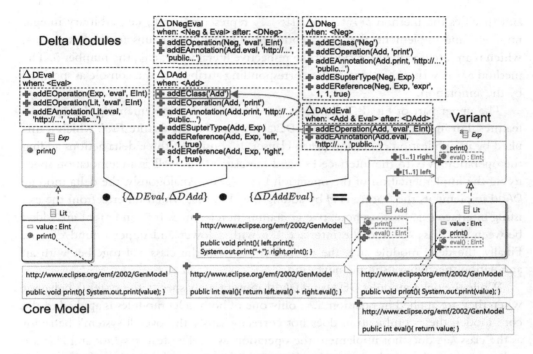

Figure 8.2 Core model of the EPL and delta modules for optional features.

The application condition maps the problem space onto the solution space by relating (combinations of) features with delta modules. Given a valid configuration, all the delta actions of all delta modules for which the application condition evaluates to *true* are applied onto the core model in a suitable order, which is implied by the after–clause of the involved delta modules.

The set of available delta operations depends on the respective modeling language and an appropriate *delta language* is needed, which specifies operations to appropriately alter a model of a given source language. An interpreter for the delta language is required to execute delta operations during variant derivation. For instance, the delta language for the EMF metamodel used in the EPL running example must specify operations to *add* and *remove* elements of the types EClass, EAttribute, EOperations, etc., and operations to *modify* such elements by setting respective properties.

8.3.2 Software product line definition

In the following, we illustrate the realization of the EPL MBSPL and give an overview of the delta modules needed to generate variants for all valid EPL configurations.

The lower left part of Fig. 8.2 shows an EMF-based implementation for the core features of the EPL which is used as core model. The interface Exp represents expressions and specifies the method `print`. The class `Lit` represents literals, i.e., arbitrary integer numbers, and implements the aforementioned interface. It contains an attribute `value`, which is an integer representation of the respective literal value, e.g., the number 5. The method `print` writes the value of the corresponding attribute to the console as specified by the annotation.

The upper part of Fig. 8.2 illustrates the delta modules implementing the optional features. For instance, the delta module **DEval** implements the feature *Eval*, which is implied by its application condition (*when*-clause). It consists of three delta actions adding the operation `eval` to the interface `Exp` and the class `Lit` as well as an annotation specifying the implementation of the operation `Lit.eval()`. Analogously, the delta module **DAdd** implements the feature *Add* by adding the class `Add`, which inherits from the existing interface `Exp`. Furthermore, two containment references `left` and `right` are added between the class `Add` and the interface `Exp` with a lower and upper bound of one. Finally, the delta module adds the operation `print` to the class `Add` together with an annotation specifying its implementation.

While the delta modules **DEval** and **DAdd** appropriately specify the system's behavior when they are applied in isolation, i.e., only one of both delta modules is applied to the core model, their combination does not correctly satisfy the overall system's behavior as the class `Add` does not implement the operation `eval`. The features *Eval* and *Add* are independent in the problem space but interact in their realization in the solution space. This general phenomenon is referred to as *optional feature problem* [14]. In delta modeling, this problem can be solved by specifying a separate delta module that implements this interaction. The delta module **DAddEval** implements the feature interaction between the features *Add* and *Eval* by adding the operation `eval()` to the class `Add` along with the specification of its implementation via annotation. The delta module's application condition consists of a conjunction of the features *Add* and *Eval*. The delta module **DAdd** inserts the class `Add` to which the operation `eval()` is added by the delta module **DAddEval**, i.e., **DAdd** must be applied first, otherwise the application of the delta module **DAddEval** would fail, which is specified by an appropriate *after*-clause in **DAddEval**. The delta modules **DNeg** and **DNegEval** implement the feature *Neg* and its interaction with the feature *Eval* similar to the previous delta modules.

8.3.3 Software product line evolution

Although SPLs are, by design, well prepared to meet diverse requirements, one cannot anticipate all future requirements which will arise throughout the entire lifetime of an SPL [5,40]. Consequently, over time, (MB)SPLs are subject to *evolution*, i.e., features along with their respective realization or valid feature combinations could be added, removed, or modified. In Fig. 8.3, the feature *Print* of the EPL becomes optional and the feature *Eval* becomes a core feature.

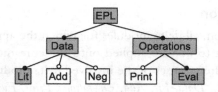

Figure 8.3 Evolved feature model for the EPL.

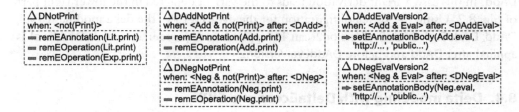

Figure 8.4 Delta modules used in the evolution of the EPL.

In addition, the implementation of the feature *Eval* is updated so that, whenever evaluating an expression with the `eval()` method, the current intermediate result is printed to the console, which effectively creates a new version of the feature *Eval*.

While making *Eval* mandatory could be implemented analogously to the previous examples by adding to the core model through delta modules, the need for removal of making *Print* optional creates a deep impact onto the overall solution space. Opposed to compositional approaches to implementing (MB)SPLs, delta modeling can deal with removals as well [30]. Fig. 8.4 shows the respective delta modules for realizing the evolution of the EPL.

The delta module **DNotPrint** removes the print operation from the interface Exp and class Lit and is applied when the configuration does not include the feature *Print*, which is implied by the corresponding application condition. The delta modules **DAddNotPrint** and **DNegNotPrint** remove the print operation and its implementation from the classes Add and Neg. The former is only applied if the configuration includes the feature *Add* but does not include the feature *Print*. Thereby, the delta module **DAdd** must be applied first, as indicated by the after–clause. Analogously, the delta module **DNegNotPrint** is applied only if the configuration includes the feature *Neg* but not *Print* where the delta module **DNeg** is applied first. The delta modules **DAddEvalVersion2** and **DNegEvalVersion2** realize the evolution of the feature *Eval* by modifying the body of the existing annotation of the operations Add.eval() and Neg.eval(). Both delta modules must be applied after their corresponding delta modules **DAddEval** and **DNegEval**.

8.3.4 Variant derivation

Given a valid configuration, all delta modules for which the application condition over selected features evaluates to *true* are applied onto the core model in a suitable order, which is implied by the after-clauses of the involved delta modules. For example, given the feature selection {*EPL, Data, Lit, Add, Operations, Print, Eval*} for the initial version of the EPL, the application condition of the delta modules DEval, DAdd, DAddEval would evaluate to *true* and they can be applied in any order in which DAddEval is applied after DAdd. In this case, variant derivation leads to the variant depicted in the lower right part of Fig. 8.2. Using the same configuration for the evolved EPL, the delta modules DEval, DAdd, DAddEval, DAddEvalVersion2 can be applied in any order in which DAddEval is applied after DAdd and before DAddEvalVersion2.

8.4. Delta modeling with DeltaEcore and SiPL

In this section, we give a short overview of the main functionality of *DeltaEcore* and *SiPL* using our running example to compare the realizations within both tool suites. We provide the artifacts for implementing the running example in an online archive[1] and direct interested readers to the webpages of DeltaEcore [33] and SiPL [25] for the latest version of the respective tool. For didactic reasons, we deliberately chose our concise running example to illustrate the tools' functionality. However, we also mention larger-scale applications for each of the tools at the end of the respective subsection.

8.4.1 DeltaEcore

The core functionality of the tool suite DeltaEcore[2] [39] lies within delta language creation, SPL definition, and variant derivation, as shown in Fig. 8.5. When using DeltaEcore to manage variability of an SPL, each of these parts is used as illustrated for the running example of the EPL.

8.4.1.1 Delta language creation

To transform a realization artifact of a particular source language through delta actions, delta modeling requires a respective delta language that supplies suitable delta operations, e.g., DeltaJava [29,47] for Java. One of the core functionalities of DeltaEcore is the generation of a delta language from the metamodel of an arbitrary source language. DeltaEcore provides the majority of a custom delta language through its *Common Base Delta Language* (*CBDL*), which contains language constructs to define, reference, and

---

[1] http://pi.informatik.uni-siegen.de/projects/sipl/mmalss2018/index.php.

[2] Note that in this chapter we are focusing on the solution space portion of DeltaEcore, which constitutes only part of the tool suite's functionality. The best use of DeltaEcore is in conjunction with an explicit feature model, specifically a Hyper-Feature Model (HFM) [40,38].

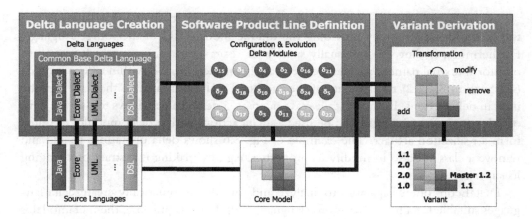

Figure 8.5 Overview of the core functionality of the tool suite DeltaEcore.

```
Ⓐ Ecore.decoredialect ⊠                                                    ⊟ ⊡
deltaDialect
{
  configuration:
    metaModel: <http://www.eclipse.org/emf/2002/Ecore>;
    identifierResolver: org.eclipse.emf.ecore.delta.resolver.EcoreIDResolver;

  deltaOperations:
    //EClass
    addOperation addEClass(EClass eClass, EPackage[eClassifiers] ePackage);
    modifyOperation setEClassName(String newName, EClass[name] eClass);
    modifyOperation setEClassAbstract(Boolean isAbstract, EClass[abstract] eClass);
    modifyOperation setEClassInterface(Boolean isInterface, EClass[interface] eClass);
    addOperation addESuperType(EClass superTypeEClass, EClass[eSuperTypes] eClass);
    removeOperation removeESuperType(EClass eClass, EClass[eSuperTypes] eSuperClass);
    detachOperation removeEClass(EClass eClass);

    //EClassifier
    customOperation setEClassifierPackage(EPackage ePackage, EClassifier eClassifier);

    //...
}
```

Figure 8.6 Excerpt from the delta dialect for Ecore models used within EPL.

require delta modules, etc., that are independent of the concrete source language. To create a custom delta language, it is necessary to interface with the respective source language. For this purpose, DeltaEcore permits the specification of so-called *delta dialects*, i.e., specifications of the delta operations that should be provided for transforming a realization artifact of the source language within a delta module. A *custom delta language* in DeltaEcore consists of the combination of the CBDL with the delta dialect for the source language. Fig. 8.6 shows an excerpt from the delta dialect for Ecore models as used within the EPL running example.

In the `configuration` part, a delta dialect has to tie to the source language via the keyword `metaModel` by providing the unique model URI of the source language. Furthermore, it allows to optionally specify an `identifierResolver`, which is used to transform textual names provided between angle brackets (<...>) within a delta module (see Section 8.4.1.2) to elements of the modified realization artifact that can then be used in delta actions. In the `deltaOperations` part, a delta dialect allows to provide a list of delta operations that can then be used as delta actions within delta modules to transform a realization artifact. The example of Fig. 8.6 shows delta operations to add and remove a class but also to modify its properties, e.g., by making it abstract or changing its super types.

DeltaEcore offers capacities to analyze and (mostly) automatically create delta languages suitable for a particular source language [39]. For this purpose, the metamodel is traversed for suitable elements and so-called *standard delta operations* to add, modify, and remove values for these elements are created automatically with predefined semantics, e.g., to set the value for nonconstant attributes. To extend a delta dialect beyond the capabilities of the atomic standard delta operations, it is possible to specify *custom delta operations* with user-defined semantics, e.g., the delta dialect for Ecore in Fig. 8.6 contains one such method to set the package of a classifier, which is only possible indirectly due to the structure of Ecore.

To facilitate execution of delta actions associated with the respective delta operations from the delta dialect, DeltaEcore generates an interpreter specific to the custom delta language. For standard delta operations, the interpreter contains a full implementation due to their defined standard semantics. For custom delta operations, the interpreter contains stub methods whose body has to be implemented by the creator of the delta language to reflect the semantics of the operation.

Note that delta language creation has to be performed only once for each source language, and in the case of Ecore a suitable delta language already exists, which is delivered with the tool suite.

8.4.1.2 Software product line definition

With a suitable delta language in place, it is possible to specify the artifacts needed by DeltaEcore to create an SPL definition. In particular, these are a set of *delta modules*, a central *mapping model* that contains all application conditions, and, optionally, a list of *application-order constraints*.

A **delta module** specifies delta actions to transform a particular realization artifact by using a delta language suitable for the source language of the realization artifact. Fig. 8.7 shows the DeltaEcore file for the delta module DAdd from the running example. In its first line, the delta module is given a descriptive name, e.g., "DAdd." The remaining lines are referred to as a *delta block*, which consists of a header and a body (within braces {...}).

```
▲ DAdd.decore ✕                                              ⚊ ☐
  delta "DAdd"

  dialect <http://www.eclipse.org/emf/2002/Ecore>
  //requires <OtherDeltaModule.decore>
  modifies <../../CoreModel/EPL.ecore>
  {
    EClass add_eClass = new EClass(name:"Add");
    addEClass(add_eClass, <epl>);
    addESuperType(<Exp>, add_eClass);

    EReference left eReference = new EReference(name:"left",
      containment:true, lowerBound:1, upperBound:1, eType:<Exp>);
    addEReference(left_eReference, <Exp>);
    EReference right eReference = new EReference(name:"right",
      containment:true, lowerBound:1, upperBound:1, eType:<Exp>);
    addEReference(right_eReference, <Exp>);

    EOperation add_eOperation = new EOperation(name:"print");
    setEOperationImplementation(add_eOperation,
      "left.print(); System.out.print(\" + \"); right.print();");
    addEOperation(add_eOperation, add_eClass);
  }
```

Figure 8.7 Delta module DAdd in DeltaEcore to realize the feature *Add*.

The header of a delta block, as first entry, specifies which delta language to use. As described in Section 8.4.1.1 only a delta dialect with suitable delta operations is needed to interface with a source language (i.e., Ecore) as the remainder of the delta language is provided by the common base delta language. Hence, using the `dialect` keyword, it is sufficient to provide the model URI used within an existing suitable delta dialect. With the `modifies` keyword, the header allows to specify which realization artifact should be transformed within the delta block, in this case, the EPL core model. For more complex scenarios, it would also be possible to provide a list of realization artifacts to be modified, e.g., when moving elements from one artifact to another. Optionally and in between the `dialect` and `modifies` keywords, it is also possible to use the `requires`[3] keyword to provide a list of delta modules that have to be applied as prerequisite for the delta module to be applicable.

The body of the delta block consists of a sequence of delta actions, i.e., calls to delta operations defined in the delta dialect with arguments to transform the targeted realization artifact(s). In the example, first, the class `Add` is added. Then, both its left and right references are created. Finally, its method `print()` is added with a suitable implementation. To target existing elements, DeltaEcore uses references, which are enclosed within angle brackets (`<...>`) and are resolved to suitable elements for the delta actions, e.g.,

---

[3] In contrast to an application–order constraint (see below), which is only enabled *if* two delta modules are selected together manually, the requires relation *forces* the required delta module(s) to be applied before the delta module.

```
[A] epl.demapping  ⊠
    Add:              <deltas/DAdd.decore>
    Eval:             <deltas/DEval.decore>
    Add && Eval:      <deltas/DAddEval.decore>
    Neg:              <deltas/DNeg.decore>
    Neg && Eval:      <deltas/DNegEval.decore>
```

Figure 8.8 Mapping in DeltaEcore to collect all application conditions for delta modules.

```
[A] epl.deapplicationorderconstraints  ⊠
    [<deltas/DAdd.decore>, <deltas/DEval.decore>]
    [<deltas/DAddEval.decore>]
```

Figure 8.9 Application-order constraints in DeltaEcore to prescribe a partial order on delta modules.

<Exp> is resolved to the interface Exp using either a generated basic reference resolver or a specialized one explicitly specified in the delta dialect (see Section 8.4.1.1).

Note that a delta module may, in principle, specify multiple delta blocks. This is sensible if a logically coherent set of transformations (e.g., to realize one particular feature) affects multiple realization artifacts of different source languages (e.g., if there also was a model specifying the concrete syntax of the EPL that would have to be subjected to variability according to the same features).

The **mapping model** relates (combinations of) features with a set of delta modules and, thus, constitutes a centralized model of application conditions for delta modules. Fig. 8.8 shows the DeltaEcore mapping model for the running example. In the general case, an entry to the mapping model consists of a left-hand side referencing a feature and a right-hand side giving a delta module within angle brackets (<...>), e.g., Add:<deltas/DAdd.decore>. However, more complex entries are possible. For one, the left-hand side may contain a complex expression over features comprising logical operators for negation (!), conjunction (&&), and disjunction (||), e.g., Add && Eval. Furthermore, the right-hand side may define a set of delta modules as comma–separated list (to be applied in an arbitrary order) if multiple delta modules are associated with the same expression over features.

Application–order constraints prescribe a partial order on delta modules to enforce constraints on their application sequence when deriving a variant from the SPL. In our running example, it has to be assured that the class Add was created and the method eval() added to its super class, i.e., the interface Exp, *before* adding the implementation of the method eval() to the class Add. However, this order only has to be obeyed *if* the selected configuration contains both features *Add* and *Eval*. Hence, the DeltaEcore delta modules **DAdd**, **DEval**, and **DAddEval** have to obey that (partial) order *if* they are selected together. Fig. 8.9 shows the respective application-order constraints file of DeltaEcore to enforce this partial order. Within brackets ([...]), DeltaEcore permits the definition of a *constrained group* containing arbitrarily many delta modules. Within a constrained group, the order of delta modules may be changed arbitrarily. However, the order of constrained groups themselves is maintained during variant derivation. Hence, a constrained group constitutes one level of a partial order specification. Note that it is also possible to nest constrained groups to realize more complex (partial) orders.

Also note that all the artifacts used in the definition of a DeltaEcore SPL are entirely model-based as they are founded in Ecore metamodels themselves. This means that DeltaEcore artifacts integrate seamlessly into an MBSPL scenario, e.g., delta modules could be generated through model transformation similar as we did in some of our work on reverse engineering SPLs [48] and pattern-based SPL development [41,42].

With the definition of delta modules, the mapping for application conditions, and the optional application–order constraints, a basic DeltaEcore SPL definition is complete and it is possible to derive variants as described in Section 8.4.1.4.

8.4.1.3 Software product line evolution

For the EPL running example, the changes for making the feature *Print* optional can be realized by delta modules that remove the respective elements from the core model. Similarly, making the feature *Eval* mandatory does not require any changes to the delta modules as the existing ones can still be applied.

However, creating a new version of a feature implementation, such as with the feature *Eval*, poses greater challenges: Not all users of an SPL may want to migrate to a new version of an SPL and its features immediately or completely, e.g., due to a dependency on an old version of a particular feature. In consequence, it may be necessary to include existing versions of features in the configuration process to be able to maintain them and combine them with other features, potentially in newer versions.

To address this challenge, DeltaEcore provides *evolution delta modules*. Evolution delta modules have the express intent of realizing versions of a feature's implementation. A version is perceived as resulting in an *incremental* change to the state of the realization of the previous version, i.e., further transformation. Hence, on the implementation level, evolution delta modules are similar to the delta modules that enable or disable a feature's functionality – which we refer to as *configuration delta modules* for disambiguation. However, conceptually, the two types of delta modules have different intents, which is relevant, e.g., in their application order where a feature has to be enabled before it can be migrated to a newer version. Fig. 8.10 shows an evolution delta module of the running example that migrates the implementation of the `eval()` method of the class `Add` to the new version of the feature *Eval*. Note that the delta module is introduced by the `evolution` keyword to signal to DeltaEcore that it is an evolution delta module.[4] In the example, the delta actions within the evolution delta module make

---

[4] In a similar manner, a `configuration` keyword may be used in the same place to make explicit that it is a configuration delta module. However, due to backward compatibility, use of this keyword is optional.

```
DAddEvalVersion2.decore  ⋇
  evolution delta "DAddEvalVersion2"

  dialect <http://www.eclipse.org/emf/2002/Ecore>
  requires <../DAddEval.decore>
  modifies <../../CoreModel/EPL.ecore>
  {
    EOperation eval_operation = <Add.eval()>;
    setEOperationImplementation(eval_operation, "int result = left.eval() + right.eval();
      System.out.println(\"Add: \" + result); return result;");
  }
```

Figure 8.10 Evolution delta module DAddEvalVersion2 in DeltaEcore to migrate the implementation of the `eval()` method.

```
epl.demapping  ⋇
  Add:                        <../deltas/DAdd.decore>
  Eval:                       <../deltas/DEval.decore>
  Add && Eval:                <../deltas/DAddEval.decore>
  Add && Eval[>="Version 2"]: <../deltas/DAddEvalVersion2.decore>

  //...
```

Figure 8.11 Excerpt of the updated mapping in DeltaEcore to collect all application conditions for delta modules including feature versions.

use of the same delta operations from the delta dialect as within configuration delta modules. However, it is possible to restrict the usage of some delta operations in a delta dialect to evolution delta modules by prepending their definition with an `evolution` keyword, e.g., especially invasive operations, such as refactorings, that should not be used for configuration.

To associate the evolution delta module with the new version of the feature *Eval*, the mapping model has to be updated. In Fig. 8.11, an excerpt from the revised mapping model shows how the new version of the feature *Eval* is used in a condition on when to apply the evolution delta module to update the implementation of the `eval()` method of the class `Add`.

When selecting a configuration that contains the feature *Eval* in the new version, the mapping model provides the appropriate evolution delta module[5] to trigger version migration of the feature.

8.4.1.4 Variant derivation

DeltaEcore provides multiple ways to select which variant should be generated: selecting a set of delta modules to be applied, defining a configuration, and, when interfacing with

---

[5] Note that evolution delta modules live up to their full potential when used with HFMs [40,38], which allow definition of features with versions on the conceptual level, automated derivation of application-order constraints, semiautomatic selection of version constellations, etc.

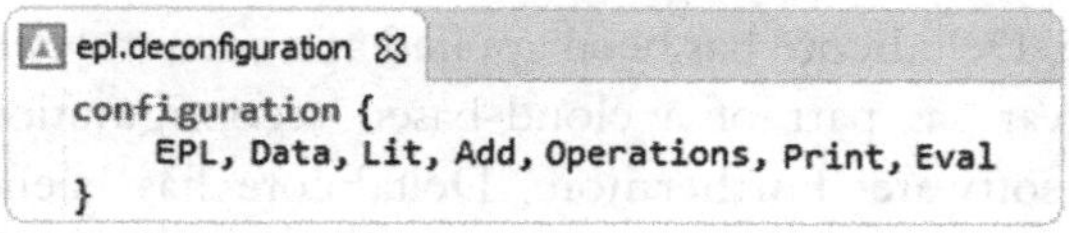

```
A epl.deconfiguration  ⊠
configuration {
    EPL, Data, Lit, Add, Operations, Print, Eval
}
```

Figure 8.12 DeltaEcore configuration for an example selection of features from EPL.

an explicit feature model [40], employing a graphical configurator. Fig. 8.12 shows an example of a DeltaEcore configuration for the EPL running example.

With the prior option, the initial set of delta modules is explicit. With the latter two options, the selection of features within the configuration is used in conjunction with the mapping model of application conditions to resolve the left-hand side of mapping entries and, if satisfied, to include the delta modules on the right-hand side in the initial set of delta modules. For the example of Fig. 8.12, the initial set of delta modules would be {DAdd, DEval, DAddEval}.

With either of these options, the determined initial set of delta modules is completed by transitively including those delta modules that are explicitly required from the header of the contained delta block(s) (see Section 8.4.1.2), which is not necessary within the running example.

With the full set of delta modules to be applied for one particular variant, DeltaEcore determines a valid application sequence. For this purpose, application-order constraints specified explicitly in the respective model and those defined implicitly by the requires relation of delta modules are considered when building a dependency graph of delta modules. Subsequently, this dependency graph is sorted topologically and, from the resulting partial order, one permissible total order is chosen as application sequence. For the example of Fig. 8.12, valid application sequences would be [DAdd, DEval, DAddEval] or [DEval, DAdd, DAddEval] due to the application-order constraints of Fig. 8.9 prohibiting DAddEval to be applied first.

For the actual variant derivation, DeltaEcore copies the core model of the SPL and then applies the relevant delta modules in the determined application sequence. For the application of delta modules, each delta block in each delta module has its delta dialect resolved so that the delta actions within the delta block body can be executed via the interpreter for the delta language (see Section 8.4.1.2). The result of variant derivation is the realization artifacts containing appropriate functionality for the selected features and, when using evolution delta modules, respective versions thereof.

8.4.1.5 Experiences

The implementation of the running example was performed by a user familiar with DeltaEcore and, with a previously existing feature and core model, consumed approximately 45 minutes of effective working time.

In larger-scale use, DeltaEcore has been applied as a corner stone technology within the EU Project HyVar[6] as part of a cloud-based reconfiguration infrastructure for variable automotive software. Furthermore, DeltaEcore has been applied in various evaluation scenarios, e.g., the TurtleBot driver software [34], a metamodel family for role-based modeling and programming languages [18], and an SPL of feature modeling notations [43]. Finally, DeltaEcore has been incorporated into multiple other tool suites, e.g., as part of reverse engineering for the generation of delta modules [48], as variant generation mechanism for context-aware SPLs [22,23], or as back-end for the new version of DeltaJava.[7]

8.4.2 SiPL

While the main functionalities of SiPL widely correspond to those of DeltaEcore, there are also some differences with respect to its underlying concepts which are illustrated using our running example.

8.4.2.1 Delta language creation

A distinguishing characteristic of *SiPL* is to derive delta modules from a model difference [26]. Therefore, SiPL is integrated with the model differencing framework SiLift [16], which provides advanced differencing and patching facilities based on graph transformation concepts. This way, the difference between an origin model and a changed version may be described as an *asymmetric difference* (also known as *edit script*) [17].

An excerpt of the conceptual structure of such a difference is illustrated in Fig. 8.13. An asymmetric difference consists of a set of *operation invocations* which, when applied onto the origin model, yield the changed model. Each operation invocation calls a parameterized *graph transformation rule* by providing actual *parameter bindings* and *mappings* of output to input parameters. A partial order is induced by the *dependencies* between the operation invocations. In the context of SiPL, an operation invocation corresponds to a delta action and a graph transformation rule to a delta operation, respectively.

That means, a delta language actually consists of a set of graph transformation rules which must be specified for the respective source language. Fig. 8.14 shows an excerpt of the set of graph-based delta operations which are typed over the Ecore metamodel. The left- and right-hand sides of a rule are merged into a single graph, following the visual syntax of the model transformation language Henshin [1,31]. The left-hand side of a rule comprises all model elements stereotyped by `delete` and `preserve`. The right-hand side contains all model elements annotated by `preserve` and `create`. The operations on

---

[6] http://hyvar-project.eu.
[7] http://www.deltajava.org.

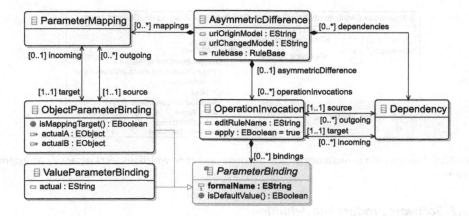

Figure 8.13 Excerpt of the conceptual structure of an asymmetric model difference used in SiPL.

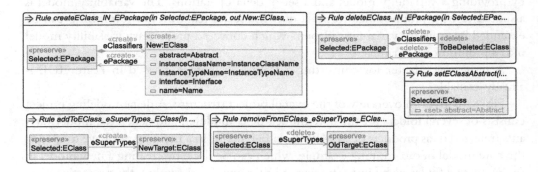

Figure 8.14 Specification of delta operations in SiPL based on graph transformation concepts.

top create/delete an `EClass`, the lower ones add/remove a `superTypes` relation between two `EClasses`. The remaining operation sets the value of the `EAttribute` called `abstract` of an `EClass`.

A basic set of operations can be automatically derived from the metamodel of the respective modeling language using the approach and supporting tool presented in [20] and [19,28], respectively. A specific capability of these operations is that they preserve elementary consistency constraints defined by the given language metamodel, e.g., multiplicities of references.

More complex delta operations can be manually specified using the Henshin editor or derived from examples following the concept of model transformation by-example [13]. Fig. 8.15 shows a delta operation creating an `EAttribute` and setting its type. Please note that the type is not declared to be mandatory by the Ecore metamodel but by a separate consistency constraint.

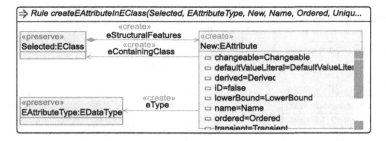

Figure 8.15 Example of a consistency preserving delta operation which respects separate consistency constraints.

8.4.2.2 Software product line definition

To create an SPL definition in SiPL, we have to create a new delta modeling project by providing a variability model and a set of delta operations. The variability model is assumed to be a propositional formula over the set of features. However, SiPL can be extended by arbitrary variability adapters which convert a proprietary variability model into a propositional formula.

A respective adapter for converting a feature diagram expressed in FeatureIDE is already available.

Fig. 8.16 gives an overview of the overall project structure. A delta modeling project consists of several folders for the core model, delta modules, and generated model variants (referred to as products of an MBSPL in Fig. 8.16). In a first step, we have to specify the core model of our running example, which can be done by creating a new model in the respective folder or by importing an existing one. Furthermore, the respective configuration must be assigned to the core model. This information is stored in a so-called *deltamodel* which manages the overall solution space, i.e., the core model, delta modules including their interrelations, and the mapping of the problem space onto the solution space. Furthermore, a delta module encapsulates an asymmetric difference, has a name, and is equipped with an application condition, i.e., a propositional formula over features defined by the feature model.

In order to create a new delta module, an origin model is generated by passing a valid feature configuration. If no configuration is given, the core model is used. An empty delta module is added to the deltamodel and the origin model is copied into the origin and modified folder of the respective delta module folder of the project. After that, a delta module can be implemented by editing the model in the modified folder and by comparing the original model with its modified version. Fig. 8.17 illustrates the derivation of the delta module **DAddEval**. Therefore, the origin model is generated by passing a configuration including the features *Eval* and *Add*.

Subsequently, we edit the model in the modified folder by adding the operation `eval` to the class `Add` and an annotation specifying its implementation. Next, an asymmetric

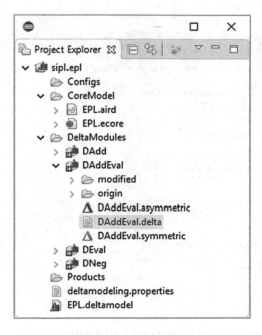

Figure 8.16 Delta modeling project structure in SiPL.

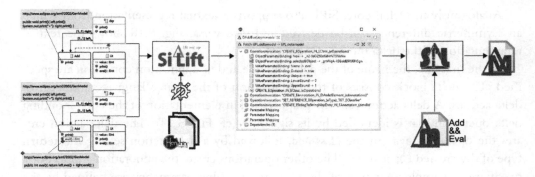

Figure 8.17 Delta module generation in SiPL.

difference is derived between the origin and modified model using SiLift. Therefore, SiLift must be configured by the respective delta operations specified as Henshin transformation rules. The asymmetric difference includes all delta actions as partially ordered set of operation invocations, according to the conceptual structure shown in Fig. 8.13. Finally, the asymmetric difference is added to the deltamodel by referencing the asymmetric difference from the respective delta module.

DAddEval.delta

```
delta DAddEval when "Add and Eval" {

    docType : "http://www.eclipse.org/emf/2002/Ecore";

    New as eval = "CREATE_EOperation_IN_EClass_(eOperations)"(
        object selectedEObject : epl.Add,
        value Name : "eval",
        value Ordered : "true",
        value Unique : "true",
        value LowerBound : "0",
        value UpperBound : "1"
    );

    "SET_REFERENCE_EOperation_(eType)_TGT_EClassifier"(
        object selectedEObject : eval,
        object NewTarget : ecore.EInt
    );

    New as New_EAnnotation = "CREATE_EAnnotation_IN_EModelElement_(eAnnotations)"(
        object selectedEObject : eval,
        value Source : "http://www.eclipse.org/emf/2002/GenModel"
    );

    New as New_EStringToStringMapEntry =
        "CREATE_EStringToStringMapEntry_IN_EAnnotation_(details)"(
        object selectedEObject : New_EAnnotation,
        value Key : "body",
        value Value : "return left.eval() + right.eval();"
    );
}
```

Figure 8.18 Textual representation of the delta module DAddEval in SiPL.

Analogously to DeltaEcore, SiPL also supports a textual representation from which an asymmetric difference can be derived and vice versa. Fig. 8.18 shows the textual representation of the delta module **DAddEval**.

In the first line, the name of the delta module and its application condition are specified. The delta block consists of the specification of the source language and a set of delta actions. A delta action calls the graph-based implementation of the corresponding delta operation that is identified by its signature (cf. Fig. 8.14). The delta action creates the operation `eval` in the class `Add`, followed by a delta action setting the return type of the created `EOperation`. The other operations create the annotation and its body specifying the implementation of the `EOperation`. Value parameters are defined by the keyword `value` and must be passed in quotation marks. Object parameters are defined by the keyword `object` and are passed using their qualified name. If several objects have the same qualified name, XMI-IDs are attached to uniquely identify the respective elements in the overall deltamodel. Created elements can be further assigned to local variables. Please note that in Henshin return values are also specified by parameters in the operation's signature. These need not to be unique in the overall set of operations. However, to uniquely identify created elements an alias can be declared using the keyword `as`.

As mentioned in Section 8.3.2, the delta module **DAddEval** depends on the delta module **DAdd**. While in DeltaEcore this relation must be managed manually as

Table 8.1 Potential relations between delta actions.

Relation	Definition	Example
Conflict	The delta actions d_1 and d_2 are in conflict if they cannot be applied together or their application in both orders would lead to different results	d_1 deletes a model element which is needed by d_2 or changes an attribute value such that a precondition of d_2 is not fulfilled anymore, or both modify the same attribute by setting different values.
Dependency	d_2 depends on d_1 if d_2 can only be applied after d_1.	d_1 creates a model element which is needed by d_2 or d_1 changes an attribute value such that an initially unfulfilled precondition of d_2 is fulfilled.
Duplicate	The delta actions d_1 and d_2 yield the same effect under the same condition.	d_1 and d_2 create or delete the same element or modify the same attribute by setting the same value.
Transient effect	The application of the delta action d_2 removes the effect of a delta action d_1.	d_1 creates a model element that is deleted by d_2.

application–order constraint (cf. Fig. 8.9 in Section 8.4.1.2), SiPL automatically detects such relations and manages them in the deltamodel.

8.4.2.3 Software product line evolution

As we will illustrate later in this section, relations such as dependency relations between delta modules can be effectively utilized to support the evolution of an MBSPL. In addition to dependencies, SiPL also supports the detection of conflicts, duplicates, and transient effects between delta actions and thus between delta modules. Table 8.1 summarizes our definition of the mentioned kinds of relations. SiPL exploits the fact that delta operations invoked by delta actions are implemented as declarative rules based on graph transformation concepts. This allows to statically reason about relations between delta actions by extending concepts presented in [17]. Relations between delta actions are further aggregated to relations between delta modules and can be validated against the feature model.

The right part of Fig. 8.19 shows the relation graph of the deltamodel for the evolution scenario of our running example. Delta modules are represented as nodes and relations as arrows. Blue arrows (dark gray in print version) represent dependencies between the connected delta modules. A delta action of the source module depends on a delta action of the target module. The red arrows (mid gray in print version) represent conflicts between the connected delta modules, i.e., the execution of a delta action of the source module would prevent the execution of a delta action of the target module.

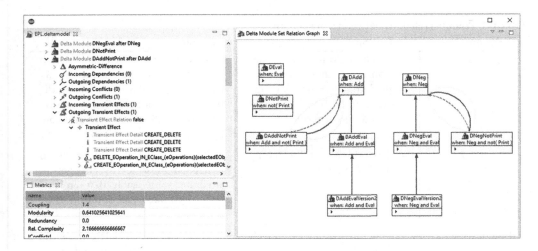

Figure 8.19 Overview of the SiPL UI components.

Transient effects are illustrated by yellow arrows (light gray in print version). The source module contains a delta action that removes the effect of a delta action of the target module. For getting more details about a relation, we can inspect the deltamodel shown in the left part of Fig. 8.19. It consists of an expandable list of delta modules and their relations.

For instance, to inspect the transient effect relation from the delta module **DAddNot-Print** to **DAdd** we can expand the item of the former and further expand the item holding the outgoing transient effects. Each transient effect item offers information about the kind of the transient effect and the involved delta actions. In this case, the operation `print` is created by one delta module and deleted by another one.

As mentioned above, the dependencies between delta modules specify the application ordering of delta modules during variant derivation. However, conflicts, duplicates, and transient effects may be a hint for a mismatch between the problem and solution space. For instance, when two features are compatible in the solution space but their delta modules are in conflict, the variant cannot be derived as expected. In our example, the detected conflicts are automatically resolved by applying the involved delta modules according to the application order defined by the dependency relation which is indicated by the red dashed arrow. However, some conflicts can only be resolved by extracting the conflicting delta actions from one of both delta modules into a new delta module which is equipped with an application condition which prevents the application of the conflicting delta actions.

Transient effects may be a hint for an inadequate variability design in the solution space. For instance, the transient effect between **DAddNotPrint** and **DAdd** is due to the

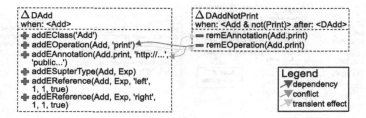

Figure 8.20 Excerpt of the delta module relation graph showing all relations between DAdd and DAddNotPrint.

Table 8.2 Restructuring operations provided by SiPL.

Operation	Resulting delta module contains ...
$extractConflicts(\Delta_1, \Delta_2)$	the delta actions of Δ_1 which are in conflict with a delta action of Δ_2 or (transitively) depend on a conflicting delta action of Δ_1.
$intersect(\Delta_1, \Delta_2)$	the common delta actions of both arguments.
$minus(\Delta_1, \Delta_2)$	the set of delta actions contained in Δ_1, but not in Δ_2.
$merge(\Delta_1, \Delta_2)$	all delta actions of both arguments and extends the partial order of the delta actions according to the dependencies between the delta modules. Furthermore, duplicates and transient effects between both arguments are eliminated.

evolution described in Section 8.3.3. Fig. 8.20 shows all occurring relations between both delta modules.

The feature *Print* has become an optional one and its implementation should be separated from the implementation of the features *Add* and *Neg* in order to correspond to the problem space.

SiPL offers a set of fundamental restructuring operations. All of them *create one new delta module* which is derived from existing delta modules; an overview of the available restructurings is given in Table 8.2.

Each restructuring operation defines a precondition which must be fulfilled, otherwise the application of the operation fails. For instance, the arguments of the operation merge must not be in conflict or the conflict must be resolved by the application order induced by the dependency. Common refactorings for delta–based SPLs as introduced by Schulze et al. [37] can be easily implemented by combining the basic restructuring operations offered by SiPL.

For instance, to eliminate the transient effects between the delta modules **DAdd** and **DAddNotPrint** of our running example, we can concatenate the operations *merge* and *minus* as illustrated in Fig. 8.21. First, we apply the operation *merge*(*DAdd*, *DAddNot Print*), which leads to the delta module **DMerge** with the manually adjusted application

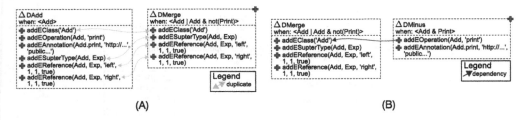

Figure 8.21 Refactoring to eliminate transient effects. (A) Result of applying *merge(DAdd, DAddNotPrint)* leading to the delta module *DMerge*. (B) Result of applying *minus(DAdd, DMerge)* leading to the delta module *DMinus*.

condition *Add|Add&not(Print)* (see Fig. 8.21A). This delta module contains all delta actions of **DAdd** and **DAddNotPrint** without those delta actions leading to a transient effect. The new delta module overlaps with the delta module **DAdd**. To eliminate the duplicates we apply the operation *minus(DAdd, DMerge)*, leading to the new delta module **DMinus** with the manually adjusted application condition *Add&Print* (see Fig. 8.21B). The delta modules **DAdd** and **DAddNotPrint** are not needed anymore and can be deleted. Now, the resulting delta module set does not contain any transient effects or duplicates. In order to perfectly correspond to the problem space, we can further merge the delta modules **DMerge** and **DAddEval**.

In general, however, finding an appropriate refactoring may be challenging. Therefore, SiPL further offers a range of metrics in order to assess the quality of the overall solution space and to recommend refactorings [27]. A metric is defined by a name, a computation function, an indicator which defines whether a high value is better or worse than a low value. For example, the *relational complexity* of a delta module measures the average number of occurrences of delta actions in relations to other delta modules. A lower value is preferable and indicates that the delta module is more easily understandable and maintainable because relations with other delta modules are less complex. The *redundancy* relates the number of duplicate delta actions and all delta actions of a nonempty delta module. A lower value indicates better reusability. Further metrics are presented in [27]. Whenever a refactoring is successfully applicable, the affected metric values are computed before and after the refactoring and presented to the developer. Regarding our running example, the initial relational complexity is about 2.2 (see lower left part of Fig. 8.19). After the elimination of the transient effects it decreases to 1.0.

8.4.2.4 Variant derivation

A variant can be derived by interactively selecting features or by creating a configuration file which is a simple list of features. In a first step the configuration can be validated against the variability model, i.e., the feature model. Next, for all delta modules whose application condition is evaluated to *true* an application order is determined according to the dependency relations. The execution of the delta operations of a delta module is performed by the patching engine of SiLift, which resolves the actual parameter bindings and delegates the actual rule invocation to the Henshin interpreter. If the application of a delta module fails due to unresolved conflicts, we can start the application again in a kind of debug mode reusing the interactive patching tool of SiLift. In this mode each delta action can be interactively applied onto the core model. Conflicting delta actions can be skipped or the patched model can be edited after each successfully applied delta action in order to resolve remaining conflicts manually.

8.4.2.5 Experiences

The implementation of the running example with SiPL was performed by a user familiar with the framework and, analogously to the implementation using DeltaEcore, with a previously existing feature and core model. It consumed approximately 80 minutes of effective working time.

In addition to the implementation of the running example, SiPL has already been demonstrated using an SPL, called *Smart Home*, which serves as a standard example domain in the SPL community [24,45], and an SPL extracted from a case study for embedded software in industrial automation, the *Pick and Place Unit* (*PPU*) [46], which serves as a standard case study within the German Priority Programme "SPP 1593 Design For Future – Managed Software Evolution".[8] Furthermore, to evaluate the larger-scale use of SiPL, we will use the bCMS-SPL, a case study having over 50,000 variants, which is used to evaluate and compare different modeling approaches [7,6].

8.5. Capabilities of DeltaEcore and SiPL

Both tool suites, DeltaEcore and SiPL, are rich in their capabilities with regard to delta modeling. While their core functionality overlaps in some areas, their fundamental concepts are based, in part, on different design decisions, and both tool suites excel in their individual areas. Table 8.3 shows an overview of the main capabilities of both tool suites and, in the following, we briefly elaborate on each of them.

Delta language infrastructure: DeltaEcore provides shared functionality for delta languages, which is independent of a concrete source language (e.g., requiring another delta module), within its *Common Base Delta Language* (*CBDL*). To create a *custom delta language*, a *delta dialect* has to be provided, which is specific to the source language and provides delta operations for modification of the respective artifacts (e.g., removing a class from an Ecore metamodel).

---

[8] http://www.dfg-spp1593.de.

Table 8.3 Comparison of capabilities of DeltaEcore and SiPL.

	Capabilities	DeltaEcore	SiPL
Solution Space	Delta Language Infrastructure	• Common Base Delta Language (CBDL) with shared functionality • Delta dialect for specific source language	• Delta Module encapsulating consistency-preserving edit script based on graph transformation concepts
	Delta Operation Syntax	• Textual (generated) • Graphical (via recorder)	• Textual (generated) • Graphical (state-based model differencing)
	Delta Operation Semantics	• Specified in Java (imperative) • Minimal standard delta operations generated • Complex/custom operations definable by implementing interpreter stubs	• Specified as graph transformation rules (declarative) • Minimal consistency-preserving delta operations generated • Complex/custom operations definable via graphical rule editor or by example
	Delta Language Creation	• Automated analysis of a language's metamodel • Generation of appropriate delta dialect with delta operations	• Automated analysis of a language's metamodel • Generation of appropriate graph transformation rules used as delta operations
	Application Condition Specification	• Centralized in a mapping model	• In individual delta modules
Problem Space	Feature Model Integration	• Plain feature models (internal) • Hyper-Feature Models (internal) • Temporal Feature Models (via DarwinSPL) • FeatureIDE Feature Models (via Composer)	• Propositional formula • Extensible via variability model adapter (FeatureIDE integration readily available)
	Configuration Contents	• Selection of features • Selection of versions	• Selection of features
	Variant Creation Trigger	• Direct selection of delta modules • Manual specification of configuration from feature model • Selection of configuration from feature model • Command line interface • Webservice for remote control	• Manual specification of configuration from feature model • Selection of configuration from feature model
Maintenance	Evolution Support	• Evolution delta modules to add/modify/remove elements for versions • Integration with Hyper-Feature Models to capture feature versions • Automatic version selection for configurations (e.g., by novelty) • Variant derivation with functionality in selected versions	• Metrics for quality assurance • Refactoring recommendation • Versioning via version control system
	Refactoring Support	• Fundamental refactoring of delta modules • Refactoring of feature models with consistency preserving operations	• Basic restructuring operations with predefined conditions: • Complex refactoring operations based on restructuring operations
	Analysis Support	• Overview of delta module dependencies and relation to modified elements for manual inspection	• Analysis of dependencies, conflicts, duplicates and transient effects • Overview and details of relations via Delta Module Relation Graph • Automatic conflict resolution strategies • Quality metrics

In SiPL a deltamodel (containing delta modules and their relations) is specified by an EMF-based metamodel. Each delta module encapsulates a consistency preserving edit script. The instantiation of deltamodels and edit scripts is independent of the concrete source language. However, we have to provide a set of graph transformation rules representing delta operations specific to the source language.

Delta operation syntax: Both DeltaEcore and SiPL provide a textual syntax for custom delta languages. However, all artifacts of DeltaEcore and SiPL are entirely model-based so that another type of concrete syntax may be used as well. For example, DeltaEcore may be integrated with the graphical syntax of a source language by providing a delta recorder that tracks operations in a graphical editor and generates a delta module from them. SiPL reuses existing state-based model differencing tools, so that the standard model editors can be used to edit a given variant and to derive a delta module from the model difference.

Delta operation semantics: DeltaEcore provides a set of *standard delta operations* to add, modify, and remove elements on a fine-grained level whose semantics are defined by a generated interpreter. DeltaEcore also supports more coarse-grained or complex user-defined operations by *custom delta operations* whose semantics are realized by implementing stub methods in the generated interpreter using the imperative programming language Java. In contrast, in SiPL delta operations are declaratively specified by graph transformation rules whose semantics are formally defined by the underlying graph transformation concepts, i.e., no additional implementation for interpreting delta operations is needed.

Delta language creation: DeltaEcore provides means for the automated analysis of a source language metamodel to determine suitable delta operations for a delta dialect, e.g., to modify values of nonconstant attributes. The resulting delta dialect has expressive names for the delta operations and can be further customized, e.g., with custom delta operations.

Likewise, SiPL offers the capability of deriving a set of basic delta operations from the metamodel of the corresponding source language. However, the generated operations are more complex than those generated by DeltaEcore, e.g., by considering multiplicity constraints defined by the metamodel. A delta operation in SiPL typically consists of several fine-grained transformation actions which are combined to form a single transaction in order to preserve the consistency of the derived model. Further complex rules can be realized using the Henshin transformation rule editor or inferred from examples.

Application condition specification: DeltaEcore collects application conditions external to the delta modules themselves in a dedicated mapping model. This has the benefit that application conditions can be created and maintained in a centralized location where logically coherent delta modules are grouped under the same application condition. On the contrary, there is no dedicated mapping artifact in SiPL but each delta module is directly equipped with an application condition. This has the benefit that application conditions are readily available with the delta modules and that a modification does not lead to unexpected side effects.

Feature model integration: DeltaEcore integrates seamlessly with plain feature models, which support the modeling of just features, as well as HFMs [40,38], which support the modeling of features and feature versions. Furthermore, the tool suite DarwinSPL [22], which has its foundation in DeltaEcore, provides Temporal Feature Models (TFMs) [23], which support the integrated modeling of evolution to the feature model structure. In addition, DeltaEcore provides a generic extension mechanism for arbitrary variability modeling notations to interface, which is also used to provide a FeatureIDE [44] integration.

In SiPL the feature model must be converted into a propositional formula which might be more complex and hard to understand. However, such formulas can be efficiently validated using existing SAT solvers. Therefore, the framework provides an extension mechanism to integrate arbitrary SAT solvers. Furthermore, the framework also provides an extension mechanism to convert proprietary variability models into such a formula, e.g., for converting feature models of FeatureIDE.

Configuration contents: DeltaEcore configurations consist of a selection of features and, in addition, an optional selection of a specific version for each of the selected features. The same holds for SiPL, except for the versioning support.

Variant creation trigger: DeltaEcore's variant creation can be triggered in multiple ways: Delta modules to be applied may be selected manually or a configuration may be specified either manually as a configuration file or, when interfacing with a feature model, by using a graphical configurator. In addition, DeltaEcore offers a command line interface for headless operation and a web service for remote operation, both of which may also trigger variant creation. SiPL also supports configuration files and the interactive selection of features as well as validating a configuration against the feature model. Furthermore, SiPL supports a kind of debug mode in order to apply delta actions and to interactively resolve occurring conflicts.

Evolution support: DeltaEcore supports versions of features through evolution delta modules which perform incremental changes based on the previous feature version. When integrating with HFMs [40,38], feature versions may be modeled explicitly so that they can be incorporated in the configuration process, with a method for automatic version selection, as well as the variant derivation procedure.

In SiPL, the actual versioning of both problem and solution space artifacts is delegated to a version control system. However, there is no conceptual limitation of integrating HFMs or Temporal Feature Models in the future.

Analysis support: DeltaEcore provides an overview of SPL projects by displaying delta modules with their dependencies and modified artifacts in various views, which may be used for manual inspection and tool-supported restructuring of the SPL on a coarse-grained level. SiPL provides a coarse-grained overview by providing a relation graph as well as fine-grained information about dependencies, conflicts, duplicates, and transient effects which can be inspected in the overall deltamodel. For each delta action its effect can be further inspected in the respective model editor. All information is provided by a generic, fully automatic, and static analysis of the delta modules. Furthermore, conflicts can be partially resolved automatically by further analyzing the relations.

Refactoring support: DeltaEcore provides fundamental support for restructuring delta modules, e.g., by providing operations to merge the contents of two delta modules or to extract a sequence of delta actions to a new delta module. Responsibility for ensuring validity of the resulting delta modules lies with the user. Furthermore, DeltaEcore provides refactoring support for feature models, with the aim of preserving the consistency of problem and solution space [35].

In contrast, SiPL provides advanced analysis and refactoring support based on concepts of metrics and a basic set of restructuring operations which can be combined to more complex refactoring operations. Each operation defines a precondition which must be fulfilled in order to be applicable. Furthermore, the impact of the refactorings to the corresponding metrics is shown to recommend an appropriate one. Finally, the analysis results can also be used to detect a mismatch between the problem and solution space which can be solved by using the proposed refactorings.

8.6. Related work

Other approaches strive for goals similar to those of this chapter by generating or providing languages to manipulate realization artifacts of various languages.

Haber et al. [10] present work that is closely related to ours as they generate a delta language for a provided source language by grammar extension. However, they merely create the concrete syntax of the language without the possibility of variant creations.

Sánchez et al. [36] present a framework to define domain-specific languages for variability management for artifacts of a particular metamodel.

Zschaler et al. [49] extend upon this work by using SPL technologies to define a family of variability modeling languages. These approaches permit the definition of languages for transformational variability; however, they are not embedded in an overall SPL methodology.

Apel et al. [2] introduce FeatureHouse as a means to generalize software composition by superimposition for realization artifacts in different source languages. However, the resulting languages utilize Feature-Oriented Programming (FOP) [3], which may merely add elements but not remove them as with delta modeling.

Only limited effort has been spent on methods and techniques to automatically detect delta module interrelations and to refactor the overall set of delta modules.

Lienhardt et al. [21] introduce a constraint-based type system for conflict detection in delta-oriented programming. Delta-oriented programming focuses on a source code-based realization of SPLs, the delta dialect is restricted to classes, attributes, and operations.

The analysis functions for SiPL are independent of the source language, i.e., the set of the delta operations. The rich structure of the transformation rules is an essential precondition for its (static) analysis functions. Most of them have no equivalent in existing work, specifically the analysis of dependencies, duplicates, and transient effects. To the best of our knowledge, only **Jayaraman et al.** [12] present a conflict and dependency analysis of deltas similar to those provided by SiPL. It is based on critical pairs of graph transformation rules. However, each feature implementation defines one special monolithic graph transformation. In contrast to this, a delta module consists of several delta actions calling the respective transformation rule. Furthermore, **Jayaraman et al.** detect only dependencies and conflicts. Duplicates and transient effects are only considered by some refactorings presented by **Schulze et al.** [37] and **Haber et al.** [11]. **Schulze et al.** introduce a catalog of refactorings in the context of delta-oriented programming and additional tool support for executing them. However, the context for refactorings, i.e., the suspicious relationships between delta modules, must be identified manually. The same limitation applies to the approach of **Haber et al.**, which focuses on architectural variability modeling. The results of SiPL's analysis functions offer the respective context and are used by restructuring operations which in turn can be exploited to apply several of the proposed refactorings. **Damiani et al.** [8] present an algorithm for refactoring in the context of delta-oriented programming. Delta modules of a monotonic implementation contain only either add and modify or remove and modify operations. Such global refactorings are not yet supported by SiPL and remain subject for future work.

In sum, to the best of our knowledge, DeltaEcore and SiPL are the only fully integrated development environments for model-based MBSPLs currently available. The approaches surveyed before only address single or small subsets of the capabilities provided by DeltaEcore and SiPL. For this reason, we deliberately abstain from including related work in Table 8.3.

8.7. Conclusion

In this chapter, we presented two tool suites for the delta-oriented development of MBSPLs; DeltaEcore and SiPL. Although sharing similar capabilities, some of them are more emphasized by one of the tools. This is not very surprising given the fact that both tools evolved largely independently of each other and emerged from different research communities. Historically, DeltaEcore is the slightly older one of both tools. Its development was strongly influenced by the software language engineering community, which is devoted to researching principles on how to systematically construct formal yet expressive languages for the sake of software development. On the contrary, the SiPL tool suite emerged from the model versioning community, which aims at establishing new techniques for differencing, patching, and merging of models with the ultimate goal of supporting professional model-based software in larger teams.

A first rather obvious difference resulting from the different historical backgrounds is the paradigm of how delta modules are being developed. DeltaEcore follows the central idea of extending an existing source language by the capability of specifying delta modules. To that end, much effort was spent on creating a sophisticated development environment which can be considered as a language workbench for engineering highly customized delta languages. On the contrary, the main motivation and thus the major paradigm shift in SiPL was to create delta modules by editing models in standard model editors and by later comparing the initial model version with its revised one. This way, a delta language is only implicitly available in terms of the set of edit operations used by the model differencing tool.

An important technological difference between both tools is that SiPL assumes delta operations to be declaratively specified by using graph transformation rules, while DeltaEcore relies on an imperative implementation of delta operations in Java. Since practitioners are generally familiar with using imperative programming languages, particularly a mainstream programming language such as Java, the latter approach is highly attractive and has the potential of a widespread adoption in industrial practice. On the contrary, a declarative approach as used by SiPL provides a formal specification of the effect of delta modules which, in turn, enables the implementation of analysis functions offering detailed insights about the relations between delta modules. The analysis support serves as a basis for the quality assurance and guided refactoring of a delta-oriented MBSPL. Although there is no empirical evidence yet with respect to the effectiveness of these techniques, the analysis and quality assurance facilities represent the most distinguishing strengths of the SiPL tool suite.

Another aspect in which both tool suites differ due to their different historical backgrounds is the way how versioning is supported in terms of the evolution of an MBSPL. In SiPL, it was a natural choice to delegate the actual versioning of both problem and solution space artifacts to a version control system. This way, all the readily available services of such a system can be simply reused, while the versioning capabilities are bound to the version space organization of the respective version control system. In contrast, DeltaEcore has the capability of supporting versioning through an integration with HFMs and Temporal Feature Models with the ultimate goal of considering SPL evolution as a primary development information. This way, there is a clear methodology of how developers can be supported in combining features of different versions instead of delegating the entire configuration to the version control system and thus to the discretion of the developer. Moreover, handling evolution as a primary development paves the way for the dynamic reconfiguration of an MBSPL at runtime.

Besides, there are some minor differences with respect to the underlying design decisions of DeltaEcore and SiPL, such as whether feature mappings are represented externally using a dedicated mapping artifact (as in DeltaEcore) or integrated with the delta modules (as in SiPL). More recently, both tool suites have started to adopt each others capabilities, for example, to provide an alternative method to specify delta modules. In SiPL, one can now synthesize a textual delta language from a set of delta operations, while DeltaEcore has adopted the idea of specifying delta modules by model editing through recording editing commands in model editors. More generally, as for future work, our aim is to closer collaborate on the development of both tool suites for the sake of learning from each other's experience and for further extending and strengthening the capabilities of each of the tools. With such a collaboration and an increasing maturity of the tools, we will be able to conduct user studies with users of different levels of experience in the future, and we are convinced that DeltaEcore and SiPL will serve as ideal experimental testbeds for a more widespread evaluation and adoption of the general idea of delta modeling in practice.

References

[1] Thorsten Arendt, Enrico Biermann, Stefan Jurack, Christian Krause, Gabriele Taentzer, Henshin: advanced concepts and tools for in-place EMF model transformations, in: International Conference on Model Driven Engineering Languages and Systems, Springer, 2010, pp. 121–135.

[2] Sven Apel, Christian Kästner, Christian Lengauer, Language-independent and automated software composition: the FeatureHouse experience, IEEE Transactions on Software Engineering 39 (1) (2013) 63–79.

[3] Don Batory, Feature-oriented programming and the AHEAD tool suite, in: Proceedings of the 26th International Conference on Software Engineering, IEEE Computer Society, 2004, pp. 702–703.

[4] Don Batory, Feature models, grammars, and propositional formulas, in: Henk Obbink, Klaus Pohl (Eds.), Software Product Lines, Springer Berlin Heidelberg, Berlin, Heidelberg, ISBN 978-3-540-32064-7, 2005, pp. 7–20.

[5] Johannes Bürdek, Timo Kehrer, Malte Lochau, Dennis Reuling, Udo Kelter, Andy Schürr, Reasoning about product-line evolution using complex feature model differences, Automated Software Engineering 23 (4) (2016) 687–733.

[6] Afredo Capozucca, Betty H. Cheng, Geri Georg, Nicolas Guelfi, Paul Istoan, Gunter Mussbacher, Requirements definition document for a software product line of car crash management systems, version: April 2013, http://cserg0.site.uottawa.ca/cma2013models/CaseStudy.pdf, Fallstudie zur Evaluation von unterschiedlichen Modellierungsansätzen, CMA Workshop at Models 2012: http://cserg0.site.uottawa.ca/cma2012/.

[7] Alfredo Capozucca, Betty Cheng, Nicolas Guelfi, Paul Istoan, OO-SPL modelling of the focused case study, in: Comparing Modeling Approaches (CMA) International Workshop Affiliated with ACM/IEEE 14th International Conference on Model Driven Engineering Languages and Systems, CMA@MODELS2011, 2011.

[8] Ferruccio Damiani, Michael Lienhardt, Refactoring delta-oriented product lines to achieve monotonicity, in: Julia Rubin, Thomas Thüm (Eds.), Proceedings 7th International Workshop on Formal Methods and Analysis in Software Product Line Engineering, in: Electronic Proceedings in Theoretical Computer Science, vol. 206, Eindhoven, The Netherlands, April 3, 2016, Open Publishing Association, 2016, pp. 2–16.

[9] Eclipse Foundation, Eclipse Modeling Framework (EMF), https://www.eclipse.org/modeling/emf/. (Accessed 27 October 2018).

[10] Arne Haber, Katrin Hölldobler, Carsten Kolassa, Markus Look, Bernhard Rumpe, Klaus Müller, Ina Schaefer, Engineering delta modeling languages, in: Proceedings of the 17th International Software

Product Line Conference, SPLC '13, ACM, New York, NY, USA, ISBN 978-1-4503-1968-3, 2013, pp. 22–31.

[11] Arne Haber, Holger Rendel, Bernhard Rumpe, Ina Schaefer, Evolving delta-oriented software product line architectures, in: Radu Calinescu, David Garlan (Eds.), Large-Scale Complex IT Systems. Development, Operation and Management, Springer Berlin Heidelberg, Berlin, Heidelberg, ISBN 978-3-642-34059-8, 2012, pp. 183–208.

[12] Praveen Jayaraman, Jon Whittle, Ahmed M. Elkhodary, Hassan Gomaa, Model composition in product lines and feature interaction detection using critical pair analysis, in: Gregor Engels, Bill Opdyke, Douglas C. Schmidt, Frank Weil (Eds.), Model Driven Engineering Languages and Systems, Springer Berlin Heidelberg, Berlin, Heidelberg, ISBN 978-3-540-75209-7, 2007, pp. 151–165.

[13] Timo Kehrer, Abdullah Alshanqiti, Reiko Heckel, Automatic inference of rule-based specifications of complex in-place model transformations, in: International Conference on Theory and Practice of Model Transformations, Springer, 2017, pp. 92–107.

[14] Christian Kästner, Sven Apel, Syed S. Rahman, Marko Rosenmüller, Don Batory, Gunter Saake, On the impact of the optional feature problem: analysis and case studies, in: Proceedings of the 13th International Software Product Line Conference, SPLC '09, Carnegie Mellon University, Pittsburgh, PA, USA, 2009, pp. 181–190.

[15] Kyo Kang, Sholom Cohen, James Hess, William Novak, A. Peterson, Feature-Oriented Domain Analysis (FODA) Feasibility Study, version: 1990 (CMU/SEI-90-TR-021), Forschungsbericht, Software Engineering Institute, Carnegie Mellon University, Pittsburgh, PA, 1990, http://resources.sei.cmu.edu/library/asset-view.cfm?AssetID=11231.

[16] T. Kehrer, U. Kelter, M. Ohrndorf, T. Sollbach, Understanding model evolution through semantically lifting model differences with SiLift, in: 2012 28th IEEE International Conference on Software Maintenance, ICSM, ISSN 1063-6773, 2012, pp. 638–641.

[17] T. Kehrer, U. Kelter, G. Taentzer, Consistency-preserving edit scripts in model versioning, in: 2013 28th IEEE/ACM International Conference on Automated Software Engineering, ASE, 2013, pp. 191–201.

[18] Thomas Kühn, Max Leuthäuser, Sebastian Götz, Christoph Seidl, Uwe Aßmann, A metamodel family for role-based modeling and programming languages, in: International Conference on Software Language Engineering, Springer, 2014, pp. 141–160.

[19] Timo Kehrer, Michaela Rindt, Pit Pietsch, Udo Kelter, Generating edit operations for profiled UML models, in: ME@ MoDELS, Citeseer, 2013, pp. 30–39.

[20] Timo Kehrer, Gabriele Taentzer, Michaela Rindt, Udo Kelter, Automatically deriving the specification of model editing operations from meta-models, in: Proceedings of the 9th International Conference on Theory and Practice of Model Transformations, vol. 9765, Springer-Verlag, Berlin, Heidelberg, ISBN 978-3-319-42063-9, 2016, pp. 173–188.

[21] Michäel Lienhardt, Dave Clarke, Conflict detection in delta-oriented programming, in: Tiziana Margaria, Bernhard Steffen (Eds.), Leveraging Applications of Formal Methods, Verification and Validation. Technologies for Mastering Change, Springer Berlin Heidelberg, Berlin, Heidelberg, ISBN 978-3-642-34026-0, 2012, pp. 178–192.

[22] Michael Nieke, Gil Engel, Christoph Seidl, DarwinSPL: an integrated tool suite for modeling evolving context-aware software product lines, in: Proceedings of the Eleventh International Workshop on Variability Modelling of Software-Intensive Systems, ACM, 2017, pp. 92–99.

[23] Michael Nieke, Christoph Seidl, Sven Schuster, Guaranteeing configuration validity in evolving software product lines, in: Proceedings of the Tenth International Workshop on Variability Modelling of Software-Intensive Systems, ACM, 2016, pp. 73–80.

[24] Klaus Pohl, Günter Böckle, Frank J. van d. Linden, Software Product Line Engineering: Foundations, Principles and Techniques, Springer-Verlag, Berlin, Heidelberg, ISBN 3540243720, 2005.

[25] Christopher Pietsch, SiPL project website, http://pi.informatik.uni-siegen.de/projects/sipl. (Accessed 27 October 2018).

[26] Christopher Pietsch, Timo Kehrer, Udo Kelter, Dennis Reuling, Manuel Ohrndorf, SiPL – a delta-based modeling framework for software product line engineering, in: 2015 30th IEEE/ACM International Conference on Automated Software Engineering, ASE, 2015, pp. 852–857.

[27] Christopher Pietsch, Dennis Reuling, Udo Kelter, Timo Kehrer, A tool environment for quality assurance of delta-oriented model-based SPLs, in: Proceedings of the Eleventh International Workshop on Variability Modelling of Software-Intensive Systems, VAMOS '17, ACM, New York, NY, USA, ISBN 978-1-4503-4811-9, 2017, pp. 84–91.

[28] Michaela Rindt, Timo Kehrer, Udo Kelter, Automatic generation of consistency-preserving edit operations for MDE tools, in: Demos@ MoDELS, vol. 14, 2014.

[29] Ina Schaefer, Lorenzo Bettini, Viviana Bono, Ferruccio Damiani, Nico Tanzarella, Delta-oriented programming of software product lines, in: Software Product Lines: Going Beyond, Springer, 2010, pp. 77–91.

[30] Ina Schaefer, Lorenzo Bettini, Viviana Bono, Ferruccio Damiani, Nico Tanzarella, Delta-oriented programming of software product lines, in: Jan Bosch, Jaejoon Lee (Eds.), Software Product Lines: Going Beyond, Springer Berlin Heidelberg, Berlin, Heidelberg, ISBN 978-3-642-15579-6, 2010, pp. 77–91.

[31] Daniel Strüber, Kristopher Born, Kanwal D. Gill, Raffaela Groner, Timo Kehrer, Manuel Ohrndorf, Matthias Tichy, Henshin: a usability-focused framework for emf model transformation development, in: International Conference on Graph Transformation, Springer, 2017, pp. 196–208.

[32] Ina Schaefer, Variability modelling for model-driven development of software product lines, in: Fourth International Workshop on Variability Modelling of Software-Intensive Systems, Proceedings, Linz, Austria, January 27–29, 2010, 2010, pp. 85–92.

[33] Christoph Seidl, DeltaEcore project website, http://deltaecore.org. (Accessed 27 October 2018).

[34] Christoph Seidl, Integrated Management of Variability in Space and Time in Software Families, PhD Thesis, Technische Universität Dresden, 2016, http://nbn-resolving.de/urn:nbn:de:bsz:14-qucosa-218036.

[35] Christoph Seidl, Florian Heidenreich, Uwe Aßmann, Co-evolution of models and feature mapping in software product lines, in: Proceedings of the 16th International Software Product Line Conference, Vol. 1, ACM, 2012, pp. 76–85.

[36] Pablo Sánchez, Neil Loughran, Lidia Fuentes, Alessandro Garcia, Engineering languages for specifying product-derivation processes in software product lines, in: Software Language Engineering, Springer, 2009, pp. 188–207.

[37] Sandro Schulze, Oliver Richers, Ina Schaefer, Refactoring delta-oriented software product lines, in: Proceedings of the 12th Annual International Conference on Aspect-Oriented Software Development, AOSD '13, ACM, New York, NY, USA, ISBN 978-1-4503-1766-5, 2013, pp. 73–84.

[38] Christoph Seidl, Ina Schaefer, Uwe Aßmann, Capturing variability in space and time with hyper feature models, in: Proceedings of the Eighth International Workshop on Variability Modelling of Software-Intensive Systems, ACM, 2014, p. 6.

[39] Christoph Seidl, Ina Schaefer, Uwe Aßmann, DeltaEcore – a model-based delta language generation framework, Modellierung 19 (2014) 21.

[40] Christoph Seidl, Ina Schaefer, Uwe Aßmann, Integrated management of variability in space and time in software families, in: Proceedings of the 18th International Software Product Line Conference, ACM, 2014, pp. 22–31.

[41] Christoph Seidl, Sven Schuster, Ina Schaefer, Generative software product line development using variability-aware design patterns, in: Proceedings of the 14th International Conference on Generative Programming: Concepts & Experiences, GPCE'15, vol. 51, ACM, 2015, pp. 151–160.

[42] Christoph Seidl, Sven Schuster, Ina Schaefer, Generative software product line development using variability-aware design patterns, in: Special Issue on the 14th International Conference on Generative Programming: Concepts & Experiences, GPCE'15, Computer Languages, Systems and Structures 48 (2017) 89–111.

[43] Christoph Seidl, Tim Winkelmann, Ina Schaefer, A software product line of feature modeling notations and cross-tree constraint languages, Modellierung 2016 (2016).

[44] Thomas Thüm, Christian Kästner, Fabian Benduhn, Jens Meinicke, Gunter Saake, Thomas Leich, FeatureIDE: an extensible framework for feature-oriented software development, Science of Computer Programming 79 (2014) 70–85.

[45] M. Voelter, I. Groher, Product line implementation using aspect-oriented and model-driven software development, in: 11th International Software Product Line Conference, SPLC 2007, 2007, pp. 233–242.

[46] Birgit Vogel-Heuser, Christoph Legat, Jens Folmer, Stefan Feldmann, Researching Evolution in Industrial Plant Automation: Scenarios and Documentation of the Pick and Place Unit, Institute of Automation and Information, Technische Universität München (TUM-AIS-TR-01-14-02), Technical Report, mediaTUM, Munich, Germany, 2014, version: 2014, https://mediatum.ub.tum.de/node?id=1208973.

[47] Tim Winkelmann, Jonathan Koscielny, Christoph Seidl, Sven Schuster, Ferruccio Damiani, Ina Schaefer, Parametric DeltaJ 1.5: propagating feature attributes into implementation artifacts, in: CEUR-WS, 2016, pp. 40–54.

[48] David Wille, Tobias Runge, Christoph Seidl, Sandro Schulze, Extractive software product line engineering using model-based delta module generation, in: Proceedings of the Eleventh International Workshop on Variability Modelling of Software-Intensive Systems, ACM, 2017, pp. 36–43.

[49] Steffen Zschaler, Pablo Sánchez, João Santos, Mauricio Alférez, Awais Rashid, Lidia Fuentes, Ana Moreira, João Araújo, Uirá Kulesza, VML★–a family of languages for variability management in software product lines, in: Software Language Engineering, Springer, 2010, pp. 82–102.

CHAPTER 9

OptML framework and its application to model optimization

Güner Orhan, Mehmet Akşit

University of Twente, Computer Science, Formal Methods & Tools Group, Enschede, The Netherlands

Contents

9.1. Introduction

Since the 2010s, there has been an increasing emphasis on Model-Driven Engineering (MDE) [1]. There has been a considerable effort in definition and implementation of models in a large category of application domains and as such many useful models are readily available for use.

Availability of models in the domains of interest, however, creates its own problems to deal with.

First, due to complexity of the domain of interest, complexity and size of models can be very large [2]. Although there have been some approaches, such as model splitting/merging/transforming [3], which can be used to deal with model complexity, generally they must be "hand-tailored" and their effects in reducing complexity can be rather limited. A number of model complexity reduction approaches has been proposed. However, as stated in Babur's article on models [4], this is an active research area and the problem of model complexity has not been solved yet satisfactorily.

Second, due to built-in variation mechanisms, models may be configured in many different ways. In addition to functional requirements, selection and configuration of models may largely depend on certain *quality attributes* and *contextual parameters*, which may not be explicitly specified as parts of models. Examples of *quality attributes* are, for example, time performance, energy reduction, and precision in computations. Examples of contextual parameters are software and hardware architectural styles of adopted platforms and their characteristics.

Last but not least, since the number of model configurations can be very large, given a set of requirements, it may be very hard for software engineers to derive the most suitable configurations in a convenient manner. Optimizing a model configuration for a single purpose is generally not satisfactory. Software engineers generally have to trade off different objectives to configure the most suitable model for a given application setting.

This chapter presents a novel tool workbench, called Optimal Modeling Language (OptML)[1] Framework, to represent certain *quality attributes* and contextual parameters, explicitly. This approach is supported by Optimal Modeling Process (OptMP) to guide the software engineer in selecting and configuring models, according to the desired optimization criteria. The framework incorporates a dedicated set of tools to compute the desired optimal model configurations. Examples of currently supported *quality attributes* are time performance, energy reduction, and precision. Furthermore, as contextual parameters, single- and multicore platforms and various distributed and/or parallel system architectures are supported. The software engineer can define new quality attributes by using the *Value* metamodel of the framework. The utility of the model and the associated processes and tools are demonstrated by a set of examples. A prototype implementation is realized using the Eclipse framework [5] and FSF application framework. FSF is a dedicated software library to implement a large category of scheduling systems [6].

This chapter is organized as follows. The following section introduces an illustrative example and explains the addressed problems. Section 9.3 presents the architecture of the framework. Section 9.4 gives a set of example models based on various architectural

---

[1] The term Optimal Modeling Language is selected for the following reason. The purpose of this framework is to compute the optimal configuration from a set of models utilized by the software engineer. We assume that each model is based on a dedicated modeling language (metamodel) in the Ecore Modeling Language tradition. In addition, to compute the optimal model, appropriate models must be introduced to specify the optimization constraints.

views. The *Model Processing Subsystem* and *Model Optimization Subsystem* of the framework are described in Sections 9.5 and 9.6, respectively. Section 9.7 briefly summarizes the related work. Section 9.8 evaluates the approach. Finally, Section 9.9 concludes the chapter.

9.2. Illustrative example, problem statement, and requirements

In this section, an illustrative example from the image processing domain is given which will be used throughout the chapter to demonstrate the problems and the proposed solutions. This example is considered illustrative for the following reasons. First, this chapter addresses the concerns where model size and complexity are considerably high. There exists a comprehensive and fully implemented software library of the example, which is considered representative for the purpose of the chapter. Second, this software library can be configured in many ways. This chapter aims at dealing with models with a large number of configurations. Third, while reusing this software library for a particular application, the software engineers are typically concerned with various quality attributes such as timeliness, energy consumption, and precision. This chapter aims to select the optimal model configuration that satisfies multiple quality constraints.

Registration is a problem of reconstructing an image output by matching two or more related images captured in different environmental conditions [7] so that the obtained image is more expressive for a particular purpose than the individual input images. This may be needed in systems where multiple sensors are used with different resolutions, positions, and imaging characteristics.

Consider, for example, the following pipeline architecture for a registration system, which is represented in five consecutive states, depicted in Fig. 9.1. This architecture is inspired from the Point Cloud Library (PCL) [8]. From the left, the state *Input* represents the data acquisition loop which gathers image data from one or more sensors. The second state *Filter* aims to reduce the data size if necessary so that only the relevant information is used for further processing. In addition, the original images are preserved. The third state *FExtract* is responsible to compute the predefined key features from the data to reason about the geometric characteristics of the images. The fourth state *Match*

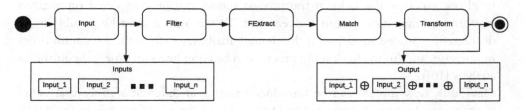

Figure 9.1 The process of a registration system to reconstruct an image from multiple inputs.

is used to correlate the extracted features with each other. The state *Transform* is used to transform the original images into a common image based on the matching process. Here, the sign $\oplus$ represents the transformation operator.

Assume that a version of the PCL library is instantiated in an Ecore MDE environment with the following models which represent the system from different architectural views [9]. More detailed information about these models can be found in Section 9.4.

- A class model, which describes the logical structure of the system.
- A feature model, which defines the variations to configure different versions of registration systems.
- A platform model, which describes the underlying computational resources of the registration system.
- A process model, which illustrates the execution flow of the processes and the necessary synchronization points among them.

From the perspective of this chapter, the following potential problems can be observed:

1. *Large configuration spaces of models*: It is a common practice that multiple related models are used in MDE environments for a given system. Each of these models may define different kinds of variations. The possible combination of all variations may potentially enable many possible instantiations of models, which can be difficult for the MDE expert to comprehend. Consider, for example, the registration system which is described by four different kinds of models. Due to the variations of each model, the design space of the registration system can be very large. In our example case, for instance, the number of variations of the defined feature model is computed as 6144 (see Section 9.5). There have been a number of proposals, such as model splitting, merging, or transforming [3,4], which can be used to deal with model complexity. However, many of these proposals provide dedicated solutions, for example, through the application of predefined rules.

2. *Lack of quality concerns in model configurations*: This problem is a natural consequence of the previous problem. An important set of criteria for creating a particular configuration from a model space is to select the configuration that fulfills the desired quality attributes.

   For example, while configuring a particular registration system, it may be necessary to check whether the tasks in the process model can be completed on a given platform configuration within a given time. To this aim, it must be possible (i) to decorate the process model with the desired attributes such as the execution times of processes and (ii) to check if the process can be completed on time (schedulability analysis [10]).

   Obviously, new models can be introduced aiming at different architectural views if necessary. Assume that we would like to extend the set of models in the MDE environment with two additional models:

Energy model: a model to define the energy demanded by processes (also called operations) to complete them on the configuration of the underlying platform in a certain time interval. The factors that affect the completion time of processes depend on the demanded energy by them and the offered energy by the platform configuration.

Precision model: a model to define the quality of the resulting image accuracy in the registration process. In the implementation, there are alternative algorithmic solutions defined with different precision. Depending on the requirements, a low-precision algorithm may be preferred to a high-precision one for the sake of time performance.

3. *Optimization of configurations*: Software engineers generally have to trade off different quality attributes to configure the most suitable model for a given application setting. For example, a particular model configuration may improve the quality attribute "reducing energy consumption" while decreasing the quality attribute "time performance." The MDE environment must provide means to optimize model configurations by considering multiple quality attributes.

Based on these observations, it is desired that the OptML Framework must support at least the following requirements.

It must be possible to:

1. evaluate configurations of models and whether there exists a configuration that can be mapped on a specified platform architecture while satisfying the time and resource constraints; and/or

2. find out the optimal model among configurations based on certain optimization criteria and objectives. Along this line, for example, it must be possible to
   - **(a)** introduce a model for each quality attribute;
   - **(b)** normalize the quality attributes;
   - **(c)** prioritize the quality attributes with respect to each other;
   - **(d)** apply the well-known logical operators on the values of attributes, such as "$<, >, <>, =$";
   - **(e)** select the models with respect to minimization or maximization of the quality attributes;

3. find out the optimal model among configurations based on certain optimization criteria and objectives; and/or

4. find out whether the introduced models are consistent with each other with respect to the predefined consistency rules.

Obviously, software engineers may demand many different kinds of models in developing their applications. The facilities provided by the OptML Framework may need to be extended accordingly. The effort that is spent in realizing the OptML Framework can only be justified if the facilities of the OptML Framework are demanded by multiple software engineers. The OptML Framework, therefore, must offer solutions to the recurring problems of software engineers. If, however, the OptML Framework is required to be extended to satisfy the emerging needs of software engineers, it must be extended accordingly. In addition, enhancements to the implementation of OptML may be necessary from time to time, for instance, to increase performance. Based on these assumptions, the following extensions are considered foreseeable:

5. supporting new models defined in the Ecore environment;

6. introducing new pruning mechanisms while extracting models from the model base;

7. introducing new value-based quality attributes;

8. introducing new value optimization algorithms where necessary;

9. adopting new search strategies for the schedulability analysis and optimization techniques.

9.3. The architecture of the framework

As shown in Fig. 9.2, the architecture of the OptML Framework consists of three subsystems. The *Model Editing Subsystem*, which is symbolically shown on the left side of the figure, can be used to define various models representing different architectural views based on the corresponding metamodels. If necessary, new metamodels can be introduced to the system using MDE facilities. We assume that this subsystem corresponds to a standard MDE framework such as the Eclipse Modeling Framework. The second process in the figure, the *Model Processing Subsystem*, is used to transform the introduced models into a representation, which can be processed by the *Model Optimization Subsystem*. The *Model Optimization Subsystem* part of the framework, which is shown symbolically at the right-hand side of the figure, automatically processes the transformed models and computes the optimal model based on the criteria provided by the model-driven engineer. The last two processing subsystems form the essential components of the OptML Framework. In our approach, we adopt the Ecore Modeling language and Eclipse Platform for the *Model Editing Subsystem*. Since this framework is well known, we do not explain it further in this chapter. Nevertheless, in Sections 9.5 and 9.6, we respectively describe the *Model Processing Subsystem* and *Model Optimization Subsystem* in detail.

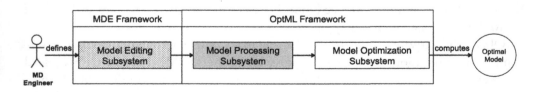

Figure 9.2 A software architecture of the OptML Framework.

9.4. Examples of models for registration systems based on various architectural views

We will now introduce six models subsequently to illustrate the contributions of this chapter and to deal with the complexity of the example design problem. It is also a common practice in MBSD that a complex design problem is decomposed into a set of models where each represents a different aspect of the system to be designed [11]. Of course, in the end, all the relevant models must be related to each other in some way to represent the system as a whole.

The modeling paradigm adopted in this chapter follows the MDE Ecore tradition, which means that first a metamodel is to be defined that conforms the Ecore metameta-model (called Ecore EMF format) [5]. A model is an instantiation of its metamodel. In the following subsections, the described models are:

- UML Class model: to depict the *logical view* [12] of the example;
- Feature metamodel: to specify the possible configurations of the example [13];
- Platform metamodel: to specify the *physical view* and the *deployment view* [12] of the example;
- Process metamodel: to specify the *process view* [12] of the example;
- Value metamodel: to specify the quality concerns of the example.

These models are selected because they are considered as fundamental models required by many applications as published by Kruchten [12]. In addition, to address the requirements of this chapter, the Value metamodel is introduced so that the optimal model can be computed accordingly.

9.4.1 UML class model

A *class diagram* is used for specifying the logical building blocks of a software system. We do not define a new metamodel for classes, but rather we adopt the standard UML class metamodel, which is one of the registered packages[2] of the Epsilon Modeling Framework in Eclipse IDE.

Fig. 9.3 shows a class model for the registration system, which is introduced in Section 9.2. Here, for brevity, the attributes and operations are not shown in the figure.

The names of classes Input, Filter, FExtract, Match, and Transform are written bold in the figure, and they correspond to the subsystems of a registration system presented in Fig. 9.1 of Section 9.2.

9.4.2 Feature metamodel

The metamodel representing feature models is shown in Fig. 9.4. The aim of the feature model is to express *commonalities* and *variabilities* in a family of software systems.

---

[2] EMF-based implementation of the Unified Modeling Language (UML).

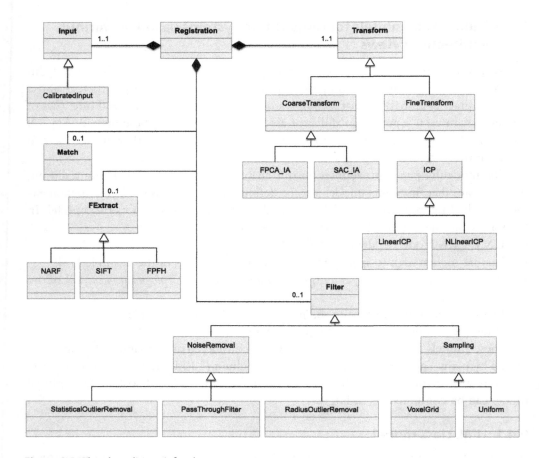

Figure 9.3 The class diagram for the registration system.

A feature model enables the model-driven engineer to express various configurations of the system. Due to various options, configuring a feature model may result in more than one software system. In the traditional MDE approach, the model-driven engineer is supposed to evaluate each configuration and choose the most suitable one based on some criteria.

Fundamentally, any feature model has exactly one **root** from which subfeature models originate. Each feature may have zero or more child features and attributes. Each attribute has a **type** and **defaultValue**, which may belong to the types **Boolean, String, Integer,** or **Object**. In addition, each feature has a **type** as **optional, alternative,** or **or** if it is a **variability**; or **mandatory** if it is a *common* asset for the product family. All the features are placed into a **Group** with **upper** and **lower**. The number of bound features inside the same group has to be between these values in any configuration instantiated from one

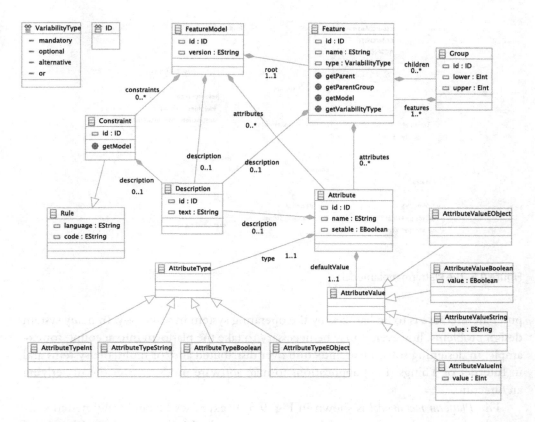

Figure 9.4 The Feature metamodel.

feature model. Finally, a feature model may have cross-tree constraints that are defined as rules in any constraint-based language.

A feature model must be consistent with the process model and class diagram. Therefore, we assume that each feature defined in a feature model must correspond to a class in the class diagram.

A feature model of the registration system, which is instantiated from the Feature metamodel, is given in Appendix 9.A.

9.4.3 Platform metamodel

As a third example, we will present the Platform metamodel, as adopted in the OptML Framework. A platform model, which is also termed deployment model [14], is an instance of this metamodel and enables the designer to express the underlying computational system. If a platform model is not specified, it is assumed that the underlying computational system is transparent and as such mapping of software modules to com-

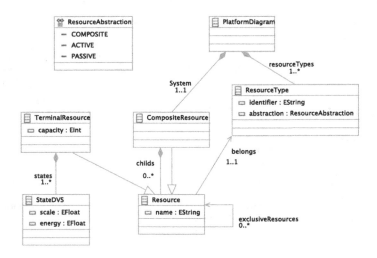

Figure 9.5 The Platform metamodel.

puter architecture is to be handled by the operating system in some way. In many system design problems, however, it may be necessary to take the platform into account, for example, in designing software architecture over distributed and/or multicore systems and in Internet of Things (IoT) applications to map software modules over the underlying architecture.

The *Platform metamodel* is shown in Fig. 9.5. It expresses hierarchically nested software/hardware architectures, which can be composed of various types of architectural components. The Platform metamodel is represented as class PlatformDiagram, which aggregates classes ResourceType and CompositeResource. The class ResourceType specifies the characteristics of the corresponding resources with a unique string type identifier and an enumeration of the literals ACTIVE, PASSIVE, and COMPOSITE. We aim to create a uniform model by considering all possible architectures as a special configuration of a composite object. To this aim the class PlatformDiagram aggregates the class CompositeResource. To create a hierarchical platform organization, the class CompositeResource uses the composite pattern format [15]. The class Resource here corresponds to an abstract representation of every architectural component, since every resource inherits its properties. The class CompositeResource may encompass zero or more terminal and/or composite resources, where composite resources may further aggregate resources, and so on. The aggregation relation from the class CompositeResource to the class Resource enables to create nested instances of classes CompositeResource and/or TerminalResource. The class TerminalResource, as the name implies, is the representation of the resources that cannot be decomposed any further. The attribute capacity of the class TerminalResource defines the maximum utilization unit which a resource can provide [16].

In recent years, mobile devices are increasingly used as a computing platform. Due to limited operational time of batteries, reducing power consumption of mobile devices has become important. To this aim, for example, the *Dynamic Voltage and Frequency Scaling* (DVFS) technique is introduced [17,18]. In Lin's article [18], the concept of operating frequency levels has been defined. The levels correspond to the frequency scaling factors varying between 0 and 1. A higher value means higher energy consumption. Due to the popularity of this approach in practice, we adopt this technique in our platform model as well; the class StateDVS is introduced for this purpose. Each level has its scaling factor and corresponding power consumption value, which are represented by the attributes scale and energy of the class StateDVS, respectively.

To avoid race conditions and simultaneous access to shared resources, the self-reference relation over the abstract class Resource is defined. It avoids any multiple resources to run at the same time.

The reference relation from Resource to ResourceType is used to denote the type of the corresponding resource.

The platform model as an instantiation of this metamodel for the registration system is given in Appendix 9.B.

9.4.4 Process metamodel

It is considered important to understand the dynamic behavior of systems, for example in allocating software to underlying architecture, verifying the operational semantics of software, or determining the time performance. In the literature, various kinds of models have been presented, such as state diagrams, process diagrams, collaboration diagrams, and activity diagrams. If the time constraints are to be considered in mapping the software system to a particular platform, we assume that a process model is defined which represents the processes and their execution flow, input–output data dependencies, and resource requirements. Consider, for example, the following Process metamodel, which is shown in Fig. 9.6.

A common practice to represent a process model is to consider it as a graph where each process is a node and the dependencies among processes are the edges of a graph [19]. In our approach, in Fig. 9.6, the class ProcessDiagram represents the root of the graph. Since there may be multiple independent processes in the system, the attribute *name* of this class is used to denote a particular process diagram.

The class ProcessDiagram aggregates zero or more nodes, where each node is represented by the class Process. The attribute *name* of this class is used to identify a particular process in a process diagram. In pure object-oriented programs, a process is associated with an object of a class. To represent this property, the attribute *namespace* is used to denote the corresponding class. In the literature, *periodic* processes are executed repeatedly at each time interval [20]. To support this characteristic of a process, the attribute

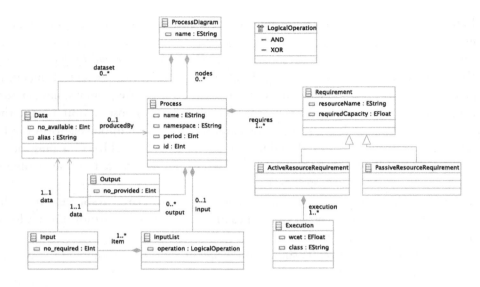

Figure 9.6 The Process metamodel.

period is defined. In case of multiple instances of a process, the attribute *id* can be used to distinguish the instances from each other.

It may be the case that the application semantics demands a process to be executed in a certain order [20,21]. To specify such conditions, we adopt the *data dependency* constraint explained in [22,23]. A data dependency constraint specifies the data that are required and/or provided by a process. If, say, the process P_1 provides the data d_1 which are demanded by the process P_2, then P_2 is eligible to be executed only after the completion of P_1. In the figure, the required and provided data for a process are represented by classes **Input** and **Output**, respectively. Both classes refer to only one instance of the class **Data**. The availability of a particular data item is indicated by the attribute **no-available** of the class **Data**. The attributes **no-required** and **no-provided** of classes **Input** and **Output** refer to the *required* and *provided* number of available data items, respectively. The class **InputList** specifies the input dependency constraints of a process. If a process requires more than one data item and is eligible to start when only one of the data items is available, the attribute **operation** of the class **InputList** should be defined as XOR. However, if the process requires the availability of all data items, then the attribute should be defined as AND, instead.

Resource requirements of a process are expressed by the class *Requirement*. The attributes **resourceName** and **requiredCapacity** refer to the necessary resource type and its capacity, respectively. To increase the utilization factor of resources, it may be profitable to divide resource types as active and passive resources [6,24]. To this aim, classes **ActiveResourceRequirement** and **PassiveResourceRequirement** are defined, which inherit

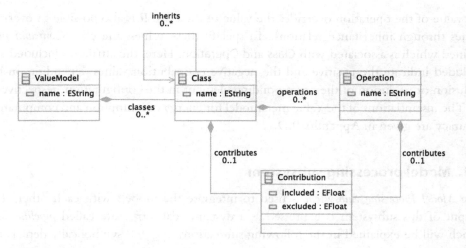

Figure 9.7 The Value metamodel.

from the class **Requirement**. For allocating active resources to processes that complete in a timely manner, worst–case execution time of a process is an important factor to consider. The attribute WCET of the class **ActiveResourceRequirement** is used for this purpose.

The process model of our example case which is created from the metamodel shown in Fig. 9.6 is presented in Appendix 9.C.

9.4.5 Value metamodel

The OptML Framework aims to optimize software models based on various quality attributes. This metamodel assumes that these attributes are expressed in *numeric values* associated with the models[3]. If there are multiple attributes, the associated values must be differentiated by the types of the qualities used. We define nevertheless a common *Value metamodel*, which can be instantiated for different quality attributes if needed. The Value metamodel is shown in Fig. 9.7. Here, the class **ValueModel** has an attribute **name** that is used to specify the type of a quality attribute. The value of a *quality attribute* is assigned to each operation within a class. As a short–hand notation, it is also possible to assign a value to a class. This means however that the values of the operations defined in that class are equal. If two different values are assigned to a class and an operation of that class,

---

[3] According to the principles of measurement theory, each quality attribute must be expressed in an appropriate value system [25]. However, the computation of values of quality attributes and their association with modeling elements can be defined in various ways. If a different value model is required than the one presented in this chapter, the tool designer must extend the OptML Framework accordingly. This is briefly discussed in Section 9.8.

the value of the operation overrides the value of the class. It is also possible to override values through inheritance relations. To specify these values, the class *Contribution* is defined which is associated with **Class** and **Operation**. Here, the attributes **included** and **excluded** indicate the positive and the negative contribution values depending on the inclusion or exclusion of the corresponding element in the configuration, respectively.

The instantiations of the Value metamodel for energy consumption and computation accuracy are given in Appendix 9.D.

9.5. Model processing subsystem

The *Model Processing Subsystem* is used to integrate the models with each other. The output of this subsystem is expressed as a dynamic data structure called *pipeline data*, which will be explained in the following subsections. Fig. 9.8 symbolically depicts the processing steps of this subsystem.

As shown in the figure, for each model that is defined, there exists a corresponding transformation unit (TU). TUs are organized in a pipeline structure and carry out the following two operations: *inputModel* and *transform*. The input and output formats of each TU conform to the *pipeline data* type. The operation *inputModel* retrieves the corresponding model from the model base.

The operation *transform* is specialized with respect to the characteristics of the corresponding model. A typical transformation operation consists of the following steps. First, it retrieves the incoming data from the pipeline. Second, it checks the consistency between the incoming data and the corresponding model. An error message is generated in case of inconsistency. Third, it transforms the structure of the corresponding model to the format of the *pipeline data*. Finally, it concatenates the incoming pipeline data with the transformed data and places it at the output. The next processing unit takes it as an incoming *pipeline data*, and so on.

An example of *pipeline data* is represented in Listing 9.1.

Listing 9.1: The structure of *pipeline data*.

```
[ ...,
  {
        "owner": <aTU>,
        "size": <aSize>,
        "data": <aData>
  },
  ...
]
```

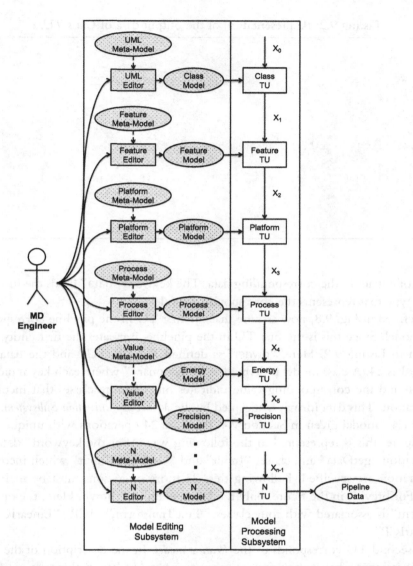

Figure 9.8 A symbolic representation of the *Model Editing* and *Model Processing* subsystems. Here, the ellipses represent data and the rectangles represent the processes.

Pipeline data consist of zero or more entities. Each TU adds its own data as an entity to the pipeline data. The three dots before and after the brackets "{" and "}" indicate a possible existence of more entities. The keyword "**owner**" denotes the identity of the TU which outputs these data. Here, the item "<aTU>" must be replaced with the name of a concrete identity of the corresponding TU. The keyword "**size**" indicates the

Listing 9.2: Representation of the output data of *Class TU*.

```
[
  {
    "owner": "Class TU",
    "size": 24,
    "data":
    {
      "getData": [ "Input", "CalibratedInput" ],
      ...
      "transform": [ "FineTransform", "ICP", "LinearICP", "NLinearICP" ]
    }
  }
]
```

number of entries in the corresponding data. The keyword "**data**" holds the instance of the data type that represents the corresponding model.

As defined in Fig. 9.8, now assume that the first TU in the pipeline corresponds to a class model. Since this is the first TU in the pipeline, it creates the first entity, which is shown in Listing 9.2. Here, "**owner**" is defined as "ClassTU," and the data size is computed as 24. A class model is defined as a "dictionary" where each key indicates an operation and the corresponding value indicates one or more classes that incorporate the operation. This data format is accepted by the *Model Optimization Subsystem*. In our example class model given in Section 9.4, there are 24 operations with unique names. In the figure, this is represented in the following way. After the keyword "**data**," first the operation "getData" and classes "Input" and "CalibratedInput" which incorporate this operation are specified. Following this, 23 more operations must be included in the list. For brevity, in the figure, only the last operation is shown. Here, the operation "transform" is associated with four classes: "FineTransform," "ICP," "LinearICP," and "NLinearICP."

The second TU corresponds to the *Feature model*. In the description of the feature model, it is assumed that each feature corresponds to a class in the class model. For this reason, *Feature TU* checks whether for each feature there exists a matching class in the class model. In case of a mismatch, an error condition is raised. In Listing 9.3, the output of our example *Feature TU* is shown. Here, we will only focus on the keywords "**size**" and "**data.**" The value of "**size**" is calculated as 6144, meaning that with the current specification of the feature model 6144 configurations are possible. *Feature TU* computes the configurations and adds these as the entries of the instance of the data type **list**, as required by the *Model Optimization Subsystem*, and stores it at "**data.**" For example, the first configuration in the list includes the features "Registration," "Input,"

Listing 9.3: Representation of the output data of *Feature TU*.

```
[ ...,
  {
    "owner": "Feature TU",
    "size": 6144,
    "data": [
        ["Registration", "Input", "Filter", "Sampling",
        "VoxelGrid", "Match", "Transform", "ICP",
        "LinearICP"],
        ...
    ]
  }
]
```

Listing 9.4: Representation of the output data of *Platform TU*.

```
[ ...,
  {
    "owner": "Platform TU",
    "size": 1,
    "data": System
  }
]
```

"Filter," "Sampling," "VoxelGrid," "Match," "Transform," "ICP," and "LinearICP." For brevity, the remaining configurations are not shown in the figure.

Platform TU is the third unit in the pipeline. The output of this TU is shown in Listing 9.4. The keyword "**size**" is set to value one, meaning that there is only a single entry in the model and that is the platform model. The keyword "**data**" refers to the instance object of the platform model, denoted by the variable name **System**, which is defined in the model base. The *Model Optimization Subsystem* directly accepts the models (objects) that conform to the Platform metamodel.

Process TU is defined as the fourth unit in the pipeline. In the process model, it is assumed that each process corresponds to an operation in the class model. Furthermore, there exists a consistency relation between the process model and the feature model, since there exists a one-to-one relation between features and classes. Therefore, *Process TU* first checks whether these conditions are satisfied. It is possible that while configuring the *Feature model* some of the optional features are not included. The operations corresponding to these excluded features must be excluded from the process model as well. Listing 9.5 shows the output data of *Process TU*. From the figure, it can be seen

Listing 9.5: Representation of the output data of *Process TU*.

```
[ ...,
  {
    "owner": "Process TU",
    "size": 966,
    "data":
      { ["Registration", "Input", "Filter", "Sampling",
        "VoxelGrid", "Match", "Transform", "ICP",
        "LinearICP"] : Process,
        ...
      }
  }
]
```

that the value of "**size**" drops from 6144 to 966 due to the elimination of the irrelevant configurations in our example. The data are organized as a dictionary type where the keys are the configurations and the values are the corresponding instances that represent the relevant portions of the process model. The *Model Optimization Subsystem* requires this dictionary data type. In Listing 9.5, only the first entry of the dictionary is shown. Here, the features "Registration," "Input," "Filter," "Sampling," "VoxelGrid," "Match," "Transform," "ICP," and "LinearICP" correspond to a relevant configuration of the feature model. This configuration is associated with the instance **Process** that includes the portion of the corresponding processes. In the figure, for brevity, the remaining 965 instances are not shown.

Energy TU is defined as the fifth unit in the pipeline. As the first step, *Energy TU* checks if the *Energy model* and the models retrieved from the *pipeline data* are consistent with each other. In this context, the consistency is specified as follows. Every class and operation defined in the *Energy model* must conform to the class model. Second, the configurations of processes are taken from the *pipeline data* and the total energy value per configuration is computed. Third, *Energy TU* creates a dictionary where the keys are the relevant configurations, and the values are the total energy value of the processes that are utilized in the corresponding configuration. Finally, this dictionary is concatenated with the incoming pipeline data and placed at the output. This data representation is required by the *Model Optimization Subsystem*. An example of output pipeline data is shown in Listing 9.6. Consider now the keyword "**data**." Here, only the first configuration is shown, which consists of "Registration," "Input," "Filter," "Sampling," "VoxelGrid," "Match," "Transform," "ICP," and "LinearICP." In this example, the total energy value consumed by this configuration is computed as 80.0.

The last unit in the pipeline is *Precision TU*. The steps carried out in this TU are the same as in the previous one, namely, checking consistency, extracting configurations

Listing 9.6: Representation of the output data of *Energy TU*.

```
[ ...,
  {
    "owner": "Energy TU",
    "size": 966,
    "data": {["Registration", "Input", "Filter", "Sampling",
      "VoxelGrid", "Match", "Transform", "ICP",
      "LinearICP"] : 80.0],
      ...
    }
  }
]
```

Listing 9.7: Representation of the output data of *Precision TU*.

```
[ ...,
  {
    "owner": "Precision TU",
    "size": 966,
    "data": {["Registration", "Input", "Filter", "Sampling",
      "VoxelGrid", "Match", "Transform", "ICP",
      "LinearICP"] : 320.0,
      ...
    }
  }
]
```

from the process model, computing the value of each configuration, and concatenating the obtained data with the incoming pipeline data. Of course, in this context, the values correspond to the precision values. The output pipeline data are shown in Listing 9.7. The first configuration associated with"**data**" is the same but the associated value of this configuration is computed as 320.0.

9.6. Model optimization subsystem

In this section, we first define the adopted optimization process. Second, we shortly describe the architecture of the subsystem. Finally, we give three example scenarios to illustrate how the subsystem computes the optimal model according to the given constraints.

9.6.1 Definition of the optimization process

The OptML Framework is used to optimize *software models*. According to our definition, the essential property of every software system is the *execution of operations* according to a certain program. The *Process metamodel* is defined to express the dynamic behavior of a software system. The model-driven engineer must specialize this metamodel to describe the dynamic behavior of the system being designed. This model is particularly useful to compute the timeliness and energy consumption properties of the models.

A *process configuration* corresponds to a program which is defined as a *valid set* of processes conforming to its process model. In general, more than one process configuration can be derived from a *Process model*. We assume that if there are inconsistent models, they are detected by the *Model Processing Subsystem* before the *pipeline data* reach the *Model Optimization Subsystem*.

The *Model Optimization Subsystem* searches for a solution of a process configuration where each process is allocated to the appropriate elements of the *Platform model*, while satisfying the constraints[4] defined by the model-driven engineer of each process. Currently, the following constraints are supported:

- The capacity. This is specified based on the units of the relevant platform elements. Examples are memory size and processing power.
- The worst-case execution time.
- The release time.
- The deadline.
- The dependency constraints.
- The preemption constraint.
- The migration constraint.
- The mutual-exclusion constraint.

The *Platform model* specifies the resources and their characteristics so that they can be matched to the constraint embedded in the process model.

While searching for an optimal configuration, the *Model Optimization Subsystem* may adopt various strategies among the set of candidate configurations, such as first-fit, *n*th-fit, and first-*n* searches. In the first-fit search approach, the first configuration that satisfies the requirements is selected and the search process is terminated. In the *n*th-fit search, as the name implies, the first *n* valid configurations are selected if existing. Finally, in the first-*n* search, the first *n* configurations are selected even if some of them are invalid.

In additional to the process requirements, the optimization process can be extended by defining constraints which can be derived from the Value metamodel, similar to the examples of the energy and precision models presented in Appendix 9.D. These additional constraints are only meaningful if more than one configuration is considered.

---

[4] These are canonical constraints taken from [20].

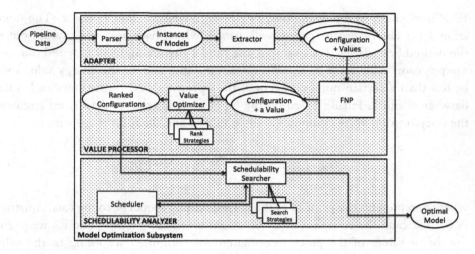

Figure 9.9 Representation of an architecture of a *Model Optimization Subsystem*.

The configurations that satisfy the process requirements and are within the boundaries of the desired value constraints are ranked according to the optimal required value. The optimal value may be either a minimum or a maximum of the values of the considered configurations.

A *Class model* is a static representation of a program, and as such it defines the bindings of processes to the operations of classes which can be overridden through inheritance. Therefore, the class model restricts the definition of configurations derived from the Process metamodel. Similarly, the *Feature model* can be seen as a restriction over the *Class model*.

9.6.2 The architecture of the model optimization subsystem

The *Model Optimization Subsystem* consists of three components: *Adapter*, *Value Processor*, and *Schedulability Analyzer*, as shown in Fig. 9.9.

Adapter has two subcomponents:

- *Parser* accepts the *pipeline data* as input and extracts the instances of models that are generated by the *transformation units*. These are a dictionary representing the class model, a list of configurations obtained from the feature model, an instance object that represents the platform, a dictionary of process configurations, and two dictionaries representing energy and precision models, respectively.
- *Extractor* processes the instances of models and generates a list of *process configurations* associated with the values to be considered for the optimization process.

The module *Value Processor* includes two subcomponents:

- *FNP* implements three operations, i.e., *Filter*, *Normalize*, and *Prioritize*. The operation *Filter* eliminates the *process configurations* which have associated values out of the desired boundaries. For example, the model-driven engineer may indicate that the precision value must be above a certain number and/or the energy value must be less than a certain number. Second, the operation *FNP* normalizes each value between 0 and 1. Finally, based on the input given by the model-driven engineer, the operation *Prioritize* computes a single value using the following formula:

$$P(\vec{v}_{m_1}, \vec{v}_{m_2}, \dots, \vec{v}_{m_n}) = \sum_{i=1}^{n} \alpha_i \times N \circ F(\vec{v}_{m_i}), \tag{9.1}$$

where α_i represents the priority of the corresponding model m_i to the total equation, $N \circ F$ indicates the composition of functions *normalize* and *filter*, and $\vec{v}_{m_i}$ represents the list of values of the process configurations computed according to the value model m_i.

- *Value Optimizer* ranks the list with respect to the selected *Rank Strategy* in an ascending or a descending order. According to the choice of the strategy, the model-driven engineer may request for the best *n* process configurations rather than delegating all of them to the following subcomponent *Schedulability Searcher*. The best configurations are computed by using an appropriate optimization algorithm. In this way, only the most promising configurations are considered first. If no quality attributes are defined, this component has no effect.

The module *Schedulability Analyzer* incorporates two subcomponents:

- *Scheduler* is based on an application framework called First Scheduling Framework (FSF) [26], which provides the necessary abstractions and mechanisms to implement schedulers. Currently, this framework supports the following abstractions as class hierarchies: Tasks, Resources, Scheduling characteristics, and Scheduling Strategy. The task and resource models of FSF are of the same type as the process and platform models of the OptML Framework, respectively. Upon defining a scheduling application using the provided models, the scheduling constraints are translated to the mathematical constraints to be solved in the generic constraint solvers provided by FSF. Since the models defined in this chapter are designed in accordance with the models in FSF, the transformation of the models is realized in the following way. Firstly, the platform model in OptML is translated to the resource model in FSF. Secondly, the *Process model* that defines the execution flow of the operations in the *Class model* is converted to the task model in FSF. Thirdly, the scheduling characteristics are defined with respect to the requirements of each process defined in Section 9.6.1, and scheduling strategy is defined as *minimizing the makespan*. Since the execution time of a process in the *Process model* changes with respect to the class in which it is defined, a separate scheduling problem is generated for each configuration. In the sense of schedulability, any feasible solution (schedule) computed by *Scheduler* makes the corresponding process configuration *valid*.

- *Schedulability Searcher* implements the search algorithm based on a certain strategy. To this aim, it retrieves the top element of the ranked configurations from the subcomponent *Value Processor* and calls on *Scheduler* to evaluate it. Depending on the result of this evaluation this process may iterate over the remaining configurations in the ranked list based on the selected strategy. For example, if the first-fit strategy is used, *Schedulability Searcher* terminates the search as soon as *Scheduler* finds a solution that satisfies the constraints. This configuration is considered to be the *optimal model*. Other related models such as the *Class model* can be reconstructed based on this result.

9.6.3 Example scenarios

In the following subsections, we will give a set of model optimization scenarios to illustrate the utility of the OptML Framework with respect to the requirements defined in Section 9.2. In practice, it is not possible to validate the correctness of a software system with the help of user-defined scenarios since the number of scenarios in any practical system can be extremely large. Therefore, we categorize the scenarios in the following way: (i) time analysis on a single- and multiprocessor architecture; (ii) model optimization based on time analysis combined with a single quality attribute; (iii) model optimization based on time analysis on multiple quality attributes; and (iv) three different search strategies to find an optimal model. We assume that these four categories represent a large number of scenarios that can be experienced in practice.

With respect to the requirements given in Section 9.2, the fulfillment of the first three requirements, finding the schedulable optimal model while satisfying the quality requirements, is demonstrated by the three categories of the scenarios that will be given in the following subsections.

9.6.3.1 Scenario 1: finding out the schedulability of the model with respect to a platform model

This scenario aims at justifying the fulfillment of the first requirement: schedulability of processes on platforms. To this aim, we have made the following assumptions about the models:

- The class, feature and process models are taken from the appendix.
- The platform model has the following characteristics: It has two terminal resources CPU0101 and MEM0101, belonging to ACTIVE and PASSIVE resource types, respectively. The object diagram for the corresponding platform model is shown in Fig. 9.10. The capacity of the active terminal resource is 1; the passive terminal resource has 1024-unit capacity. Each of these terminal resources has one running state. In this scenario, we choose a platform model with single processing unit that

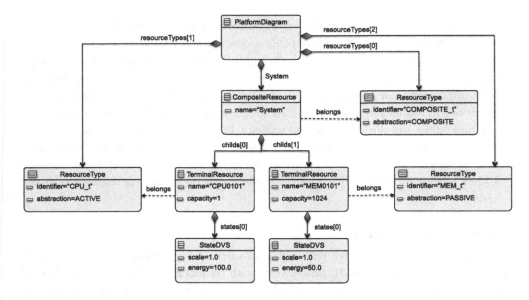

Figure 9.10 An object diagram of the platform model.

is more restricted than the one given in the appendix to demonstrate the effect of platform capacities on the schedulability process.

- The criteria of the scheduling objective is set to "minimizing the maximum makespan" [21].
- The scheduling window for the processes is set to 50.
- The search strategy is chosen as *first-fit*.
- The process configurations are not ranked by the subcomponent *Value Optimizer*.

With these given assumptions, *Scheduler* returns a solution that is depicted in Fig. 9.11. The figure consists of two rectangles. The top horizontal rectangle depicts the schedule computed by *Scheduler*. This figure only includes the processes that are selected by the optimizer. The horizontal axis corresponds to *Time*, and the vertical axis corresponds to the capacities of the two *terminal resources*, CPU0101 and MEM0101. Each unit in the vertical coordinate corresponds to total amount of capacity for each resource. The larger rectangle shows all processes, some of which may not be included in the schedule. The legend of each process is indicated by a different color.

9.6.3.2 Scenario 2: finding the schedulable optimal model with respect to a single quality attribute

The second scenario is defined to illustrate the first three requirements in Section 9.2: introducing new quality attributes and finding out the optimal model that satisfies both the schedulability and the quality requirements.

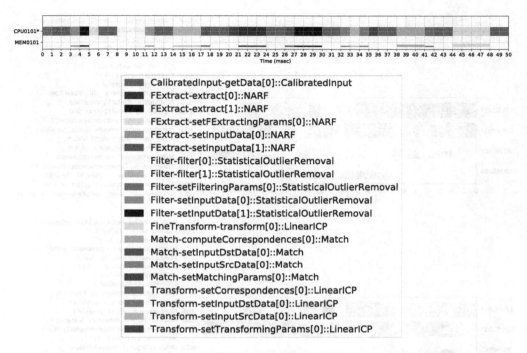

Figure 9.11 A schedule produced by *Scheduler* using the first-fit strategy. The representation given in this figure conforms to the presentation that is commonly used in scheduling and timetabling research [27]. The top horizontal rectangle depicts the schedule computed by *Scheduler*. The colors correspond to the tasks and they do not have a specific meaning; they are selected by the tool randomly.

Along this line, the following assumptions are made:

- The class, feature, platform, and process models are taken from the appendix.
- A single quality model *energy consumption* is introduced, which is defined in the appendix.
- The criteria of the scheduling objective is set to "minimizing the maximum makespan."
- The scheduling window for the processes is set to 23.
- The search strategy is chosen as *third-fit*.
- The process configurations are ranked by the subcomponent *Value Optimizer* with respect to the ascending energy values.

In this scenario, the multiprocessor architecture is selected. The result of the *optimization process* based on the given assumptions is shown in Fig. 9.12. This figure consists of three subfigures, where each subfigure shows the evaluation of a particular process configuration. The top subfigure shows a schedule with the lowest energy

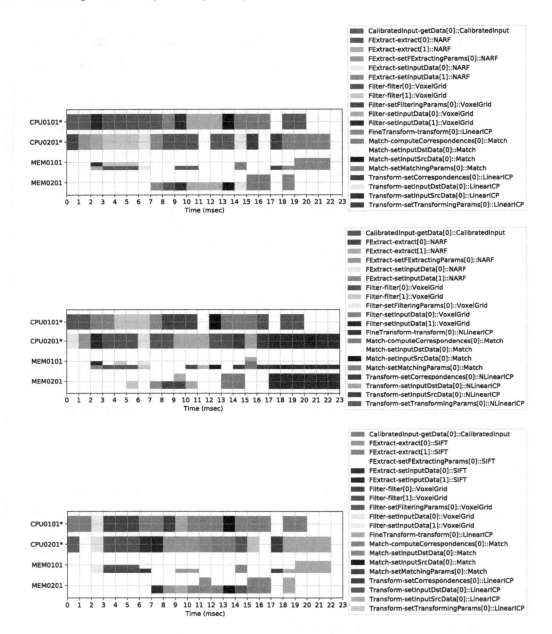

Figure 9.12 The three schedules ordered according to their energy consumption values.

consumption; the bottom subfigure shows one with the highest consumption among the ones that are evaluated. Therefore, the top subfigure is considered as the optimal model.

9.6.3.3 Scenario 3: finding the schedulable optimal model with respect to multiple quality attributes

This scenario extends the previous one with an additional quality attribute *precision*. To this aim, the assumptions are the same as the previous scenario, except the following:

- The scheduling window size is set to 21.
- The search strategy is set to first–25.
- A new quality attribute *precision* is defined in the appendix.
- The priorities of the quality attributes *energy* and *precision* are defined as 0.2 and 0.8, respectively.
- The process configurations are ranked by the subcomponent *Value Optimizer* with respect to the ascending energy and precision values. To preserve uniformity, a normalized value of 1 corresponds to lowest precision, whereas 0 is the highest.

The result is presented in Fig. 9.13. Here only two configurations are shown. The second configuration found in the ranked configurations happens to be not schedulable,

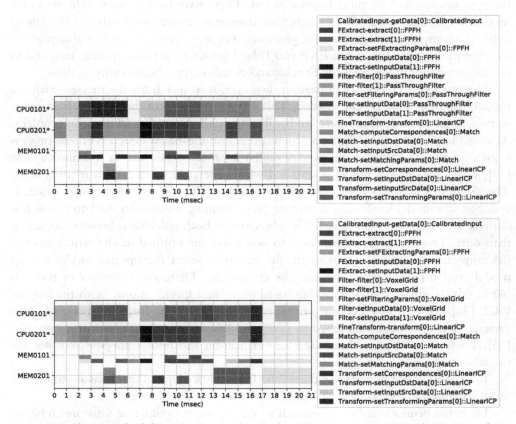

Figure 9.13 The two schedules ordered according to their energy consumption and precision values.

because the operation **transform** exceeds the specified scheduling window size if it is implemented with a high precision value.

The figure consists of two subfigures, where the top subfigure corresponds to a configuration with a higher quality value. This configuration is selected as the optimal model.

9.7. Related work

Model-Driven Architecture (MDA) aims at separating platform-independent and platform-dependent models from each other [28]. MDE extends MDA with metamodels and model transformations [29]. In MDE, not only models but also metamodels and model transformations are the core assets of software development. The research activities in MDE are very broad, including domain-specific models, model building, model verification, model reuse, model transformation, and code generation [30,31].

In the literature, the terms model and optimization are used in two ways: (i) models for optimization and (ii) model optimization. There have been considerable works on models for optimization where researchers investigate mostly mathematical models to define and implement optimization processes. For our approach, such techniques are adopted in the subcomponents *FNP* and *Value Optimizer* for value optimization and in the subcomponents *Schedulability Searcher* and *Scheduler* for schedulability analysis.

The purpose of model optimization, however, is to search for the models within a model base that satisfy certain criteria. This is the main focus of the chapter. In contrast to models for optimization, there are hardly any publications that address this problem. As stated by Chenouard [32], a constraint programming-based design synthesis process is presented using MDE techniques. A similar approach is adopted in Joachim's article [33], where an optimal model is searched within the context of certain requirements. The difference between these two articles is that in the former, the optimal model is searched at a model level using constraint programming, whereas in the latter, search is defined as a model transformation. The objectives of both articles are, however, different than ours. The aim in these articles is to synthesize the optimal model which satisfies the constraints, whereas in our work, the aim is to select the optimal model among model configurations which satisfies the constraints. There are a number of research works which aim at verifying models based on certain specifications. With the help of OCL [34], for example, certain properties of models can be formally specified. Also various tools have been developed to verify models decorated with such specifications [35]. These tools in general are used for verification and testing purposes but not for model optimization.

To the best of our knowledge, there is no framework proposal that aims at optimizing models defined in the context of MDE, as proposed in this chapter.

There has been a number of research works aiming at optimizing software architectures according to a set of quality attributes. In [36], algorithms are proposed to optimize

TV architectures for the qualities availability, reduced memory usage, and time performance. Multiobjective optimization techniques are proposed in [37] with respect to certain quality attributes such as production speed, reduced energy usage, and print quality. A design method for balancing quality attributes energy reduction and modularity of software is proposed in [38].

9.8. Evaluation

We will now evaluate this chapter with respect to the nine requirements given in Section 9.2. The fulfillment of the first three requirements given in Section 9.2 is explained in Section 9.6.3.

The fourth requirement, checking consistency among models, is realized by the TUs defined in the *Model Processing Subsystem*, as explained in Section 9.5.

The fifth requirement, supporting new models in the Ecore environment, is provided with the following condition: For each new model, the *tool engineer* must define the corresponding TU with the necessary model extraction, consistency checking, and model transformation functions.

The sixth requirement, pruning models, is supported in the TUs by redefining the operation *inputModel* with different retrieval strategies so that only the relevant parts of the model are selected. For example, in the current implementation of the framework, we support the following strategies: (a) retrieve the complete model (default); (b) include only the classes denoted by the *model-driven engineer*; and (c) include all the classes with a query. The architecture allows introducing new strategies modularly. However, if not all model elements are selected, the pruned model can be inconsistent with the other models. In our framework, for example, if some classes, which are eliminated from the class model, are included in the feature model, the feature TU will give an error message. For brevity, this chapter does not focus on the pruning techniques. An interested reader may refer to the following report [39]. As explained in Section 9.5, due to the pipeline architecture of the *Model Processing Subsystem*, the framework allows the tool designer to introduce such extensions to the existing utilities.

As demonstrated in Section 9.6.3, the seventh requirement, introducing new quality attributes, can be realized by specializing the *Value metamodel* with the following restrictions: (a) the quality attributes are expressed in numbers and associated with classes and their operations; and (b) if necessary, the quality attributes per configuration are computed by the corresponding TU. If this way of representing the desired quality attribute is not appropriate, a new Value metamodel and the corresponding TU must be introduced. The pipeline architecture of the *Model Processing Subsystem* makes this possible without changing the other TUs.

The eighth requirement, introducing new value optimization algorithms, is supported in the following way: Currently, the OptML Framework implements a rather

Table 9.1 The performance values of the scenarios. The scenarios are implemented on a Mac-Book Pro with 2.6 GHz Intel Core i5 processor and 8 GB 1600 MHz memory.

Parameter	Scenario		
	Scenario 1	Scenario 2	Scenario 3
Model pruning	No	No	No
Normalization	NA	[0 1]	[0 1]
Rank strategy	NA	Ascending order	Ascending order
Search strategy	First-fit	Third-fit	First-25
Utilized solver	Gurobi [40]	Gurobi	Gurobi
Average execution time of the solver per configuration	305 sec	30 sec	26 sec

straightforward value optimization based on filtering, normalizing, prioritizing, and priority-based ranking with the help of the subcomponents *FNP* and *Value Optimizer*. Different quality values are merged into a single value. As the next step, the schedulability of the ranked configurations is analyzed. One may adopt, however, different value optimization techniques depending on the needs, and the feasibility of time and space complexity of the optimization algorithms. For example, one may aim at reducing the time of optimization processes by using heuristic rules. In the literature, various optimization algorithms are presented, such as hill climbing, exhaustive search, and pareto-front multiobjective optimization [36]. These changes can be encapsulated in the subcomponents *FNP* and *Value Optimizer*.

The last requirement, adopting new search strategies for scheduling, can be introduced by defining a new search strategy for the subcomponent *Schedulability Searcher*.

The time performance of the model optimization process is considered important for the usability of the framework. The current architecture allows performance improvement in the following ways: (a) introducing effective model pruning strategies in the corresponding TUs; (b) applying different optimization algorithms in the *Model Optimization Subsystem*; (c) using efficient schedulability search strategies; and (d) using efficient solvers in the subcomponent *Scheduler*.

The average execution times of the solver per configuration are shown in Table 9.1. In the figure, the bottom row shows the results of the evaluation. The top five rows define the parameters of the implementations. In none of these scenarios, model pruning is used. Normalization of the quality values is only applied to the second and third scenarios where the values are mapped into the range of 0 to 1. Here, NA means not applicable. The definitions of rank and search strategies are self-explanatory. The row "Utilized solver" defines the constraint solver that is used in the subcomponent *Scheduler*. The execution time of the solver for Scenario 1 is much higher due to the longer schedulability window size. The time that is required to search among the possible solutions by the subcomponent *Schedulability Searcher* is negligible when compared with the time performance of the solvers. Of course, to determine the total time performance, one needs to multiply the last column value with the number of iterations in the search strategy.

9.9. Conclusion

This chapter identifies the following problems: (a) model complexity; (b) the lack of quality-based model selection; and (c) the lack of support of multiple quality-based evaluations. To address these problems, the OptML Framework is presented. This framework accepts various models defined in the Ecore MDE environment, processes them according to user preferences and model properties, and computes the optimal schedulable models based on value optimization and constraint-based scheduling algorithms. To the best of our knowledge, this is the first generic MDE framework that is suitable for model optimization.

The utility of the framework is demonstrated by implementing three scenarios inspired from *registration systems* used in the area of image processing. The scenarios show that the required objectives of model optimization as defined in this chapter are fulfilled. With the help of design patterns and various architectural styles, the architecture has a modular structure and thereby allows for the introduction of new strategies for model extraction and transformation, value optimization, and schedulability analysis. The framework is fully implemented and tested. It integrates a number of third-party software such as Eclipse EMF [41], pyecore [42], pyuml2 [43], FSF [26], Numberjack [44], matplotlib [45], Gurobi [40], SCIP [46], and Mistral [47].

Appendix 9.A. Feature model

The tool designer instantiates the Feature metamodel shown in Fig. 9.4. The instantiated model is shown in Fig. 9.14. The number of systems that can be configured from this model is computed as 1288.

In this section, we present a feature model for the example registration system. This model conforms to the Feature metamodel presented in Fig. 9.4 of Section 9.4. This model is used in the scenario implementations presented in Section 9.6.3. By definition, in our framework, each feature corresponds to a class in the class model. The descriptions of the names of the adopted features are presented in Section 9.4.2. To avoid repetition, the features are not described here again. The number of systems that can be configured from this model is computed as 1288.

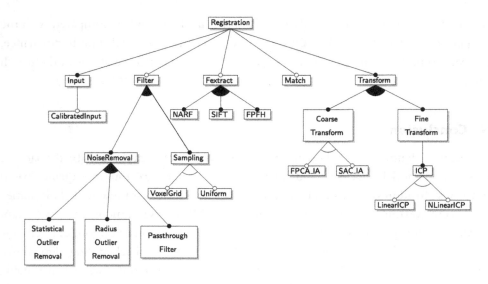

Figure 9.14 A feature model of the registration system derived from the feature metamodel.

Appendix 9.B. Platform model

The metamodel shown in Fig. 9.5 is instantiated according to the registration system given in Section 9.2. To this aim, we define a platform model depicted in Fig. 9.15 in which the system has two composite resources, each of which consists of one active and one passive resource. Each terminal resource has one state except the active resource of the second composite resource. There exist three resource types: processing unit (ACTIVE), memory (PASSIVE), and computation node (COMPOSITE). Unlike the other terminal resources, the processing unit of the second node has two states: half- and full-speed running modes. Each processing unit has a unit capacity. The memory components have 512- and 256-unit capacity on the first and second computation nodes, respectively.

Appendix 9.C. Process model

In Figs. 9.16 and 9.17, the process model is presented which is created from the Process metamodel presented in Fig. 9.6 in Section 9.4.4. Since the process model is rather large, in the figure, we will only elaborate a selected set of processes. As explained previously, the input data are acquired by utilizing the class Input. For this purpose, the operation getData is defined. In some cases, the accuracy of acquired data may be crucial. To this aim, the class CalibratedInput is used instead of using the superclass of it. The operation calibrate aims to increase the quality of the data if it is called before calling getData.

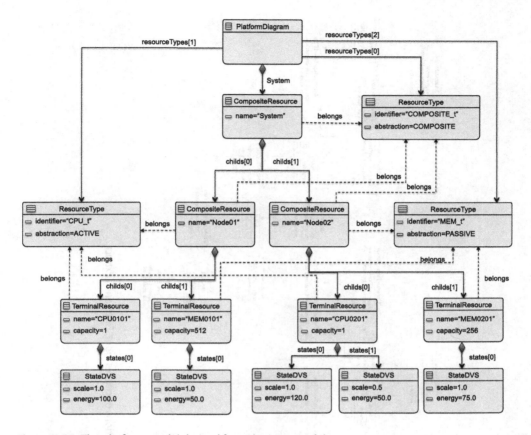

Figure 9.15 The platform model derived from the metamodel.

To reduce the size of the input data, the class **Filter** is defined. The operation **set-InputData** is used to set the interested data. To set the filtering-related parameters, the operation **setFilteringParams** is defined. The operation **filter** is, finally, called to gather the filtered data. The class **Filter** is specialized further into classes **NoiseRemoval** and **Sampling**, which are responsible for eliminating the erroneous data and getting a part of the data to reduce the size, respectively. Classes inheriting to these classes, such as **StatisticalOutlierRemoval** or **VoxelGrid**, correspond to the different algorithmic approaches.

The class **FExtract** is responsible for computing predefined key features of the data to reason about the geometric properties. Similarly, the operation **setInputData** of the class **FExtract** is utilized to set the interested data, and the operation **setFExtractingParams** is responsible to adjust the settings of the class. To gather a combination of computed key features and the given data, the operations **extract** is called. Classes **NARF**, **SIFT**, and **FPFH** represent the definitions for computing various predefined features.

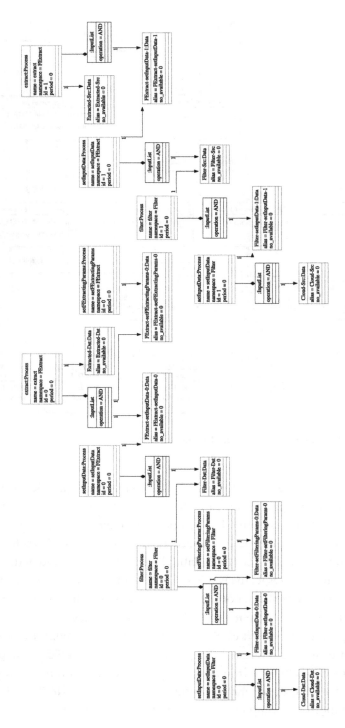

Figure 9.16 The partial process model including the functions of classes *Input*, *Filter*, and *FExtract*.

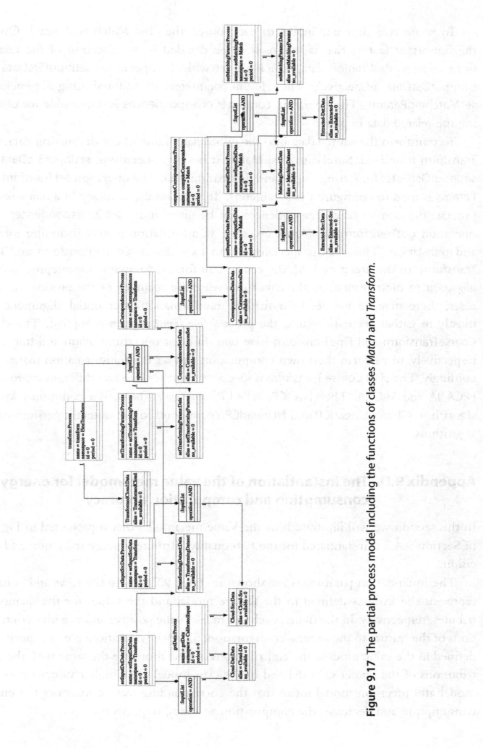

Figure 9.17 The partial process model including the functions of classes *Match* and *Transform*.

To relate two different inputs to each other, the class **Match** is defined. One of the important factors that is supposed to be decided is the direction of the relation from *source* to **destination** data, which are set with the operations **setInputSrcData** and **setInputDstData**, respectively. The relevant parameters are adjusted using the operation **setMatchingParams**. The operation **computeCorrespondences** is responsible for obtaining the related data in pairs.

To transform the source data into the coordinate frame of the destination data, class **Transform** is utilized. Similar to class **Match**, it has two operations, **setInputSrcData** and **setInputDstData**, for setting source and destination data. The operation **setTransformingParams** is used to configure the parameters. To increase the accuracy of a *transformation* process, the computed correspondences can be given using **setCorrespondences**. The operation **getTransformationMatrix** gives the transformation matrix including rotation and translation. This class is specialized into two classes, **CoarseTransform** and **FineTransform**. In the literature [48], the coarse transform is known as the preprocess *initial alignment* to pretransforming the data to increase the accuracy of the process. In some cases, there may be no need for further transformation after initial alignment, but mostly to gather accurate results, the process *fine transformation* is applied. The classes **CoarseTransform** and **FineTransform** have two different operations, **align** and **transform**, respectively, to perform their own computation and set the transformation matrix, accordingly. The class **Coarse Transform** is specialized further into two different approaches, **FPCA_IA** and **SAC_IA**. The class **ICP**, called *iterative closest point*, is a commonly known algorithm. Classes **LinearICP** and **NLinearICP** correspond to the different versions of **ICP** algorithms.

Appendix 9.D. The instantiation of the value metamodel for energy consumption and computation accuracy

In this section we will illustrate how the Value metamodel that is presented in Fig. 9.7 of Section 9.4.5 is instantiated for the two quality attributes energy reduction and precision.

The instantiation parameters are shown in Table 9.2, where the rows and columns represent the features defined in the feature model and the values for the quality attributes, respectively. In the figure, each cell includes the positive and negative contributions of the feature to the process configuration. Since the operations are not specifically defined in the value models, the contributions of operations are the same with the contributions of the owner class defined in the class model. The higher values for energy model and precision model mean that the corresponding feature increases the energy consumption and decreases the computation accuracy, respectively.

Table 9.2 The values of energy and precision models for each feature are defined. The positive and negative contributions of the value models, which are explained in Section 9.4.5, are shown using the notation "+/−" in the value cells.

	Energy value	Precision value
Registration	0 / 0	0 / 0
Input	0 / 0	0 / 0
CalibratedInput	50 / 30	20 / 50
Filter	5 / 0	5 / 10
NoiseRemoval	10 / 0	10 / 20
PassthroughFilter	15 / 0	15 / 30
RadiusOutlierRemoval	15 / 0	15 / 30
StatisticalOutlierRemoval	15 / 0	15 / 30
Sampling	10 / 0	10 / 20
VoxelGrid	15 / 0	15 / 30
Uniform	15 / 0	15 / 30
FExtract	10 / 0	5 / 10
SIFT	30 / 0	10 / 30
NARF	30 / 0	10 / 30
FPFH	30 / 0	10 / 30
Match	20 / 0	15 / 50
Transform	5 / 0	5 / 5
CoarseTransform	10 / 0	10 / 30
FPCA_IA	12 / 0	5 / 30
SAC_IA	12 / 0	5 / 30
FineTransform	15 / 0	2 / 90
ICP	15 / 0	3 / 90
LinearICP	20 / 0	1 / 110
NLinearICP	20 / 0	1 / 110

References

[1] F. Buschmann, K. Henney, D. Schimdt, Pattern-Oriented Software Architecture: On Patterns and Pattern Language, vol. 5, John Wiley & Sons, 2007.
[2] L.A. Belady, M.M. Lehman, A model of large program development, IBM Systems Journal 15 (3) (1976) 225–252.
[3] G. Brunet, M. Chechik, S. Easterbrook, S. Nejati, N. Niu, M. Sabetzadeh, A manifesto for model merging, in: Proceedings of the 2006 International Workshop on Global Integrated Model Management, GaMMa '06, ACM, 2006, pp. 5–12.
[4] Ö. Babur, L. Cleophas, M. van den Brand, B. Tekinerdogan, M. Aksit, Models, more models, and then a lot more, in: M. Seidl, S. Zschaler (Eds.), Software Technologies: Applications and Foundations, Springer International Publishing, Cham, 2018, pp. 129–135.
[5] D. Steinberg, F. Budinsky, M. Paternostro, E. Merks, EMF: Eclipse Modeling Framework 2.0, 2nd edition, Addison–Wesley Professional, 2009.

[6] G. Orhan, M. Akşit, A. Rensink, A formal product-line engineering approach for schedulers, in: 22nd International Conference on Emerging Trends and Technologies in Convergence Solutions, The Society for Design and Process Science (SDPS), 2017, pp. 15–30.

[7] L.G. Brown, A survey of image registration techniques, ACM Computing Surveys 24 (4) (1992) 325–376.

[8] R.B. Rusu, S. Cousins, 3D is here: point cloud library (PCL), in: 2011 IEEE International Conference on Robotics and Automation, 2011, pp. 1–4.

[9] P. Clements, D. Garlan, R. Little, R. Nord, J. Stafford, Documenting software architectures: views and beyond, in: Proceedings of the 25th International Conference on Software Engineering, IEEE Computer Society, 2003, pp. 740–741.

[10] E. Fersman, L. Mokrushin, P. Pettersson, W. Yi, Schedulability analysis of fixed-priority systems using timed automata, Theoretical Computer Science 354 (2) (2006) 301–317.

[11] P. Clements, D. Garlan, L. Bass, J. Stafford, R. Nord, J. Ivers, R. Little, Documenting Software Architectures: Views and Beyond, Pearson Education, 2002.

[12] P. Kruchten, The 4+1 view model of architecture, IEEE Software 12 (1995) 42–50, https://doi.org/10.1109/52.469759.

[13] K.C. Kang, J. Lee, P. Donohoe, Feature-oriented product line engineering, IEEE Software 19 (4) (2002) 58–65, https://doi.org/10.1109/MS.2002.1020288.

[14] J. Odell, Extending UML for agents, 2000.

[15] E. Gamma, R. Helm, R. Johnson, J. Vlissides, Design Patterns: Elements of Reusable Object-Oriented Software, Addison–Wesley Longman Publishing Co., Inc., Boston, MA, USA, 1995.

[16] M. Holenderski, R.J. Bril, J.J. Lukkien, Parallel-task scheduling on multiple resources, in: Real-Time Systems (ECRTS), 2012 24th Euromicro Conference on, IEEE, 2012, pp. 233–244.

[17] J.-J. Chen, T.-W. Kuo, Multiprocessor energy-efficient scheduling for real-time tasks with different power characteristics, in: 2005 International Conference on Parallel Processing, ICPP'05, 2005, pp. 13–20.

[18] X. Lin, Y. Wang, Q. Xie, M. Pedram, Task scheduling with dynamic voltage and frequency scaling for energy minimization in the mobile cloud computing environment, IEEE Transactions on Services Computing 8 (2) (2015) 175–186.

[19] S. Azizi, V. Panahi, Formal specification of semantics of UML 2.0 activity diagrams by using graph transformation systems.

[20] G.C. Buttazzo, Hard Real-Time Computing Systems: Predictable Scheduling Algorithms and Applications, vol. 24, Springer Science & Business Media, 2011.

[21] M.L. Pinedo, Scheduling: Theory, Algorithms, and Systems, 4th edition, Springer, New York, Dordrecht, Heidelberg, 2010.

[22] E.A. Lee, D.G. Messerschmitt, Static scheduling of synchronous data flow programs for digital signal processing, IEEE Transactions on Computers 36 (1) (1987) 24–35.

[23] W. Ahmad, R. de Groote, P. Holzenspies, M. Stoelinga, J. van de Pol, Resource-constrained optimal scheduling of synchronous dataflow graphs via timed automata, in: Proceedings of the 14th International Conference on Application of Concurrency to System Design, ACSD 2014, IEEE Computer Society, United States, 2014, pp. 72–81.

[24] W. Zhao, K. Ramamritham, J.A. Stankovic, Preemptive scheduling under time and resource constraints, IEEE Transactions on Computers C-36 (8) (1987) 949–960.

[25] N. Fenton, J. Bieman, Software Metrics: A Rigorous and Practical Approach, 3rd edition, CRC Press, Inc., Boca Raton, FL, USA, 2014.

[26] G. Orhan, M. Aksit, A. Rensink, Designing reusable and run-time evolvable scheduling software, in: Proceedings of the 12th International Conference on the Practice and Theory of Automated Timetabling, PATAT 2018, 2018, pp. 339–373.

[27] L. Abeni, G. Buttazzo, Resource reservation in dynamic real-time systems, Real-Time Systems 27 (2) (2004) 123–167, https://doi.org/10.1023/B:TIME.0000027934.77900.22.

[28] U. Assmann, M. Aksit, A. Rensink, Model driven architecture: European MDA workshops: foundations and applications, MDAFA 2003 and MDAFA 2004, Twente, The Netherlands, June 26–27, 2003 and Linköping, Sweden, June 10–11, 2004. Revised selected papers, 2005, https://doi.org/10.1007/11538097.

[29] A.R. da Silva, Model-driven engineering: a survey supported by the unified conceptual model, Computer Languages, Systems and Structures 43 (2015) 139–155.

[30] F. Jacob, A. Wynne, Y. Liu, J. Gray, Domain-specific languages for developing and deploying signature discovery workflows, Computing in Science & Engineering 16 (1) (2014) 52–64.

[31] Y. Sun, J. Gray, J. White, A demonstration-based model transformation approach to automate model scalability, Software & Systems Modeling 14 (3) (2015) 1245–1271.

[32] R. Chenouard, C. Hartmann, A. Bernard, E. Mermoz, Computational design synthesis using model-driven engineering and constraint programming, vol. 9946, 2016, pp. 265–273.

[33] J. Denil, M. Jukss, C. Verbrugge, H. Vangheluwe, Search-based model optimization using model transformations, in: D. Amyot, P. Fonseca i Casas, G. Mussbacher (Eds.), System Analysis and Modeling: Models and Reusability, Springer International Publishing, Cham, 2014, pp. 80–95.

[34] J. Warmer, A. Kleppe, The Object Constraint Language: Getting Your Models Ready for MDA, 2nd edition, Object Technology Series, Addison–Wesley, Reading, MA, 2003.

[35] M. Gogolla, J. Cabot, Continuing a benchmark for UML and OCL design and analysis tools, in: P. Milazzo, D. Varró, M. Wimmer (Eds.), Software Technologies: Applications and Foundations, Springer International Publishing, Cham, 2016, pp. 289–302.

[36] H. Sözer, B. Tekinerdogan, M. Aksit, Optimizing decomposition of software architecture for local recovery, Software Quality Journal 21 (2) (2013) 203–240, https://doi.org/10.1007/s11219-011-9171-6.

[37] A. de Roo, H. Sözer, L. Bergmans, M. Aksit, Moo: an architectural framework for runtime optimization of multiple system objectives in embedded control software, Journal of Systems and Software 86 (10) (2013) 2502–2519, https://doi.org/10.1016/j.jss.2013.04.002, eemcs-eprint-24550.

[38] S. te Brinke, S. Malakuti Khah Olun Abadi, C. Bockisch, L. Bergmans, M. Akşit, S. Katz, A tool-supported approach for modular design of energy-aware software, in: SAC '14, Association for Computing Machinery, United States, 2014, pp. 1206–1212.

[39] B. Kuyucu, On the Design of a User-Interface for Optimal Modeling Language (OptML) Framework, University of Twente, Drienerlolaan 5, 7522NB, Enschede, The Netherlands, report on practical training, 2018.

[40] L. Gurobi, Optimization, Gurobi optimizer reference manual, http://www.gurobi.com, 2018.

[41] E. Foundation, Eclipse modeling framework, http://eclipse.org/emf/, 2016.

[42] V. Aranega, A python(nic) implementation of emf/ecore (eclipse modeling framework), https://github.com/pyecore/pyecore, 2016.

[43] V. Aranega, A python implementation of the UML2 metamodel based on pyecore, https://github.com/pyecore/pyuml2, 2018.

[44] E. Hebrard, E. O'Mahony, B. O'Sullivan, Constraint Programming and Combinatorial Optimisation in Numberjack, Springer, Berlin, Heidelberg, 2010, pp. 181–185.

[45] J.D. Hunter, Matplotlib: a 2D graphics environment, Computing in Science & Engineering 9 (3) (2007) 90–95.

[46] G. Gamrath, T. Fischer, T. Gally, A.M. Gleixner, G. Hendel, T. Koch, S.J. Maher, M. Miltenberger, B. Müller, M.E. Pfetsch, C. Puchert, D. Rehfeldt, S. Schenker, R. Schwarz, F. Serrano, Y. Shinano, S. Vigerske, D. Weninger, M. Winkler, J.T. Witt, J. Witzig, The SCIP Optimization Suite 3.2, Tech. Rep. 15-60, ZIB, Takustr. 7, 14195 Berlin, 2016.

[47] I. Dillig, T. Dillig, A. Aiken, Cuts from Proofs: A Complete and Practical Technique for Solving Linear Inequalities over Integers, Springer Berlin Heidelberg, Berlin, Heidelberg, 2009, pp. 233–247.

[48] R.B. Rusu, N. Blodow, M. Beetz, Fast point feature histograms (FPFH) for 3D registration, in: Robotics and Automation, 2009, ICRA'09, IEEE International Conference on, Citeseer, 2009, pp. 3212–3217.

PART 3

Industrial applications

CHAPTER 10

Reducing design time and promoting evolvability using Domain-Specific Languages in an industrial context

Benny Akesson[a], Jozef Hooman[a,b], Jack Sleuters[a], Adrian Yankov[c]
[a]ESI (TNO), Eindhoven, The Netherlands
[b]Radboud University, Nijmegen, The Netherlands
[c]Altran, Eindhoven, The Netherlands

Contents

Model Management and Analytics for Large Scale Systems
https://doi.org/10.1016/B978-0-12-816649-9.00020-X

10.1. Introduction

Development of contemporary systems is becoming increasingly complex, time consuming, and expensive. This happens in response to a number of trends. Firstly, more and more dependent software and hardware components are being integrated to realize a wider range of functionality. Increased integration results in systems with complex behaviors that are difficult to design and validate, increasing development time. This problem is exacerbated by an increasing *system diversity* due to recent trends towards mass-customization of systems [1], which increasingly creates situations where every manufactured system has a unique hardware configuration and feature set. Lastly, system requirements frequently change as new technology is being introduced, or because of new expectations from the market. This means that substantial effort goes into reengineering systems to ensure they match customer needs throughout their life cycle.

These trends in development of complex systems result in three key challenges:

C1) Development time needs to be shortened to reduce cost and time-to-market.

C2) Systems must be quick and easy to customize for a particular customer to manage increasing diversity.

C3) System functionality must be evolvable to ensure that it continuously matches the needs of the customer during its life cycle.

Model-Based Engineering (MBE) is an engineering approach where models play an important role in managing complexity by providing abstractions of the system that separate the problem domain from the implementation technologies of the solution space. This has helped bringing development closer to domain experts, enabling them to express their ideas using familiar notations from their domain and automatically generate system artifacts, such as documentation, simulation models, and production code [2–4]. MBE can take many forms as there is a plethora of development methodologies used in industry with as much as 40 modeling languages and 100 tools being reported as commonly used [5]. The most commonly used modeling languages, at least in the embedded systems domain, are UML and SysML for software engineering and system engineering, respectively [6,7]. However, Domain-Specific Languages (DSLs) are becoming increasingly prevalent in narrow and well-understood domains [4].

This chapter is an experience report about addressing the increasing system complexity in an industrial context using an MBE development approach based on DSLs. First, Section 10.2 discusses how DSL technology addresses the three complexity challenges mentioned above, as well as stating five technical research questions related to this approach. Section 10.3 then discusses the five research questions in the context of the published state of the art to determine the extent to which they are recognized in literature and identify the range of available solutions. We continue in Section 10.4 by presenting our approach to practically investigate the research questions in the context of a case study from the defense domain. The design of a DSL ecosystem developed for this case study is discussed in Section 10.5, after which we explain how the research questions were addressed by the case study in Section 10.6. Section 10.7 presents an intermediate evaluation of the work before we end the chapter by discussing conclusions in Section 10.8.

10.2. Domain-Specific Languages

Determining whether a particular design methodology is a good fit for a given problem is not easy. This problem has also been recognized in the context of MBE [8,9]. This section discusses how an MBE methodology based on DSLs addresses each of the three challenges outlined in Section 10.1 and presents five research questions related to the approach that will be investigated through a literature study (Section 10.3) and a case study (Section 10.6).

10.2.1 Reducing development time (C1)

Compared to general-purpose programming, DSL-based development approaches require an initial investment in terms of effort [6,7]. This investment involves defining the abstract and concrete syntaxes of the language and implementing model validation, as well as model-to-text transformations that can generate artifacts for all supported variants. However, once this investment has been done, development time is ideally reduced, resulting in return on investment in the longer term if sufficiently many model instances are created [6]. Note that this assumes that the considered variants are not so different that they constantly require the DSL and its transformations to be extended, limiting reuse and increasing development effort. For this reason, DSLs are particularly well suited in the context of (mass-)customization, since a potentially large number of variants are needed that fit within the confines defined by product lines. In this context, which is the context of this work, DSLs efficiently address Challenge C1.

10.2.2 Improved customization (C2)

Customization of a system or component can also be addressed using feature models [10]. However, a limitation of feature models is that they are context-free grammars that can only specify a bounded space that is known a priori. This means that feature models are only suitable for a restricted form of customization, i.e., selecting a valid combination of features that are known up front [11]. However, it is not possible to use a feature model to specify new features that were not previously considered at an abstract level. If this is necessary, an alternative approach is to specify variability using general-purpose programming languages, which are fully flexible but expose low-level implementation details and do not separate the problem space and solution space. DSLs bridge the gap between feature models and general-purpose programming languages, as they are recursive context-free grammars that can specify new behavior from an unbounded space, while keeping problem space and solution space separate [11]. DSL technology is hence a good fit for systems with a high degree of variability, addressing Challenge C2.

10.2.3 Improved evolvability (C3)

Since DSLs make it quick and easy to customize systems or components by modifying model instances of DSLs and then generate artifacts, it also follows that the instance can be easily modified and artifacts regenerated if the system is evolved, e.g., due to changing requirements. In contrast, a change in the underlying implementation technology does not require DSL instances to be changed, as the specification in terms of domain concepts has not changed. Instead, changes to implementation technology imply a change only in the model-to-text transformations that express the semantics of the models, e.g., in terms of code. This provides a rather clean separation of concerns, which is reflected in surveys [12,9] and case studies [6] listing improved flexibility, reactivity to changes, and portability as benefits of DSLs. DSL technology hence also addresses Challenge C3.

10.2.4 Industrial research questions

The above reasoning suggests that a design methodology based on MBE and DSLs might be suitable to address all three challenges stated in Section 10.1. However, any design methodology has its drawbacks and it is essential to make sure that these do not offset the benefits [5]. Although there are surveys suggesting that the benefits of MBE often outweigh the drawbacks [7], leading to adoption of the approach, there are also examples of the opposite [4,13]. A credible business case hence has to be built on a case-by-case basis. It is widely recognized that this involves not only technical, but also organizational and social considerations [4,5,14,13,15]. A list of 14 industrial research questions, or challenges, can be found in [16]. These research questions are based on experiences from our partner companies that are active in different application domains in the high-tech industry, e.g., defense, healthcare, and manufacturing. In this chapter, we limit the scope to discuss a subset of five research questions that are relevant in the context of our case study.

(RQ1) How do you achieve modularity and reuse in a DSL ecosystem?

(RQ2) How do you achieve consistency between model and realizations?

(RQ3) How do you manage an evolving DSL ecosystem?

(RQ4) How do you ensure model quality?

(RQ5) How do you ensure quality of generated code?

10.3. State of the art

This section continues by discussing the five research questions and the extent to which they are recognized in the state of the art. We choose to focus on state of the art work in an industrial context, i.e., empirical studies, case studies, and best practices in industry, and review the proposed solutions. We choose this focus to limit the discussion to relevant industrial problems and proven solutions. A broader exploration including more academic solutions is highly relevant, but is left as future work. Note that many of the research questions correspond to broad research areas and that an exhaustive discussion is outside the scope of this chapter.

10.3.1 Modularity and reuse (RQ1)

Software engineering has seen great increases in productivity by enabling software to be decomposed into reusable modules that can be used as building blocks. This practice allows commonly used functionality to be implemented only once and then gradually mature as it is gradually reused, extended, and maintained. This same development is also desirable in language engineering. As DSLs evolve to cover a broader and broader domain, they inevitably reach the point where they need to be split into multiple modules or sublanguages to create a separation of concerns and reduce complexity. Since multiple languages describing aspects of the same domain are likely to share common concepts, further modularization is often beneficial to enable reuse and improve maintainability [17]. Examples of this can be found in industrial case studies from a variety of domains [18–20]. In [20], it was reported that modularizing a large DSL into a number of sublanguages incurred an overhead of approximately 10% in terms of grammar rules and 5% in terms of lines of code. However, a great reduction of complexity was reported by separating concerns, as well as improvements in maintainability.

The widespread use of small DSLs that can serve as modules in a larger DSL ecosystem, even within a single project, results in an integration challenge [5]. The available features for language composition vary significantly between different language workbenches and the metametamodels they support [21]. For example, composition features such as language extension/restriction where a base language is extended/restricted without modifying its implementation are quite common. In contrast, language unification that allows the implementation of both languages to be reused by only adding glue code is relatively rare [22]. It is hence clear that the problem of modularity and reuse is recognized and features to address it are considered differentiating features of existing language workbenches. We continue by briefly describing the language composition features available in Xtext, which is the language workbench used our in case study.

Xtext has quite limited and heavy-weight support for DSL modularity [20]. Each module is created as a DSL in its own right and results in five Eclipse projects being created. A DSL ecosystem hence quickly contains tens to hundreds of Eclipse projects. Xtext supports *single inheritance* at the level of grammars, which works similarly to the concept of inheritance in many object-oriented programming languages. This feature enables language extension or specialization by overriding concepts in the inherited grammar. It also supports a feature called *mixin*, which allows the metamodel defined by another grammar to be imported and its elements referenced. However, it is not possible to use the imported *grammars* by referring to its rules, and the including language can thereby not use its syntax to create objects. This is only possible through inheritance. Lastly, Xtext also has a feature called *fragments* that allows frequently occurring rule fragments to be factored out and reused, reducing duplication and improving reuse and maintainability. However, this feature is limited to reuse within the particular grammar in which it was defined. In addition to the features supported directly by Xtext, it is shown in [20] how to creatively combine Xtext features to create a notion of interface-based modularity, where unassigned rule calls in Xtext can be used to create abstract rules that are later implemented by languages importing the grammar.

The language composition features offered by the language workbench affect to which extent and at what granularity modularity and reuse occur. The limited language composition features provided by Xtext are sufficient to enable fine-grained reuse within a single grammar (fragments) and coarse-grained reuse between languages of the same DSL ecosystem (inheritance and mixins). However, an implication of these features is that there is very limited reuse, at any granularity, between languages in different domains, i.e., different ecosystems. While this may sound natural, since DSLs are domain-specific, not even common language concepts such as expressions or concepts for date and time are typically reused. This shows that the equivalence of libraries in regular software engineering is missing from Xtext. Instead, reuse between languages in different domains often happens by copying and pasting rules and generator fragments from previous languages. Although it is stated in [23] that this type of reuse already goes a long way, we believe that further improvements to Xtext are necessary to achieve the required productivity and maintainability benefits offered by DSLs. *We hence conclude that for industrial cases where advanced language composition features are required, it may be worthwhile to consider other mature language workbenches than Xtext.* A suitable candidate in this case may be JetBrains MPS,[1] where modularization and language composition are fundamental design concepts [24].

10.3.2 Consistency between model and realizations (RQ2)

The problem of inconsistencies between software artifacts is mentioned as a current challenge for MBE in [25,26]. A concrete example of this is that software designs, modeled in languages like UML, often quickly become forgotten and inconsistent once development starts. This problem may occur for multiple reasons, one being that many practitioners do not take diagrams seriously and see them as doodles on the back of a napkin before the real implementation work starts with textual languages [27]. Another reason for inconsistencies is that many tools are not able to keep models at different levels of abstraction synchronized. This problem is recognized in a survey about MBE practices

---

[1] https://www.jetbrains.com/mps/.

in industry [4], which suggests that 35% of respondents spend significant time manually synchronizing models and code. The problem of manually synchronizing artifacts is also explicitly mentioned in a case study at General Motors [28]. In this case, the lack of tool support for merging and diffing models resulted in tedious and error-prone manual workarounds that would lead to inconsistent artifacts.

Consistency between models and realizations (or other artifacts) can be bridged by generating all artifacts from a single source. In fact, this way of working is considered a best practice of MBE [15] and is a key benefit of MDE approaches that easily and efficiently support generation, which is a core purpose of DSLs. This benefit was explicitly highlighted in [6], where both code and documentation were generated from models specified using DSLs. This means that the model was always consistent with the generated artifacts. Similarly, [19] generates a simulation model, C++ code, visualizations, run-time monitoring facilities, and documentation that is consistent with an interface description based on a family of DSLs. *These works suggest that DSL technology is appropriate for ensuring consistency between model and realizations.*

Generation of artifacts can ensure that they are always up-to-date with respect to the model, but this does not necessarily mean that they all correctly and consistently implement the semantics of the DSL. This is because the semantics of the DSL is typically hidden inside the generators and there are no simple ways to ensure that these semantics are consistent with each other [17]. This problem is addressed in [29], which combines formalizing (parts of) the semantics of a DSL with conformance testing to validate that these semantics are correctly implemented by generated artifacts, in this case code and an analysis model. The approach is demonstrated through case study using a DSL for collision prevention developed by Philips. A drawback of this approach is that it requires substantial effort (possibly years) and very particular expertise to formalize two nontrivial languages to the point where equivalence can be proven. Proofs furthermore often become (partially) invalid as models or generators change, making software evolution more costly and time consuming. Using this approach to address RQ2 may hence exacerbate problems related to RQ3. For this reason, *formally proving semantic equivalence between realizations is not considered practical for complex industrial systems.*

10.3.3 Evolving DSL ecosystems (RQ3)

Just like regular software, DSL ecosystems evolve over time. This may be in response to required changes in syntax, semantics, or both [30] as domain concepts are added, removed, or modified. While evolution is often positive and helps the DSL stay relevant in a changing world, it creates a legacy of old artifacts, such as models, transformations, and possibly editors, that may no longer conform to the evolved metamodel and cannot be used unless they co-evolve [31]. This problem is well recognized in the literature and is explained with examples from popular metamodels, such as UML and Business Process Model and Notation (BPMN), in [31]. Although it is possible to manually co-evolve models and transformations to reflect changes in the metamodel, this manual process becomes tedious, error-prone, and costly when the legacy is large [31,32]. For example, the Control Architecture Reference Model (CARM) ecosystem [33] developed at ASML consists of 22 DSLs, 95 QVT transformations, and 5500 unit test models to support development of those transformations. Co-evolving a DSL ecosystem is more difficult than a single language, due to dependencies between its constituent parts [32]. Manually co-evolving a large industrial ecosystem like CARM is hence not feasible in terms of time and effort, but requires extensive automation.

Co-evolution of metamodels and artifacts has been an active research topic for many years. A list of 13 relevant aspects that can be used to classify co-evolution approaches, such as the type of artifact they consider or the technique used to determine the evolution specification for the metamodel, is presented in [31]. Furthermore, an overview of five existing representative co-evolution approaches and a classification using the 13 aspects is presented. Together, the five presented approaches cover co-evolution of all artifacts, i.e., models, transformations, as well as editors. No precise conclusions are drawn about the state of existing tools. However, it is suggested that there is no single tool that adequately considers all cases of co-evolution and that dealing with the problem requires modelers to learn to use different tools and techniques to co-evolve their artifacts. It is mentioned that co-evolution of transformations is intrinsically more difficult than models, which is reflected in the availability of mature approaches. In the rest of this section, we focus on co-evolution of models, an easier problem for which industrial-strength tools exist. For example approaches for co-evolution of model transformations, refer to, e.g., [34,35]. Other interesting aspects of evolution, such as its impact on code generation, are relevant and challenging, but outside the scope of this work.

Apart from manual co-evolution of artifacts, there are four (semi)automated approaches for obtaining an evolution specification [36,37]. (1) In *operator-based* approaches, evolution of the metamodel is manually specified in terms of reusable operations representing frequently occurring patterns of evolution. Based on the specified sequence of operators, a co-evolution specification for artifacts can be automatically derived. The usability of this approach is to a large extent determined by the completeness of libraries with reusable operators. Edapt,[2] the standard co-evolution tool for the Eclipse Modeling Framework (EMF), previously known as COPE [38], is a prominent example of a well-known tool in this category. (2) In *recording* approaches, modifications are recorded to the metamodel and a specification reflecting the performed changes is automatically created. This approach is also supported by Edapt. (3) In *state-based differencing* approaches, the original and evolved version of the metamodel are compared and an approximate specification of the changes is derived. Example approaches in this category include EMFMigrate [39] and EMFCompare.[3] (4) In *by-example* approaches [40],

2 https://www.eclipse.org/edapt/.
3 https://www.eclipse.org/emf/compare/.

the user manually migrates a number of model instances and the specification is derived by looking at the changes. The strengths and weaknesses of these four approaches are further discussed in [36].

The mentioned methods for co-evolution apply to co-evolution of models that have been manually specified by a user. However, another method applies to models that have been automatically created using static or dynamic techniques for model inference, e.g., using the Symphony process [41]. In this case, it may be faster to simply update the software creating the models to comply with the new metamodel and just rerun it to infer the models again. Of course, this method assumes that the data from which the models are inferred are stored.

It is clear that several methods and tools exist to address the co-evolution problem, although there are only limited studies that evaluate their applicability in the context of industrial DSL ecosystems. The extent to which Edapt could be used to perform DSL/model co-evolution in the CARM ecosystem was investigated in [42]. It was concluded that the standard operators could fully support 72% of the changes, with another 4% being partially supported. Implementing a set of model-specific operators increased the supported changes to around 98%. With these extensions, the authors conclude that Edapt is suitable for maintenance of DSLs in an industrial context. Further extensions to improve the usability of Edapt in industry have also been proposed in [43]. *Based on this evidence, we consider Edapt and its extensions relevant candidates for managing evolution of EMF-based metamodels in cases where the workflow used to modify the metamodels supports the usage of such tools.*

Determining the required operators by means of case studies, even on a large DSL ecosystem like CARM, is not necessarily sufficient to make statements about the suitability for other ecosystems. This was shown in [37], where a theoretically complete operator library for specifying any sequence of evolutionary steps for the EMF metametamodel was derived. This investigation showed that state of the art operator libraries could only specify 89% of DSL evolutions and that most of the remaining deficiencies could not be identified using a case study of the CARM ecosystem.

10.3.4 Ensuring model quality (RQ4)

If models are used as the sole source of all generated artifacts, it is essential to validate models to ensure their correctness. In addition, it is frequently stated as a best practice to test and find defects as early as possible [44], since this has been shown to increase quality and reduce the total time and effort required to develop or maintain software [6,45].

There are several ways to improve the quality of models and ensure correctness. For DSLs, a good starting point is to use the validation features of the language workbench. Features for model validation exist in all language workbenches, although the supported validation features vary [21]. Validation of structure and naming in model instances are relatively common features, while built-in support for type checking is less commonly supported. Many language workbenches have a programmatic interface allowing domain-specific validation routines to be implemented to make sure the model makes sense in the domain where it will be used, which is considered a best practice [44].

A more refined approach to model validation may involve tools and methods external to the language workbench. In [14], the quality and correctness of models is established by simulating the models against an executable test suite. The methodology proposed in [46] generates POOSL [47] simulation models connected via a socket to custom-made visualization tools for the considered system. The main benefit of this is that it helps make interactions between components and the behavior of the system explicit to reach an early agreement between stakeholders. Another example is to generate formal models to validate domain properties. For this purpose, the work in [48,49] generates satisfiability modulo theories problems that were solved by an external solver. The results from this solver are then fed back into the validation framework of the language workbench to interactively notify the user directly in the development environment. Lastly, one best practice is to review models, just like source code [44]. This is currently done by many practitioners to build confidence in the quality of code generators and the generated code [45,50,51]. Based on this brief review, *we conclude that suitable validation methods are available both internally in language workbenches and through external tools.* The exact choice of method, as well as criteria for validation, is highly problem-specific and should be determined on a case-by-case basis.

10.3.5 Quality of generated code (RQ5)

Quality of generated code is a very broad research question, since software quality can mean a lot of different things [52]. A tertiary study, i.e., a study of literature surveys, in the area of quality in MBE is presented in [53]. The study considers as many as 22 literature surveys, many of which choose maintainability as the quality metric of choice. They conclude that the field is not yet fully mature as most surveys target researchers and focus on classifying work, rather than targeting industry practitioners and aggregating quantitative evidence according to established quality metrics. We proceed by discussing a few relevant primary studies, most of which conclude that code generation leads to quality improvements.

A case study [6] in the Dutch IT industry showed that introducing MBE in the maintenance phase of a software project improves software quality. More specifically, they showed that a lower defect density was achieved using modeling, although at the expense of increasing time to fix a defect. However, the total result of these effects was a decrease in the total effort spent on maintenance of versions of the software. A reduction of defects is also claimed in [23,54], although the latter does not substantiate this with any quantitative evidence. A similar observation was made by Motorola in [14], which states that it is sometimes faster and sometimes slower to find the root cause of a software defect using MBE. They also provide quantitative estimates suggesting a reduction in the time to fix defects encountered during system integration, overall reduction of defects, and improvements in phase containment of defects (i.e., that defects are more likely to be detected and fixed in the development phase in which they are introduced) and productivity.

Motorola also points out a problem related to code quality using MBE. They state that code generation using off-the-shelf code generators can become a performance bottleneck unless it is possible to customize the generation [14]. The problem of generated code not being of desirable quality is also recognized in surveys with more than 100 participants [7,26]. In the most recent of the two surveys, 21% of participants is negative or partially negative about the quality of the resulting code [7]. While this number shows that there are practitioners that are not satisfied with the quality of generated code, the number of practitioners that are neutral (30%) or (highly) positive (49%) is much higher. From this, *we conclude that the quality of generated code is a problem worth investigating further, especially for performance-critical applications designed with tools that provide little or no control over code generation.*

Another aspect of generated code quality is the extent to which it is readable by humans. Best practices state that generated code should follow acceptable style guides. This may seem like a waste of time, since other best practices suggest that generated code should not be modified [44]. However, people still benefit from readable code in several ways. (1) Just like for any other code, generated code is inspected by developers trying to track down the root cause of a defect and this goes faster if the meaning of the code is clear. (2) Manual reviews of generated code are part of the development practice in many places to ensure correctness of the code and its generators [45,50,55]. (3) Readable code provides an exit strategy in case the company decides to stop using MBE by simply checking in the generated code and continue using it manually [50]. A case where this did not work out was reported in [2], where generated Simulink code was not human-readable, making the adoption of MBE hard to roll back.

Testing is an essential way to ensure the quality of software. However, code generation complicates testing, since there are often many possible paths through the code generator. There are two fundamental approaches to address this issue. The first approach implies *testing the code generator* itself. The challenge with this approach is to achieve sufficient coverage of the possible paths. Testing all possible (combinations of) paths through the generator is typically not feasible, due to the combinatorial explosion of possibilities. However, it may be possible to exercise all possible control flows in the code generator (i.e., every outcome of every single if statement) or use Pairwise Independent Combinatorial Testing[4] to exercise pairwise combinations of control flows. If the desired number of test models is too large to generate manually, a generator can be implemented to generate models that trigger the appropriate paths through the code generator.

---

[4] https://github.com/Microsoft/pict.

The second approach is to ignore the code generator and *test the generated code*. In this case, all existing testing practices remain valid, but testing needs to be repeated for each generated variant. Although this suggests that the overall testing effort is increased, it is important to recognize that the quality of the generated code increases over time as the generator matures. This approach is common for software in safety-critical domains, such as healthcare, automotive industry, and avionics, since it is often more practical and cost-effective to certify generated code than trying to qualify the code generator itself [55].

10.4. Approach to practical investigation

Having discussed the relevant state of the art for the five research questions related to our MBE approach based on DSLs, we proceed by explaining the organization of the practical investigation into these questions. We start by motivating our choice of modeling technology, before presenting our case study from the defense domain.

10.4.1 Modeling technology

The five research questions in this work are practically investigated using an MBE approach based on DSLs. This is because Section 10.2 suggested that DSL technology is an intuitive fit with the three challenges stated in Section 10.1, assuming the five stated research questions could be answered. The review of the state of the art in Section 10.3 suggests that there are promising solutions that answer many of those questions. In addition, we have many years of experience of transferring DSL technology to industry and applying it in different domains, e.g., [19,56,57,18]. There are many approaches [58] and tools [21] for developing DSLs. This work uses Xtext, which is a mature language workbench that has been around since 2006 and has relatively high coverage in terms of important features for language development [21]. Additionally, it is open source and available as a plugin for the Eclipse IDE, one of the most commonly used tools for MBE [7]. Generators are defined in the Xtend language, which is a DSL built on top of Java that can be combined with regular Java code. Details on how to develop DSLs and generators based on Xtext and Xtend can be found in [59].

10.4.2 Case study

A suitable case study is needed to drive the practical investigation into the five industrial research questions stated in Section 10.2. We start by presenting the general context of our case study from the defense domain. This study is centered around the engagement chain of a Combat Management System, shown in Fig. 10.1. This work considers a single ship, referred to as the *own ship*, with a number of sensors, e.g., surveillance radars and tracking radars, and a number of effectors to counteract possible threats.

Figure 10.1 Overview of the engagement chain.

The engagement chain consists of a number of steps that execute periodically, e.g., every few seconds. The input to this chain is the current state of the world. In the first step, surveillance radars are observing the world and produce sensor tracks, which can be intuitively understood as a radar blip with a position and speed corresponding to, e.g., another ship, a missile, or a jet. The sensor track is then passed on to a track management process that fuses sensor tracks from multiple sensors to generate a single, more accurate, system track. The system tracks are sent to the threat evaluation process, which determines the types of threats, investigates their intentions, and produces a ranking that indicates the relative threat level. A sorted list of threats is then sent to the engagement planning process, which determines the combinations of sensors and effectors that should be used against each threat and at what time. Depending on the choice of the planning algorithm, it may plan actions to counteract the threats strictly following threat ranking, or it may plan more flexibly using the ranking as a guideline. The generated engagement plan is then executed, and the actions of the sensors and effectors close the loop by interacting with the world, affecting its state.

For this case study, we have implemented two DSLs (partially) corresponding to two steps of the engagement chain in Fig. 10.1. The first DSL covers threat ranking, which is a part of threat evaluation, and the second covers engagement planning.

10.5. DSL ecosystem design

This section presents the design of the DSL ecosystem that was developed for the case study driving our investigation into the five research questions. First, we explain the rationale behind the design of this ecosystem, followed by a description of the Threat Ranking DSL. This particular DSL was chosen because it is the smallest and conceptually simplest language to discuss, allowing us to describe it in limited space, yet give a feeling for the level of abstraction chosen in this work. A complete and detailed description of the entire ecosystem is left as future work.

10.5.1 Design rationale

In terms of the classification of DSL development patterns in [58], this work used informal domain analysis, primarily based on discussions with relevant domain experts and architects, to identify suitable domain models for the different components. The design of the languages followed the language invention pattern, i.e., new DSLs were designed from scratch. The two DSLs were developed one at a time, starting with Threat Ranking, to incrementally build trust in the overall approach and evaluate its benefits and drawbacks [16].

The DSL design process was incremental and iterative through a series of meetings with domain experts and architects, being the main technical stakeholders. The meetings discussed relevant domain concepts and possible variation points in the languages. After the meetings, there was a formal design phase where we prototyped the DSL by specifying the abstract and concrete syntaxes through a grammar in Extended Backus–Naur Form (EBNF), which is the starting point for DSL design in Xtext. The proposed grammar and a few example instances were then discussed in the following meeting along with new possible concepts and variation points that could be introduced in the next iteration. This process was repeated until the languages were considered sufficiently expressive. Only at this point, generators with model-to-text transformations were implemented. In our experience, this incremental way of working with frequent prototypes helps drive development forward, as well as mitigate analysis paralysis [60].

In terms of the implementation pattern, we used the compiler/application generator approach to translate constructs of our DSL to existing languages. This choice of implementation pattern was motivated by the desire to enable analysis and validation of DSL instances, as well as to be able to tailor the notation to the specific domain. In that sense, the choice of implementation pattern is consistent with recommendations in [58]. Since the intended users of the language are domain experts and system engineers, rather than software developers, it was decided that the language should look and read more like text than code. This means some extra keywords have been added to make it easier to read and understand, at the expense of slightly longer specifications. This is not expected to be an issue as specifications are quite short. It was also decided to give the languages a common look and feel by using the same structure and notation, wherever possible.

10.5.2 Threat ranking DSL

This section aims to give a feeling for the DSLs created in this work by discussing the concepts of the Threat Ranking DSL in the context of an example instance. The basic idea behind our Threat Ranking DSL is to assign priority levels to each threat and to use a tiebreaker metric to resolve the order in which threats with the same priority level are ranked. As seen in Fig. 10.2, instead of using numbers to indicate priority, we use six threat levels, going from higher to lower: CRITICAL, SEVERE, SUBSTANTIAL, MODERATE, LOW, and NONE. The first five levels (CRITICAL to LOW) indicate threats that will appear in the output threat ranking, while threats with the last level (NONE) are filtered out and are not considered for engagements. The benefit of this use of

```
JET assign level SEVERE
MISSILE assign level MODERATE
OTHER assign level NONE

If JET isInbound then INCREASE level
If ANY distance < 1 km then assign level CRITICAL

Weight a = 1.5
Weight b = 0.9
Metric custom = a * keepOutRange + b * lethality
Tiebreaker: custom higherIsMoreDangerous

Objective: protectOwnShip
```

Figure 10.2 Example instance of the Threat Ranking DSL.

threat levels over priority levels represented by numbers is that it ties into an existing classification that is used in the domain.

Threat levels are assigned in two ways in the language: (1) statically per threat type (e.g., JET and MISSILE) and (2) dynamically per individual threat. The static assignment associates each threat type with a threat level that initially applies to all threats of that type. The proposed DSL requires all threat types to have a statically assigned threat level and is hence a common feature among all instances. To facilitate this in a simple way without explicitly listing all 10 currently supported threat types, the types OTHER and ANY have been introduced. ANY covers all types, whereas OTHER captures all threat types that have not been listed (i.e., neither explicitly nor by an ANY).

The static threat level assignment can be dynamically modified per threat during each execution of the Threat Ranking algorithm based on properties of the threat at that particular time, e.g., kinematic information or the distance to the own ship. This is done using optional if-statements, making this a variable feature of the language. Values representing distances, speeds, or times are required to have an appropriate unit to improve readability and remove ambiguity that can lead to incorrect implementation. A number of units are available in each category, allowing the user to choose whatever feels more natural. Behind the scenes, the generators convert all values into common units, i.e., meters for distances, seconds for time, and meters per second for speed.

The DSL instance in Fig. 10.2 contains two examples of dynamic threat level modifications. First, it states that any inbound jet, i.e., a jet flying towards the own ship, should have its threat level increased by one step, i.e., from SEVERE to CRITICAL in this case. This is an example of a *relative threat level assignment*, as the resulting threat level depends on the level before this assignment. Secondly, it states that any threat that is less than 1 km from the own ship should have its level reassigned to CRITICAL. This is an *absolute threat level assignment* as it is independent of the previous threat level. It is possible to have any number of if-statements and they are executed in order. If a relative

INCREASE or DECREASE of the threat level is done on a threat with the highest or lowest threat level, respectively, the level remains unchanged.

All threats will be assigned a final threat level based on the combination of static and dynamic threat level assignments. To arrive at a final ranking, the order in which to rank threats with the same threat level must be decided. This is done by either choosing any of nine predefined tiebreaker metrics or by specifying a custom metric as an expression consisting of different threat properties, such as kinematic information (e.g., speeds and distances). The latter possibility vastly increases the possibilities for how to rank threats with the same threat level, as custom metrics can specify arbitrarily complex expressions, which as discussed in Section 10.2 form an unbounded space of behaviors that cannot be captured by feature models. This is a key argument for modeling the Threat Ranking algorithms using DSLs. The example instance in Fig. 10.2 defines a new metric as a weighted combination of the specified keep out range and the lethality of the threat type. For each metric, it is possible to indicate whether a higher or a lower value is more dangerous.

Lastly, there is the concept of a High-Value Unit (HVU), which is a critical unit, e.g., a cargo ship or an aircraft carrier, that may require protection by the own ship. The DSL is extended with the ability to specify an objective related to an HVU, i.e., to protect the HVU, protect the own ship, or protect both.

In conclusion, the presented Threat Ranking DSL defines the Threat Ranking algorithm at a high level of abstraction using terminology from the application domain, which is commonly considered a best practice [44,61,13]. The DSL is furthermore so narrowly defined that it is impossible to use it to model different domains, which is a useful test to determine if the right balance between generic and specific has been found [60].

10.6. Results of practical investigation

This section explains the techniques employed and lessons learned from applying our MBE approach using DSLs to the case study with the goal of addressing the five industrial problems highlighted in Section 10.2. Note that a less complex DSL ecosystem than, e.g., the CARM [33] ecosystem suffices for our case study, which may impact some of the conclusions in this section. We proceed by discussing each research question in turn.

10.6.1 Modularity and reuse (RQ1)

The structural design of the DSL ecosystem developed for the case study is shown in Fig. 10.3. The ecosystem comprises three DSLs, one for each of the two considered functions in the Combat Management System, Threat Ranking and Engagement Planning, and an additional language that factors out common domain concepts that are

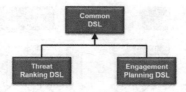

Figure 10.3 DSL ecosystem for parts of a Combat Management System.

shared among the other two languages. Examples of concepts that are shared between the languages are expressions, units, objectives, metrics, threat types, and threat properties. As suggested by the figure, the languages are composed by means of Xtext's single inheritance mechanism, where Threat Ranking DSL and Engagement Planning DSL both inherit the grammars of Common DSL. Our ecosystem is making use of neither Xtext's mixin feature nor the fragment feature. Mixins are not used because there is no need for either Threat Ranking DSL or Engagement Planning DSL to refer to the metamodel of the other language. Fragments are not used as there are no repetitive patterns in the rules of any of the individual grammars.

Although the language composition features of Xtext are limited, we conclude that they are sufficient for the needs encountered during our case study. However, it is easy to see the world through the limitations of tools [60] and it is possible that another language workbench with more and lighter-weight language composition features would have encouraged us to modularize at finer granularity to enable reuse of, e.g., expressions and units in other languages outside this work.

10.6.2 Consistency between model and realizations (RQ2)

The proposed DSL-based approach to MBE generates both simulation models and code, reflecting the observation in [7] that code generation and simulation are the most common uses of models in the embedded domain. The DSL instance is the sole source of truth from which both simulation models and code are generated, following best practices from [15]. Code generators have been implemented for the relevant programming and modeling languages and their execution environments (env), corresponding to different simulators at different levels of abstraction, or the system itself. This ensures that simulation models and production code are always consistent with their corresponding DSL instance, as illustrated in Fig. 10.4. The fact that multiple artifacts are generated from a single DSL instance means that our approach is consistent with "the rule of two," i.e., that DSL instances should be used for at least two different purposes to fully benefit from a model-based design approach [15]. We do currently not generate any documentation from the DSL instance, but an additional generator could be implemented to generate documentation using LaTeX. There are also available tools for automatic gener-

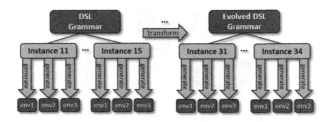

Figure 10.4 Generation of models for multiple simulation environments, as well as production code. Model-to-text transformations migrate DSL instances to newer versions as the DSL grammars evolve.

ation of Word documents, e.g., Gendoc[5] and m2doc.[6] However, we leave this as future work.

Generation of artifacts from a single source does not ensure that the semantics of the DSL is consistently implemented in the generators. This means that simulation models and production code may be consistent with the model, but have inconsistent views on what the model actually means. In our particular case, a consistent interpretation of the semantics implies that generated simulation models and production code always produce the same ranking, given the same input. Ensuring consistent semantics hence boils down to validating that this is really the case. The challenge is that the models and code execute in different environments that use different languages and model some system components at different levels of abstraction. As a result, even if the exact same Threat Ranking algorithm is used in all environments, the inputs of the Threat Ranking component are not expected to be the same. For example, a threat may be detected slightly earlier or later, impacting the set of threats to rank at a particular point in time, which in turn affects the scheduled engagements and the set of threats later in the scenario. For this reason, it is not always possible to compare results across environments and draw meaningful conclusions about consistency of semantics between generators.

Our solution to mitigate this concern is to remove the differences in environment and execute all implementations in a single execution environment. This is achieved by wrapping the generated production code and run it in one of the simulation environments as software-in-the-loop, which ensures that all generated implementations have the same inputs and that all other components are implemented identically. This in turn enables us to establish the consistency of semantics between generators by extensive regression testing through comparison of results, following the recommendation in [17]. Over time, through extensive testing and use, this approach builds confidence that the different generators implement the same semantics of the DSL and hence that the Threat Ranking component works correctly.

[5] https://www.eclipse.org/gendoc/.
[6] http://www.m2doc.org/.

There are two main drawbacks of this approach. (1) The semantics are implemented in each generator and any semantic change must hence be consistently implemented in all generators manually, which is error-prone. (2) It is limited to cases where the exact same result is expected from all realizations, or slightly more generally, where results maximally differ by some known maximum bound, which is often not the case. An interesting option could be to specify the semantics on an abstract level and generate implementations that are consistent by construction. This would ensure that changes in semantics would only be made in a single place, which could be advantageous for highly evolvable systems. This direction is considered future work.

10.6.3 Evolving DSL ecosystems (RQ3)

The DSL ecosystem in our case study consists of three DSLs (Threat Ranking DSL, Engagement Planning DSL, and Common DSL), four model-to-text transformations that generate simulation models and production code in a variety of languages, as well as an analysis model for custom metrics, described in Section 10.6.4. Our DSL ecosystem is hence considerably smaller than the CARM ecosystem [33], discussed in Section 10.3.3, resulting in a smaller legacy as metamodels evolve. The evolution of our ecosystem has taken place entirely during the development phase of the DSLs, which has been approximately two years.

We considered using Edapt [42], since it is relatively mature and has been positively evaluated for other industrial DSL ecosystems. However, Edapt works directly with the EMF metamodel, which Xtext generates from the specified grammars. The grammars will hence become inconsistent with the evolved metamodel, unless a formal link is established that propagates the changes to the grammar. To the best of our knowledge, there is currently no tool implementing this link, removing Edapt from further consideration in this work. Instead, we focused on solutions available within Xtext itself. We only needed to consider co-evolution of models and transformations, since Xtext regenerates the editor based on changes to the metamodel. Due to the limited evolvability burden in our case study, we have opted for the simplest option that satisfied our needs. This involved manually implementing a generator with a model-to-text transformation whose input was a model conforming to the nonevolved grammar and output a textual representation of that instance conforming to the evolved grammar, as shown in Fig. 10.4. This corresponds to a largely manual approach, as the mapping between concepts in the nonevolved and evolved metamodel, as well their semantics, was done manually. However, the implemented transformation could quickly and easily be applied in a batch run to evolve all existing model instances. This simple approach is hence not limited to ecosystems with a few models, but is primarily restricted by the number and complexity of the languages in the ecosystem and the complexity of their dependencies.

10.6.4 Ensuring model quality (RQ4)

In the proposed DSL-based approach to MBE, the model is the sole source of truth from which both simulation models and code are ultimately generated, following the best practice from [15]. It is hence important that the quality of these models is high and that any problems are detected as early as possible. To this end, we experimented with three ways to improve model quality:

1) The Eclipse-based IDE for the Threat Ranking DSL, which is automatically generated from the DSL grammar by Xtext, ensures syntactic correctness and immediately validates that the syntax of an instance complies with the grammar.

2) A number of model validation rules have been implemented that exploit knowledge about the domain to detect problems with instances. These validation rules can either lead to warnings, which only alert the user but still allow generation of artifacts, or to errors, which prevent the generators from running altogether until the problem is resolved. This is generally a good place to address deprecation issues as the DSL is evolving. A warning can be triggered when a deprecated construct is encountered in a model, assuming an appropriate model transformation is available to map it to an equivalent construct in the evolved DSL. In contrast, if a model transformation is not available (anymore), an error is triggered.

   More specifically for our Threat Ranking DSL, one validation rule triggers a warning if there are multiple static threat level assignments to a single threat type to alert the user that only the last assignment is useful. In contrast, another rule throws an error in case not all threat types have a static threat level assignment, since this violates a fundamental assumption of the ranking algorithm. Yet another validation rule checks the correctness of units, i.e., that metrics related to time or distance are only compared to values whose units relate to time and distance, respectively. This prevents comparing apples to pears, or more literally, seconds to meters, by raising an error. For many of these validation rules, quick fixes were built into the editor to help the developer resolve violations quickly and reliably.

3) An analysis tool was also implemented in a generator that immediately produces a report providing visibility on the results provided by custom metrics, previously introduced in Section 10.5.2, without having to run the simulator. The generated report is based on a single given list of threats to be ordered. Representative lists of threats are easily obtained by recording inputs to the Threat Ranking components during simulation. The report, shown in Fig. 10.5, demonstrates how the custom tiebreaker metric is computed for each threat. This immediately shows the user an example outcome when applying the metric and gives insight into what caused that outcome. For example, it could show that a particular parameter is typically dominating the metric and that weights should be adjusted to make the metric achieve the desired goal. This is particularly helpful when experimenting with complex custom metrics.

```
Analysis of custom metric:                    Example: 5-MISSILE

Weights: smallNumber := 0.000001              Parameters:
Expression: timeToOwnShip * timeToKOR +          CPADistance : 48.30 m
keepOutRangeViolated * smallNumber / speed       altitude : 19.86 m
                                                 speed : 799.93 m/s
Ranking by custom metric                         timeToKOR : 22.82 s
(lower is more dangerous):                       timeToOwnShip : 0.06 s

   1) [1.37] 5-MISSILE                         Substituted: 0.06 * 22.82 +
   2) [2.07] 3-MISSILE                            0.0 * 0.000001 / 799.93
   3) [2.08] 1-MISSILE
   4) [2.29] 4-MISSILE                         Evaluated: 1.37
   5) [2.56] 2-MISSILE
```

Figure 10.5 Generated analysis showing results of applying a custom metric to a particular set of threats. The numbers in the example are not indicative of any real systems.

10.6.5 Quality of generated code (RQ5)

Most of our practical work related to quality of generated code is related to testing, which is done at three different levels: (1) unit testing, (2) component testing, and (3) integration testing. Unit testing of the DSLs follows the method described in [59] and performs low-level validation of the generators by asserting that particular model constructs result in the expected code being generated. In contrast, the component-level testing focuses on the semantics of the generated code and validates that this is consistent across implementations, as described in Section 10.6.2. Note that comparing results from multiple implementations is useful to validate consistency, but it does not necessarily imply that any implementation is correct. However, following this approach, all implementations must provide the same incorrect result in order for it to pass the test, which is rather unlikely. Lastly, we perform integration testing in the complete system to verify that components communicate correctly and that system-level results, such as when and where threats are neutralized in a particular scenario for a given DSL instance, do not unexpectedly change during development. For this purpose, golden reference results have been generated for relevant threat scenarios and DSL test instances and are used for comparison during testing. To further increase confidence in the results, different generators can be implemented by independent developers based on a common specification, following requirements for certification of software components in safety-critical avionics systems [62].

Manual validation is tedious and time consuming labor, especially when software is being developed in parallel on many different branches. Following the GitFlow work-flow,[7] our repository has two main branches, i.e., master and development. New features are developed on feature branches that are integrated into development after passing tests

---

[7] https://www.atlassian.com/git/tutorials/comparing-workflows/gitflow-workflow.

at all three above-mentioned levels. Despite passing all tests, the development branch contains newly integrated experimental features and is not considered perfectly stable. Once it is time to make a new release, additional manual validation is done on this branch and once it is considered to be sufficiently stable for users, it is integrated into the master branch.

To reduce the manual effort of all commits on these branches, testing has been automated to make it possible to run all combinations (or a chosen subset) of DSL instances and scenarios by pushing a single button. As recommended in [44], the DSL instances used for testing have been designed in such a way that they exercise as many constructs of the DSLs as possible to improve coverage. Since we are preparing for a situation where the DSL itself evolves over time, it is important that integration testing is always done with the latest versions of the ecosystem and its generators. However, Xtext does not support automatic generation of a command line DSL parser and generator that can be used for integration testing after each commit. As a contribution of this work, we have defined a method for automatic generation of such a command tool that can be used with any Xtext project without even requiring an Eclipse installation. A description of this method and an example project is available online.[8]

To automate all aspects of testing and enable *continuous integration*, we have set up a Jenkins Automation server that checks out the latest version of the code after each commit on any branch, builds the compiler, and runs all tests. General benefits of this setup include enabling defects to be caught early, improving phase containment of defects, and ensuring that only the latest changes must be reviewed and debugged when a bug is detected. There are also specific benefits with respect to Challenge C3. For example, it allows DSL instances of deployed systems to be stored in branches and continuously validates that evolved versions of the DSL do not accidentally change their behavior. This makes it easier to maintain and upgrade systems after deployment.

The automation server also provides *continuous deployment* by automatically generating Eclipse plugins based on both the development branch and master branch for Threat Ranking DSL and Engagement Planning DSL as soon as a commit to either of these branches has passed all tests. These plugins are then made available on an internal update site. Experienced users or developers can hence subscribe to the latest development version and experiment with the latest features, while regular users can subscribe to the latest stable version. This setup ensures that the plugins used by both of these communities are always up-to-date.

Unlike Motorola [14] and some of the survey participants in [7,26], discussed in Section 10.3.5, we have not experienced problems with the performance of generated code during our case study. The two main reasons for this are the following: (1) our models are relatively small, making them less prone to performance problems, and (2) DSL development using Xtext gives full control over the model-to-text transformations used

---

[8] https://github.com/basilfx/xtext-standalone-maven-build

for code generation, which means that the differences with hand–written code are typically small. When these differences do occur, it is mostly easy to simplify the structure of the generators and avoid complex control flows that slow down development and complicate testing.

10.7. Evaluation

As a part of an informal evaluation of an intermediate version of the DSL ecosystem, an event was organized on the premises of the industrial partner where about 20 employees with various functions ranging from software and system engineers to domain experts and even a director participated. Some of the participants were familiar with the domain from before, but many of them were not. The event consisted of a 30-minute introduction after which participants were divided into four groups that experimented with the DSL ecosystem. After a short tutorial that explained the basics of the DSLs and the associated tooling for simulation and visualization, the teams were tasked with using the DSL ecosystem to solve a particular assignment. It turns out that a short tutorial was sufficient to get three out of the four groups to productively experiment with making their own model instances to solve the assignment, at which point we only needed to answer a few simple questions, e.g., about the definitions of keywords in the language. The last of the four groups completed the tutorial, but did not get off the ground with making their own instances. This was due to a combination of lacking motivation, insufficient domain knowledge, and group dynamics.

The feedback from the participants was largely positive. Some participants had domain knowledge and suggested features that could be included in future versions of the DSL. It was also reported that the participants found experimenting with the DSL an effective way to learn about the domain, as the DSL and associated tooling made it and easy to customize, deploy, and evaluate model instances. This suggests that our DSL was on its way towards delivering on Challenges C2 and C3. This feedback also resonates with the claim that MBE empowers users without software background, e.g., domain experts and system engineers, by enabling them to work productively without having to rely on software engineers to implement their ideas [2–4].

As the work was concluded, a final evaluation was organized to assess the potential of the DSL–based development methodology. The goal of the evaluation was to let a number of intended DSL users experience with the DSL way of working and assess its potential within the organization. The means to achieve this was to let them experiment with the DSL ecosystem. These experiments took place in the intended application context, in this case together with suitable tools for simulation and visualization. Although this is a specific example of a DSL in context, the participants were asked to assess the general potential of DSLs and not limit themselves to the particular DSL or the domain of engagement planning. Ten participants considered representative for the potential users of DSLs were asked to join the evaluation. Some participants in the evaluation were system engineers/architects with only limited software development experience, corresponding to the primary audience of the developed DSL ecosystem. Others had experience with implementing algorithms directly in general-purpose programming languages and could hence provide a complementary perspective. The setup of the final evaluation was nearly identical to that of the intermediate evaluation, but featured newer versions of the DSL ecosystem and the tutorial to reflect improvements made during the six months between the two events.

The participants identified a number of classic gains of DSLs during the session, e.g., (1) the demonstrated DSL-based environment was easy to use and accessible to nontechnical people, (2) the DSL hides the implementation technology, allowing the problem to be decoupled from its implementation, (3) DSLs enable faster customization and prototyping, at least of variants that fit within the boundaries of the language, and (4) DSLs may improve communication within a group of people, but also with the outside world. These observations relate to known benefits of DSLs, discussed in Section 10.2, and together they address all three challenges identified in Section 10.1. This is an encouraging result! A number of challenges were also identified during the evaluation, e.g., DSLs require higher upfront investment compared to traditional software development, modeling requires different skills, and adopting a DSL-based methodology requires organizational support. These issues have been previously identified and mitigation techniques have been documented and shared with the industrial partner.

10.8. Conclusions

This chapter addressed the problem of reducing design time and improving evolvability of complex systems through an MBE approach based on DSLs. Five research questions raised by our industrial partner related to the approach were investigated by means of a literature study and a practical case study from the defense domain, namely, how to (RQ1) achieve modularity and reuse in a DSL ecosystem, (RQ2) achieve consistency between model and realizations, (RQ3) manage an evolving DSL ecosystem, (RQ4) ensure model quality, and (RQ5) ensure quality of generated code.

A DSL ecosystem with two DSLs inheriting common concepts from a third language was developed for the case study. The ecosystem also features four model-to-text transformations to generate an analysis report and code for a variety of programming and modeling languages. Further transformations have been developed to support migration of models as the DSL ecosystem evolves. We discussed how the generated analysis report and model validation rules help ensuring correctness of models and how the quality of code generated by the ecosystem is improved using continuous integration and continuous deployment practices.

Both intermediate and final evaluation results suggest that the proposed DSL-based development methodology and example DSL ecosystem deliver on their design goals and address the aforementioned challenges for complex systems. A number of relevant issues related to DSL-based development were explicitly identified by the users of the ecosystem, but they were already known and had been discussed along with existing mitigation techniques within the company. Based on this work, next steps will involve the industry partner deciding whether the gains of DSL-based development outweigh the pains, and for what application domains the gains are maximized.

References

[1] H. Geelen, A. van der Hoogt, W. Leibbrandt, F. Beenker, HTSM roadmap embedded systems, 2018.
[2] J. Aranda, D. Damian, A. Borici, Transition to model-driven engineering, in: International Conference on Model Driven Engineering Languages and Systems, Springer, 2012, pp. 692–708.
[3] H. Burden, R. Heldal, J. Whittle, Comparing and contrasting model-driven engineering at three large companies, in: Proceedings of the 8th ACM/IEEE International Symposium on Empirical Software Engineering and Measurement, ACM, 2014, p. 14.
[4] J. Hutchinson, J. Whittle, M. Rouncefield, Model-driven engineering practices in industry: social, organizational and managerial factors that lead to success or failure, Science of Computer Programming 89 (2014) 144–161.
[5] J. Whittle, J. Hutchinson, M. Rouncefield, The state of practice in model-driven engineering, IEEE Software 31 (3) (2014) 79–85.
[6] N. Mellegård, A. Ferwerda, K. Lind, R. Heldal, M.R. Chaudron, Impact of introducing domain-specific modelling in software maintenance: an industrial case study, IEEE Transactions on Software Engineering 42 (3) (2016) 245–260.
[7] G. Liebel, N. Marko, M. Tichy, A. Leitner, J. Hansson, Model-based engineering in the embedded systems domain: an industrial survey on the state-of-practice, Software & Systems Modeling 17 (1) (2018) 91–113.
[8] J. Whittle, J. Hutchinson, M. Rouncefield, H. Burden, R. Heldal, A taxonomy of tool-related issues affecting the adoption of model-driven engineering, Software & Systems Modeling 16 (2) (2017) 313–331.
[9] J. Hutchinson, J. Whittle, M. Rouncefield, S. Kristoffersen, Empirical assessment of MDE in industry, in: Proceedings of the 33rd International Conference on Software Engineering, ACM, 2011, pp. 471–480.
[10] D. Beuche, H. Papajewski, W. Schröder-Preikschat, Variability management with feature models, Science of Computer Programming 53 (3) (2004) 333–352.
[11] M. Voelter, E. Visser, Product line engineering using domain-specific languages, in: Software Product Line Conference (SPLC), 15th International, IEEE, 2011, pp. 70–79.
[12] M. Torchiano, F. Tomassetti, F. Ricca, A. Tiso, G. Reggio, Relevance, benefits, and problems of software modelling and model driven techniques – a survey in the Italian industry, Journal of Systems and Software 86 (8) (2013) 2110–2126.
[13] D. Wile, Lessons learned from real DSL experiments, Science of Computer Programming 51 (3) (Jun. 2004) 265–290.
[14] P. Baker, S. Loh, F. Weil, Model-driven engineering in a large industrial context – Motorola case study, in: Model Driven Engineering Languages and Systems, 2005, pp. 476–491.
[15] P.F. Smith, S.M. Prabhu, J. Friedman, Best Practices for Establishing a Model-Based Design Culture, SAE Technical Paper, Tech. Rep., 2007.
[16] B. Akesson, J. Hooman, R. Dekker, W. Ekkelkamp, B. Stottelaar, Pain-mitigation techniques for model-based engineering using domain-specific languages, in: Proc. Special Session on Model Management and Analytics, MOMA3N, 2018, pp. 752–764.

[17] M. Voelter, Architecture as language, IEEE Software 27 (2) (March 2010) 56–64.

[18] J. Verriet, L. Buit, R. Doornbos, B. Huijbrechts, K. Sevo, J. Sleuters, M. Verberkt, Virtual prototyping of large-scale IoT control systems using domain-specific languages, in: Proceedings of the 7th International Conference on Model-Driven Engineering and Software Development, MODELSWARD 2019, 2019.

[19] I. Kurtev, M. Schuts, J. Hooman, D.-J. Swagerman, Integrating interface modeling and analysis in an industrial setting, in: Proceedings of the 5th International Conference on Model-Driven Engineering and Software Development, MODELSWARD 2017, 2017, pp. 345–352.

[20] C. Rieger, M. Westerkamp, H. Kuchen, Challenges and opportunities of modularizing textual domain-specific languages, in: Proceedings of the 6th International Conference on Model-Driven Engineering and Software Development, MODELSWARD 2018, 2018, pp. 387–395.

[21] S. Erdweg, T. Van Der Storm, M. Voelter, L. Tratt, R. Bosman, W.R. Cook, A. Gerritsen, A. Hulshout, S. Kelly, A. Loh, et al., Evaluating and comparing language workbenches: existing results and benchmarks for the future, Computer Languages, Systems and Structures 44 (2015) 24–47.

[22] S. Erdweg, P.G. Giarrusso, T. Rendel, Language composition untangled, in: Proceedings of the Twelfth Workshop on Language Descriptions, Tools, and Applications, ACM, 2012, p. 7.

[23] F. Hermans, M. Pinzger, A. Van Deursen, Domain-specific languages in practice: a user study on the success factors, in: International Conference on Model Driven Engineering Languages and Systems, Springer, 2009, pp. 423–437.

[24] M. Voelter, Language and IDE Modularization and Composition with MPS, Springer, Berlin, Heidelberg, 2013, pp. 383–430.

[25] G. Mussbacher, D. Amyot, R. Breu, J.-M. Bruel, B.H.C. Cheng, P. Collet, B. Combemale, R.B. France, R. Heldal, J. Hill, J. Kienzle, M. Schöttle, F. Steimann, D. Stikkolorum, J. Whittle, The Relevance of Model-Driven Engineering Thirty Years from Now, Springer International Publishing, Cham, 2014, pp. 183–200.

[26] A. Forward, T.C. Lethbridge, Problems and opportunities for model-centric versus code-centric software development: a survey of software professionals, in: Proceedings of the 2008 International Workshop on Models in Software Engineering, ACM, May 2008, pp. 27–32.

[27] D. Harel, B. Rumpe, Meaningful modeling: what's the semantics of "semantics"?, Computer 37 (10) (2004) 64–72.

[28] A. Kuhn, G.C. Murphy, C.A. Thompson, An exploratory study of forces and frictions affecting large-scale model-driven development, in: International Conference on Model Driven Engineering Languages and Systems, Springer, 2012, pp. 352–367.

[29] S. Keshishzadeh, A.J. Mooij, Formalizing and testing the consistency of DSL transformations, Formal Aspects of Computing 28 (2) (2016) 181–206.

[30] J. Mengerink, L. van der Sanden, B. Cappers, A. Serebrenik, R. Schiffelers, M. van den Brand, Exploring DSL evolutionary patterns in practice: a study of DSL evolution in a large-scale industrial DSL repository, in: 6th International Conference on Model-Driven Engineering and Software Development, MODELSWARD 2018, 2018.

[31] D. Di Ruscio, L. Iovino, A. Pierantonio, Coupled evolution in model-driven engineering, IEEE Software 29 (6) (2012) 78–84.

[32] J. Mengerink, R. Schiffelers, A. Serebrenik, M. van den Brand, DSL/model co-evolution in industrial EMF-based MDSE ecosystems, in: ME@ MODELS, 2016, pp. 2–7.

[33] R.R. Schiffelers, W. Alberts, J.P. Voeten, Model-based specification, analysis and synthesis of servo controllers for lithoscanners, in: Proceedings of the 6th International Workshop on Multi-Paradigm Modeling, ACM, 2012, pp. 55–60.

[34] J. Di Rocco, D. Di Ruscio, L. Iovino, A. Pierantonio, Dealing with the coupled evolution of metamodels and model-to-text transformations, in: ME@ MoDELS, 2014, pp. 22–31.

[35] J. García, O. Diaz, M. Azanza, Model transformation co-evolution: a semi-automatic approach, in: International Conference on Software Language Engineering, Springer, 2012, pp. 144–163.

[36] L.M. Rose, R.F. Paige, D.S. Kolovos, F.A. Polack, An analysis of approaches to model migration, in: Proc. Joint MoDSE-MCCM Workshop, 2009, pp. 6–15.

[37] J. Mengerink, A. Serebrenik, R.R. Schiffelers, M. van den Brand, A complete operator library for DSL evolution specification, in: 2016 IEEE International Conference on Software Maintenance and Evolution, ICSME, IEEE, 2016, pp. 144–154.

[38] M. Herrmannsdoerfer, S. Benz, E. Juergens, COPE – automating coupled evolution of metamodels and models, in: European Conference on Object-Oriented Programming, Springer, 2009, pp. 52–76.

[39] J. Di Rocco, L. Iovino, A. Pierantonio, Bridging state-based differencing and co-evolution, in: Proceedings of the 6th International Workshop on Models and Evolution, ACM, 2012, pp. 15–20.

[40] G. Kappel, P. Langer, W. Retschitzegger, W. Schwinger, M. Wimmer, Model transformation by-example: a survey of the first wave, in: Conceptual Modelling and Its Theoretical Foundations, Springer, 2012, pp. 197–215.

[41] A. Van Deursen, C. Hofmeister, R. Koschke, L. Moonen, C. Riva, Symphony: view-driven software architecture reconstruction, in: Software Architecture, 2004. WICSA 2004. Proceedings. Fourth Working IEEE/IFIP Conference on, IEEE, 2004, pp. 122–132.

[42] Y. Vissers, J.G.M. Mengerink, R.R.H. Schiffelers, A. Serebrenik, M.A. Reniers, Maintenance of specification models in industry using Edapt, in: 2016 Forum on Specification and Design Languages, FDL, Sept 2016, pp. 1–6.

[43] J.G.M. Mengerink, A. Serebrenik, M. van den Brand, R.R.H. Schiffelers, Udapt: Edapt extensions for industrial application, in: Proceedings of the 1st Industry Track on Software Language Engineering, ITSLE 2016, ACM, New York, NY, USA, 2016, pp. 21–22.

[44] M. Voelter, Best practices for DSLs and model-driven development, Journal of Object Technology 8 (6) (2009) 79–102.

[45] M. Broy, S. Kirstan, H. Krcmar, B. Schätz, J. Zimmermann, What is the benefit of a model-based design of embedded software systems in the car industry?, in: Emerging Technologies for the Evolution and Maintenance of Software Models, 2012, pp. 343–369.

[46] J. Hooman, Industrial application of formal models generated from domain specific languages, in: Theory and Practice of Formal Methods, Springer, 2016, pp. 277–293.

[47] B.D. Theelen, O. Florescu, M. Geilen, J. Huang, P. van der Putten, J.P. Voeten, Software/hardware engineering with the parallel object-oriented specification language, in: Proceedings of the 5th IEEE/ACM International Conference on Formal Methods and Models for Codesign, IEEE Computer Society, 2007, pp. 139–148.

[48] A.J. Mooij, J. Hooman, R. Albers, Early fault detection using design models for collision prevention in medical equipment, in: International Symposium on Foundations of Health Informatics Engineering and Systems, Springer, 2013, pp. 170–187.

[49] S. Keshishzadeh, A.J. Mooij, M.R. Mousavi, Early fault detection in DSLs using SMT solving and automated debugging, in: International Conference on Software Engineering and Formal Methods, Springer, 2013, pp. 182–196.

[50] A.J. Mooij, J. Hooman, R. Albers, Gaining industrial confidence for the introduction of domain-specific languages, in: Computer Software and Applications Conference Workshops (COMPSACW), 2013 IEEE 37th Annual, IEEE, 2013, pp. 662–667.

[51] A.J. Mooij, G. Eggen, J. Hooman, H. van Wezep, Cost-effective industrial software rejuvenation using domain-specific models, in: International Conference on Theory and Practice of Model Transformations, Springer, 2015, pp. 66–81.

[52] International Organization for Standardization, ISO-IEC 25010: 2011 Systems and Software Engineering-Systems and Software Quality Requirements and Evaluation (SQuaRE)-System and Software Quality Models, ISO, 2011.

[53] M. Goulão, V. Amaral, M. Mernik, Quality in model-driven engineering: a tertiary study, Software Quality Journal 3 (24) (2016) 601–633.

[54] P. Mohagheghi, V. Dehlen, Where is the proof? – a review of experiences from applying MDE in industry, in: Lecture Notes in Computer Science, vol. 5095, 2008, pp. 432–443.

[55] M. Voelter, B. Kolb, K. Birken, F. Tomassetti, P. Alff, L. Wiart, A. Wortmann, A. Nordmann, Using language workbenches and domain-specific languages for safety-critical software development, Software & Systems Modeling (2018) 1–24.

[56] A.J. Mooij, M.M. Joy, G. Eggen, P. Janson, A. Rădulescu, Industrial software rejuvenation using open-source parsers, in: International Conference on Theory and Practice of Model Transformations, Springer, 2016, pp. 157–172.

[57] R. Doornbos, B. Huijbrechts, J. Sleuters, J. Verriet, K. Sevo, M. Verberkt, A domain model-centric approach for the development of large-scale office lighting systems, in: Complex Systems Design & Management (CSD&M) Conference, IEEE, 2018.

[58] M. Mernik, J. Heering, A.M. Sloane, When and how to develop domain-specific languages, ACM Computing Surveys 37 (4) (2005) 316–344.

[59] L. Bettini, Implementing Domain-Specific Languages with Xtext and Xtend, Packt Publishing Ltd., 2016.

[60] S. Kelly, R. Pohjonen, Worst practices for domain-specific modeling, IEEE Software 26 (4) (2009).

[61] G. Karsai, H. Krahn, C. Pinkernell, B. Rumpe, M. Schindler, S. Völkel, Design guidelines for domain specific languages, preprint, arXiv:1409.2378, 2014.

[62] RTCA, Inc., RTCA/DO-178C, U.S. Dept. of Transportation, Federal Aviation Administration, 2012.

CHAPTER 11

Model analytics for industrial MDE ecosystems

Önder Babur[a], Aishwarya Suresh[a], Wilbert Alberts[b], Loek Cleophas[a,c], Ramon Schiffelers[a,b], Mark van den Brand[a]

[a]Eindhoven University of Technology, Eindhoven, The Netherlands
[b]ASML N.V., Veldhoven, The Netherlands
[c]Stellenbosch University, Matieland, Republic of South Africa

Contents

Model Management and Analytics for Large Scale Systems
https://doi.org/10.1016/B978-0-12-816649-9.00021-1

11.1. Introduction

The increased use of Model-Driven Engineering (MDE) techniques leads to the need to address issues pertaining to the growing number and variety of MDE artifacts, such as Domain-Specific Languages (DSLs) and the corresponding models. This is indeed the case when large industries adopt MDE for multiple domains in their operation. ASML, the leading producer of lithography systems, is an example of such a company where multidisciplinary teams work on various MDE ecosystems involving tens of languages and thousands of models [1]. Automated analyses of those artifacts can potentially aid in the maintenance and evolution of those ecosystems. One example issue with these ecosystems is that of duplication and cloning in those artifacts. The presence of clones might negatively affect the maintainability and evolution of software artifacts in general, as widely reported in the literature [2]. In the general sense, when multiple instances of software artifacts (e.g., language or model fragments in our case) exist, a change required in such a fragment (to fix a bug, for instance) would also have to be performed on all other instances of this fragment. Inconsistent changes to such fragments might also lead to incorrect behavior. Therefore, eliminating such redundancy in software artifacts might result in improved maintainability. While not all cases of encountered clones can be considered negative [3], as some might be inevitable or even intended, it is worthwhile to explore what types of clones exist and what their existence might imply for the system.

The growing number of DSLs in the variety of ecosystems, on the other hand, also demands ways to automatically analyze those languages, e.g., to give an overview of the domains and subdomains of the enterprise-level ecosystem (i.e., system of ecosystems). Other interesting analyses would include the similarities, conceptual relatedness, and clone fragments among the various languages both within and across the ecosystems.

In this work, we explore a variety of model analytics approaches using our framework SAMOS (Statistic Analysis of MOdelS) [4,5] in the industrial context of ASML ecosystems. We perform case studies involving clone detection on ASML's data models and control models of the ASOME ecosystem, cross-DSL conceptual analysis and language-level clone detection on three ecosystems (ASOME, CARM2G, wafer handler), and finally architectural analysis and reconstruction, using a technique called topic modeling [6], on the CARM2G ecosystem DSLs. We provide insights into how model analytics can be used to discover factual information on MDE ecosystems (e.g., what types of clones exist and why) and opportunities such as refactoring to improve the ecosystems.

The rest of the chapter is structured as follows. In Section 11.2 we introduce our main objectives for analyzing MDE ecosystems. In Sections 11.3 to 11.5, we give some background information on our SAMOS analysis framework as the basis of our studies, ASML ecosystems, and the concept of model clones, respectively. We detail how we used and extended SAMOS for the clone detection tasks on ASML's ASOME ecosystem models in Section 11.6. We provide extensive case studies in Section 11.7: clone detection in ASOME data models and control models, cross-DSL conceptual analysis and language-level clone detection, and finally architectural analysis of the CARM2G ecosystem. We continue in Section 11.8 with a general discussion and threats to validity, with related work on important topics such as model clone detection and topic modeling in Section 11.9, and finally with conclusions and pointers for future work in Section 11.10.

11.2. Objectives

This section presents the objectives that we pursued to analyze the MDE ecosystems at ASML. First, we would like to point out that we used and extended our model analytics framework, SAMOS, to perform various analyses on the MDE artifacts. Since SAMOS already provides a means to detect clones for Ecore metamodels (representing the DSLs in the ecosystems), we explore how this framework can be extended (1) to analyze models adhering to the domain-specific metamodels used at ASML and (2) to incorporate additional techniques, e.g., for architectural analysis.

ASML uses the ASOME modeling language [7] to model the behavior of its machines. To analyze ASOME models in SAMOS, we first need to understand the elements involved in these models, based on the metamodels they adhere to. This is necessary to extend the feature extraction part, determining, e.g., which model parts to extract (and in which specific way) or to ignore. Moreover, while SAMOS defines comparison schemes for the comparison of features extracted from, e.g., Ecore metamodels, it has yet to be examined if these comparison schemes are suitable for ASOME models.

Our analysis of ASOME models in this work, namely, clone detection, also needs to be evaluated with respect to accuracy and relevance. The goal of clone detection in this context is to find a way to use this information to investigate, and if possible reduce, the level of cloning in the models. The largest part of the analysis done in this chapter is clone detection on ASOME models at ASML. We consider three aspects: (1) applying and extending SAMOS to detect clones in ASOME models, (2) assessing the accuracy and relevance of the clones found, and (3) improving the maintainability of the MDE ecosystems at ASML based on the discovered cloning information.

Given the variety of MDE ecosystems at ASML, each of which consists of several languages represented by metamodels, we have a few additional objectives related to language-level analyses. First of all, we would like to investigate what overview and high-level insights we can gain by clustering the metamodels of multiple ecosystems. Similarly, we are also interested in the cloning at the language-level within and among the ecosystems, along with their relevance, nature, and actionability for improving the quality of the ecosystems. Finally we also consider a focused study on the CARM2G ecosystem only and reconstruct it in terms of the conceptual and architectural layers for architectural understanding and conformance.

The related analyses, addressing the objectives presented above, are discussed in various sections of the chapter. The extension of SAMOS for clone detection on ASOME models is addressed in Sections 11.6.1 and 11.6.2. The actual clone detection and the interpretation of the results are discussed in the first case studies in Sections 11.7.1 and 11.7.2. The case studies in Sections 11.7.3–11.7.5 on the MDE ecosystems address the rest of the objectives on the language level.

11.3. Background: SAMOS model analytics framework

The SAMOS framework is a tool developed for the large-scale analysis of models using a combination of information retrieval, natural language processing, and statistical analysis techniques [8]. The model analytics workflow of SAMOS is shown in Fig. 11.1. The process starts with an input of a collection of models that adhere to a particular

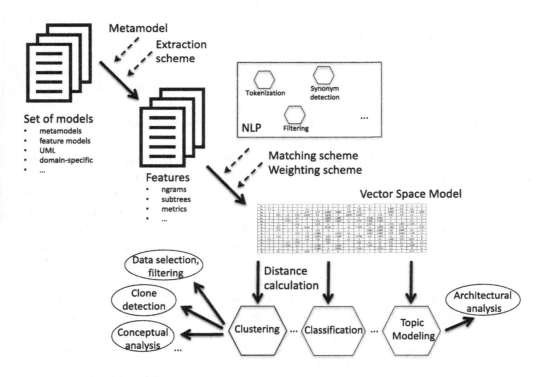

Figure 11.1 SAMOS workflow.

metamodel. SAMOS has so far been used for the analysis of, e.g., Ecore metamodels [9] and feature models [10].

Given a collection of models,[1] SAMOS first applies a metamodel-specific extraction scheme to retrieve the features of these models and store them in feature files. Features can be, for instance, singleton names of model elements or larger fragments of the underlying graph structure such as n-grams [8]. Once the features have been extracted, the following steps are independent of the type of the input models. SAMOS computes a term frequency-based Vector-Space Model (VSM), using comparison schemes (for instance determining whether to match metaclasses or ignore them), weighting schemes (for instance classes weighted higher than attributes), and natural language processing (NLP) techniques such as stemming and synonym checking. After choosing the suitable schemes, a VSM is constructed where each model is represented as a vector comprised of the features that occur in these models. Applying various distance measures suitable to the problem at hand, SAMOS applies different clustering algorithms (using R statistical software [11]) and can output automatically derived cluster labels, for instance for clone detection, or diagrams for visualization and manual inspection and exploration.

The workflow as detailed above can be modified to include scopes. By identifying meaningful scopes for models (such as treating classes and packages separately in a class diagram, in contrast to the whole model as a single entity), the settings in SAMOS allow for an extraction of features at the level of the defined scope. This allows to extract model fragments, effectively mapping a model into multiple data points for comparison among as well as within the models.

11.4. MDE ecosystems at ASML

The development of complex systems involves a combination of skills and techniques from various disciplines. The use of models allows one to abstract from the concrete implementation provided by different disciplines to enable the specification, verification, and operation of complex systems. However, shortcomings or misunderstandings between the disciplines involved at the model level can become visible at the implementation level. To avoid such shortcomings, it is essential to resolve such conflicts at the model level. To this end, MultiDisciplinary Systems Engineering (MDSE, used synonymously with MDE in our work for simplicity, although strictly speaking it is a broader domain) ecosystems are employed to maintain the consistency among interdisciplinary models.

ASML is developing such MDE ecosystems by formalizing the knowledge of several disciplines into one or more DSLs [12]. The separation of concerns among the different

---

[1] In the model analytics context, we use metamodels synonymously with models, in the sense that they are models too (adhering to the corresponding metametamodel).

disciplines helps with handling the complexity of these concerns. Clear and unambiguous communication between the different disciplines is facilitated to enable not only the functioning of the complex system, but also its ability to keep up with the evolving performance requirements. Furthermore, the design flow is optimized, resulting in a faster delivery of software products to the market [7,12].

In such an ecosystem, concepts and knowledge of the several involved disciplines are formalized into of one or more DSLs. Each MDE ecosystem has its own well-defined application domain. Examples of developed MDE ecosystems at ASML are:

- *ASOME*, from ASML's Software application domain. It enables functional engineers from different disciplines to define data structures and algorithms, and it allows software engineers to define supervisory controllers and data repositories [7].
- *CARM2G*, from ASML's Process Control application domain. It enables mechatronic design engineers to define the application in terms of process (motion) controllers (coupled with defacto standard Matlab/Simulink), providing a means for electronic engineers to define the platform containing sensors, actuators, the multiprocessor, multicore computation platform and the communication network, and means for software engineers to develop an optimal mapping of the application onto the platform; see [13,14].
- *Wafer handler* (*WLSAT*), from ASML's Manufacturing Logistics application domain. It provides a formal modeling approach for compositional specification of both functionality and timing of manufacturing systems. The performance of the controller can be analyzed and optimized by taking into account the timing characteristics. Since formal semantics are given in terms of a (max, +) state space, various existing performance analysis techniques can be reused [15–17].

11.4.1 ASOME models

The ASOME MDE ecosystem is a software development environment that supports the DCA architecture, which separates Data, Control, and Algorithms. A motivation to employ this architecture pattern is to avoid changes in the control flow of a system based on a change in data. Using techniques of MDE, ASOME provides metamodels to create data and control models independently of each other.

In the context of DCA and ASOME, data are one of the aspects. Similarly, we also talk about the "control," "algorithm," and (overall) "system" aspects. Within this data aspect, several kinds of systems, interfaces, and realizations can be recognized. Domain interfaces and system realizations are just a few examples; other operational examples would include data shifters and services. We further limit our studies on domain interfaces in the data models.[2]

---

[2] Although the concept of a data model, in the strict sense, actually does not exist, we simply refer to any such model originating from the data realm as *data model* in this work for simplification.

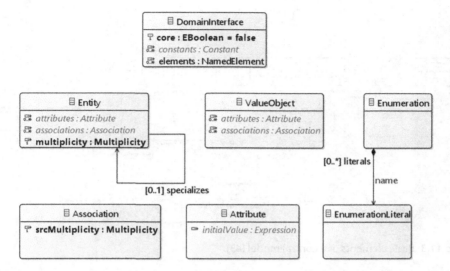

Figure 11.2 Basic elements in a data model [43]. Images from this thesis were reproduced with permission from Technische Universiteit Eindhoven.

Data elements of an ASML component are represented using one or more data models adhering to several metamodels. Data models contain the following (Fig. 11.2):

- *Domain Interfaces:* Any kind of interface in ASOME can express a dependency on another interface. Interfaces allow dependencies among elements originating from other models, hence reuse across models. Within these domain interfaces, several model elements reside including enumerations, entities, and value objects.
- *Attributes and Associations:* Simple elements with a name, *type*, and multiplicities representing how many instances of these elements can exist at run-time. While attributes are involved strictly in a containment structure, associations can refer to other elements without containing them. Associations additionally distinguish between source and target multiplicities. For collections, the order of the elements contained might be relevant (i.e., list) or not (i.e., set), indicated by the optional flag *order*.
- *Entities:* Model elements that contain value objects, attributes, and associations to other entities (within the same model or from different models). Entities additionally allow a user to define properties such as deletability, mutability, and persistency.
- *Value Objects:* Model elements that contain only attributes of primitive types enumerations or other value objects, but no associations. The concept of value objects has been introduced to be able to avoid repetition.
- *Enumerations:* Collections of constant values called *Enumeration Literals*.
- *Primitive Types:* Simple types to act as basic building blocks for the ASOME language.

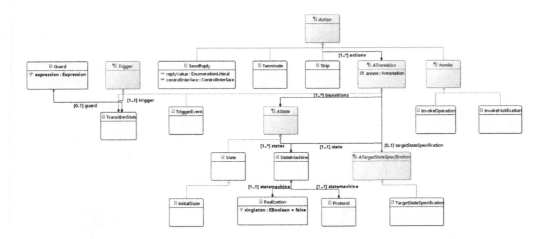

Figure 11.3 Basic elements of a control model [43].

Control models, on the other hand, allow a user to model the flow of control of different components of the system at hand. This is done using state machines. Control models can be of three different types – composite, interface, and design.[3] The construction of complex systems in ASOME control models is done using instances of some smaller systems. Composite models contain a decomposition defining what system instances are made up of, along with how they are connected through ports and interfaces. An interface model provides a protocol for a state machine along with a definition of how the system and its interfaces can be defined. A design model uses this protocol to define a concrete realization of this system. Fig. 11.3 represents the elements of interest in a control model. These elements are the following:

- *State Machines:* Elements defined in the protocols of interface models or realization design models. A state machine consists of states, transitions, and variables used within these states.
- *States:* Elements used to represent different states of the system being modeled using control models. Every state machine consists of a number of states, one of which is indicated as the initial state.
- *Transition States:* Elements that are contained within the state to model the behavior of the system based on different triggers. Each transition state is associated with such a trigger, followed by one or more actions (see below) and optionally a guard (an expression to be solved) and a target state specification.[4]

---

[3] Composite is strictly a part of the "system" aspect and not the "control" aspect. As with data models, the concept of a control model as such does not exist within ASOME. In the scope of this work, we group these types of systems or interfaces and call them control models.

[4] Note that it might be slightly confusing to have a concept mixing the terms *transition* and *state*. In the newer versions of the ASOME language, the name has been changed to *Transition* avoid confusion.

- *Actions:* Elements specifying the activity that follows when a triggering event occurs. A sequence of one or more actions is defined in each transition state. These actions could include sending a reply to an interface of a model, terminating the control flow, invoking an operation or a notification, etc.

We will refer to the above basic concepts within ASOME models, when discussing our approach for clone detection in Section 11.6. While ASOME also facilitates the specification of Algorithm models, these are not considered for the purpose of finding clones in this work. This is due to the fact that there is ongoing effort at ASML to model algorithms and as a consequence there are no models that contain sufficient algorithmic aspects to analyze.

11.5. Model clones: concept and classification

Before detailing the process of clone detection, it is essential to consider what defines a clone. Model clone detection is a relatively new area of exploration as compared to code clone detection [18]. While there are clear definitions of what constitutes a clone for code, such a definition is not as clear for models. The first step to approach the problem of clone detection for ASML models using SAMOS was to define what model clones are. A model fragment (a part of a model) is considered to be a clone of another fragment if the fragments are considered to be highly similar to each other. Therefore, the idea of model clones boils down to *groups of model fragments that are highly similar to each other* in the general sense.

Another aspect of model clone detection is the categorization of the types of clones that can be detected. For the purposes of this work, the classification used in [8] has been used, i.e., the following:
- *Type A.* Duplicate model fragments except secondary notation (layout, formatting) and internal identifiers, including *cosmetic changes* in names such as case.
- *Type B.* Duplicate model fragments with a *small percentage* of changes to names, types, and attributes with little addition or removal of parts.
- *Type C.* Duplicate model fragments with a *substantial percentage* of changes, addition, or removal of names, types, attributes, and parts.

For the ASOME data models, the names of elements are considered relevant (argument being that they are similar to conceptual domain models) and the classification of clone types takes changes in the name of model fragments into account. However, for the ASOME control models, since the behavior of these models is analyzed and the structure of the models represents behavior, the classification of clones takes into account the addition or removal of components that modify the structure of the model (in the sense of finding *structural* clones). This is partly in line with the clone category of *renamed clones*, as investigated in the model clone detection literature (e.g., in [19] for Simulink model clones).

11.6. Using and extending SAMOS for ASOME models

SAMOS is natively capable of analyzing certain types of models, such as Ecore metamodels. However, it needs to be extended and tailored to the domain-specific ASOME models; this can be considered an extended implementation rather than a conceptual extension. The current section discusses the applicability and extension of SAMOS for clone detection on the ASOME models at ASML. The workflow of SAMOS, as represented in Fig. 11.1, involves the extraction of relevant features from the models. This extraction scheme is metamodel-specific and, therefore, an extension to SAMOS is first required, to incorporate a feature extraction scheme based on the ASOME metamodels. As addressed in Section 11.3, SAMOS already uses a customizable workflow for extracting and comparing model elements, e.g., for clone detection. The first step to do this is the metamodel-based extraction of features, i.e., via a separate extractor for each model type, which is addressed in the following sections.

11.6.1 Feature extraction

The first step for detecting model clones is to determine the information that is relevant for comparing model elements. In feature extraction, first, the collection of metamodels which jointly define what the Data and Control models adhere to were inspected. Along with input from a domain expert, we gained insight into the features for each model element that could be considered relevant for clone detection. These include, among others, names and types of the model elements, depending on the particular model element involved. Separate extraction schemes were developed for the Data and Control models.

The above settings describe how a model element (i.e., the vertex in the underlying graph) should be represented as a feature. Next, SAMOS allows a structure setting for feature extraction: unigrams, effectively ignoring the graph structure; n-grams, capturing structure in linear chunks; and subtrees, capturing structure in fixed-depth trees [5]. These have implications on the comparison method needed (as will be explained in the following sections; see [5] for details) and on the accuracy of clone detection overall.

The extraction in SAMOS can be specified to treat models as a whole (i.e., map each model to the set of its model elements). In addition, the extraction scope can be narrowed to smaller model fragments, such as extracting features per class in a class diagram. In such cases the analysis done in SAMOS is performed on a model fragment level rather than at the model level, effectively allowing SAMOS to compare and relate model fragments at the chosen scope. For the ASOME models, a number of scopes were investigated. The relevant ones used in the scope of this work are the following.

Scopes for data models

Fig. 11.2 is a basic representation of the elements contained in the data models. The extraction scopes are listed below:

- *Structured Type and Enumerations:* Similar to the EClass scope we used for metamodel clone detection [5], we treat each entity, value object, and enumeration within a model separately.
- *LevelAA:* This lowest-level scope treats each attribute or association separately; hence the name *AA* denoting attributes and associations. These could be considered as one-liner microclones considered in the code clone detection literature [20].

Scopes for control models

Fig. 11.3 represents the basic elements of ASOME control models. For those models, we considered the following scope:

- *Protocol:* A Control Interface, defined in an interface model, uses a state machine to specify the allowed behavior, i.e., *Protocol*, along with its interface. A Control Realization, defined in a design model, needs to provide a specification that adheres to the Control Interfaces it provides and requires. Similarly, this specification takes the form of a state machine.

11.6.1.1 Domain specific concerns for extraction

A direct (and nonfiltered) treatment of the models as their underlying graphs might lead to inaccurate (and noisy) representations, and in turn inaccurate comparison results. We had several domain-specific adaptations for feature extraction of the new model types.

Redundant information in the model graphs

Fig. 11.4 represents the structure of attributes and associations, respectively, as modeled in the ASOME language. A blind extraction of features along the tree structure for these model fragments would lead to redundant representation of features. For instance, consider a tree-based comparison of any two attributes based on this representation. Since the tree nodes of *Collection* and *Multiplicity* would by definition exist in any attribute, the tree comparison would always detect some minimum similarity (2/7 tree nodes matching). In the extreme case, all attributes with matching multiplicities would have a too high similarity (at least 5/7 tree nodes matching). This would lead to unfair

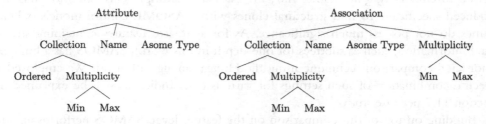

Figure 11.4 Containment structure for *Attribute* and *Association* [43].

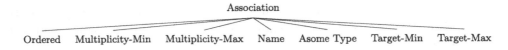

Figure 11.5 Modified containment structure for Association [43].

similarities between those model fragments, and this is against the *fine-tuned distances* policy of SAMOS [5].

To solve this problem, we appended the multiplicity bounds and ordering flag into the attribute or association. Fig. 11.5 depicts the new flattened representation for Association. This allows us to have a more meaningful comparison, and in turn more accurate clone detection.

Filtering out some model elements

With MDE systems, maintaining traceability between models and eventually derived or generated artifacts, such as code, is important. ASOME uses annotations in Control models to provide this traceability between systems. In Control models, for transition states within a state, such annotations are introduced. During the extraction of features from models, annotations are also extracted. However, the behavior of the model does not depend on these annotations and therefore, including these annotations hampers the accuracy of detecting relevant clones for our interest. To avoid this, the extraction of model features excluded the extraction of these annotations.

11.6.2 Feature comparison, VSM calculation, and clustering

While SAMOS has the basic building blocks for the next steps in clone detection, namely, feature comparison and VSM construction (see Section 11.3 for a summary, and [5] for details), we need to specify and extend the comparison needed for our case studies. The feature comparison setting on the vertex or unit level in SAMOS involves, e.g., whether to consider domain type (i.e., metaclass) information of model elements for comparison, and whether and how to compare names using NLP techniques such as tokenization and typo and synonym checking. For this work, we introduced a new option to effectively *ignore* names (i.e., the *No Name* setting). This extension was introduced specifically to find structural clones within ASOME control models, where names do not possess much significance. As for aggregate features containing structural information, such as subtrees (of one–depth in this work), SAMOS has a built-in unordered comparison technique using the Hungarian algorithm [5]. We employed a specific combination of such settings for various case studies, as will be explained in Section 11.7 per case study.

Building on top of this comparison on the feature level, SAMOS performs an all-pairs comparison to compute a VSM, representing all the models (or model fragments, depending on the extraction scope) in a high-dimensional space. In the case of clone detection, by selecting distance measures (specifically masked Bray–Curtis) and clustering methods (density-based clustering), SAMOS performs the necessary calculations to identify clone pairs and clusters [5].

11.7. Case studies with ASML MDE ecosystems

We have performed a wide range of case studies on the models and languages/metamodels used at ASML. In the first two case studies we have detected and investigated the clones in ASOME data and control models, while the others contain language-level analyses on various ecosystems.

11.7.1 Clone detection in ASOME data models

This section discusses the results of the case studies performed using the different settings of SAMOS on the ASOME data models.

11.7.1.1 Dataset and SAMOS settings

The dataset consists of 28 data models, containing one domain interface each. These domain interfaces in total contain 291 structured type and enumeration model fragments and 574 attributes and associations. Our preliminary runs with the scopes *Model* and *Domain Interface* did not yield significant results, and therefore we report here only the lower-level scopes. The settings of SAMOS for this case study are as follows:

- *Scopes:* Structured Type and Enumerations, LevelAA.
- *Structure:* Unigrams. For the model fragments at the chosen (low-level) scopes, there is no deep containment structure. So, a unigram representation suffices for this case study.
- *Name Setting:* Name-sensitive comparison, as model element names are important parts of the data models.
- *Type Setting:* A relaxed type comparison (standard setting of SAMOS) for the scope Structured Types and Enumerations and strict type comparison for LevelAA. For the latter, we are interested in strictly similar microclones, facilitating easy refactoring.

On the given set of data models, using the settings above, we discuss the results we found in the next section.

11.7.1.2 Results and discussion

This section discusses, per scope, the results obtained through the chosen settings. The discussion is structured as follows. First, the model fragments considered to be clones are discussed; second, the proposition for reducing the level of cloning is presented, and finally, the opinion of a domain expert on this proposition is presented.

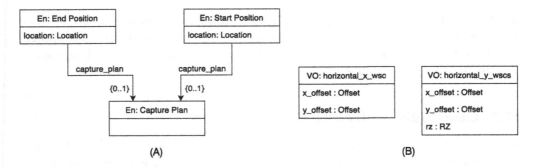

Figure 11.6 Examples for clone pairs in the Structured Types and Enumerations scope for data models: (A) Type B clone pair example; (B) Type C clone pair example [43].

We found the following clone clusters in the scope of Structured Type and Enumerations:

- *Type A Clones.* Only one clone cluster was found for this category consisting of two Value Objects, named *XYVector*, representing coordinates. This is a small example of duplication in two models and can be easily eliminated by reuse. The domain expert has commented that in fact one of the models is actually called *core*, with the intention that it contains the commonalities, while other models import and reuse it.
- *Type B Clones.* We found four clone clusters, each having a single clone pair. We show an example of such Type B clone pairs in Fig. 11.6A, consisting of model fragments with partly similar names (e.g., End Position vs. Start Position) and otherwise the same content. The domain expert's remark was that such cases of redundancy can be considered candidates for elimination via inheritance abstraction. However, due to specific design constraints and the additional effort to integrate this abstraction in the existing practice, the expert told it is difficult and unlikely that such improvements will be realized.
- *Type C Clones.* We found 23 Type C clone clusters. Fig. 11.6B shows an example clone pair, with a slight name difference and an additional attribute. Other pairs included changes in names and attributes. For such clones, redundancy can be eliminated by creating an abstract class with commonalities and extending it. The domain expert commented it is in any case useful to discover such variants in the modeling ecosystem, and it can be investigated which ones can be refactored (as in the discussion above for Type B clones).

As for the microclones at scope LevelAA, we have the following results.

- *Type A Clones.* We found 53 Type A clone clusters. The most interesting result proved to be the Association *task*, found in a cluster of nine items. The target of this association is an entity *Task* which belongs to a core data model (a specific instance of a data model which other models depend on). This pattern, along with the fact that these associations were all named the same, is an indication of consistency and good design, as confirmed by the domain expert. This is an example of the fact that not all clones are harmful and in this case, the clones are an indication of good design; outliers (if any) can be investigated as an indication of violation of the common practice. As for the duplication in Attributes, the idea of refactoring by lifting these attributes up to a common superclass was considered by the domain expert with suspicion, due to the additional complexity of introducing inheritance. Duplicate associations in some cases cannot even be eliminated at all, especially in entities from different models.

- *Type B and C Clones.* We found 65 Type B clones clusters, consisting with very similar elements with small differences in, e.g., multiplicities. Clones of these types are not candidates for elimination or refactoring however, as remarked by the domain expert. Similarly, among the 81 Type C clone clusters with a higher percentage of differences, the domain expert could not find any good candidate for refactoring. This might indicate that only exact duplicate microclones should be considered as useful and actionable, and therefore studied.

Overall discussion

We have provided separate discussions above for our results on different scopes and clone types. A general remark is to be made about the NLP component of SAMOS. In the current setting, due to the tokenization and stopword removal, SAMOS considers model elements with names *element_m_1* and *element_m_2* as identical; numbers and short tokens are omitted. Moreover, the lemmatization and stemming steps lead SAMOS to consider the following as identical or highly similar names: *changed, unchanged, changing*. In the future we might consider further fine-tuning (and partly disabling) several NLP components considering the problem at hand, when looking for exact clones under the scope of LevelAA.

11.7.2 Clone detection in ASOME control models

This section discusses the case studies performed on control models as well as the results of these case studies.

11.7.2.1 Dataset and SAMOS settings

The approach taken to detect clones within control models is different compared to the one for data models. This is due to the importance of structure in these models. However, the tree-based setting in SAMOS is still considerably expensive for large datasets. On the other hand, a structure-agnostic unigram-based detection with SAMOS [4] would not be accurate enough. Therefore, we follow an iterative approach (similar to [5]). We first narrow down the number of elements for comparison using a cheaper unigram-based analysis. On each cluster found in this first step, we perform a more accurate clone detection separately, thereby reducing the total complexity of the problem. In our previous work [5], we showed that this iterative process leads to only minor drops in recall, but we leave the assessment of its accuracy in this work for future work.

The dataset of control models for this case study contained 691 models, 531 protocols, and realizations. A preprocessing step excluded 10 protocols and realizations because these protocols and realizations were very large compared to the other models. Excluding these for the comparison was justified considering it was less likely to find models similar to these based on their size. Moreover, these models would slow down the comparison significantly while constructing the VSM. The following settings were chosen for the comparison of control models.

- *Scope:* Protocol level. On this level of comparison, one can compare models based on their behavior, as defined using the state machines residing in these protocols or realizations.
- *Structure:* For the first round of comparisons, the unigram setting was used to find clusters of similar model elements. Fifty such clusters of models were found. The second round of comparison involved inspecting some of the clusters found in the first round. For this round, one-depth subtrees were extracted and compared.
- *Name Setting:* A no-name setting was used for the two rounds of comparison of control models. This was done so we could find models that were structurally equivalent ignoring names.
- *Type Setting:* A strict type setting was used for both rounds of comparisons for control models. In a no-name comparison, to find structurally similar models, this setting allows one to detect model elements with similar types and other attributes.

11.7.2.2 Results and discussion

In this section we provide a detailed discussion and qualitative evaluation on some exemplary control model clone clusters found by SAMOS. Based on the first round, which results in a number of *buckets* with potential candidates for clones, we ran SAMOS with the more accurate subtree setting for a second round of clustering. Fig. 11.7 represents the hierarchical clustering of elements contained in one bucket. Note that this hierarchical clustering of elements is used for clarification and discussion purposes only; SAMOS employs a threshold-based automatic cluster extraction technique. The dendrogram represents the Protocol-scope model fragments at each leaf represented by a number, and the vertical axis and the joints in the tree denote the distance, i.e., dissimilarity, of the fragments.

The models inspected in this cluster were quite large. These contained a single state with a variation in the number and type of transition states, representing an *all accepting* state. A combination of patterns found in the models is shown in Fig. 11.9. The state X

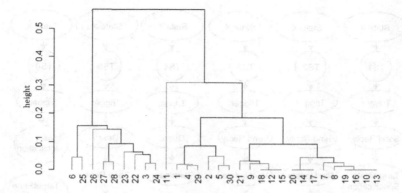

Figure 11.7 A dendrogram representation of cluster 1 with potential clones [43].

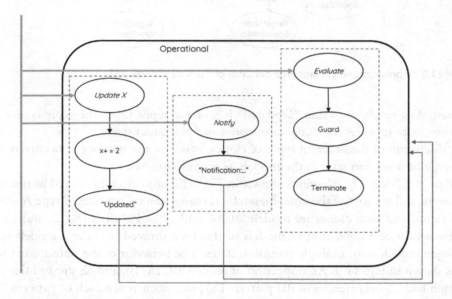

Figure 11.8 Example visualization of some transition states in cluster 1 [43].

contains a number of transition states. The patterns of the different types of transition states found in the models are represented by TS1 through TS6.

Fig. 11.8 is an example of a visualization of a few of the transition states in the single state models found in this cluster. The figure shows a single state *Operational* which defines behavior using three transition states. A trigger exists for each transition state. The triggers here are *Update X*, *Notify*, and *Evaluate*. Depending on the trigger that has been received, the corresponding transition state is executed. For example, the *Update X* trigger is followed by the action of a State Variable Update where the variable *x* is

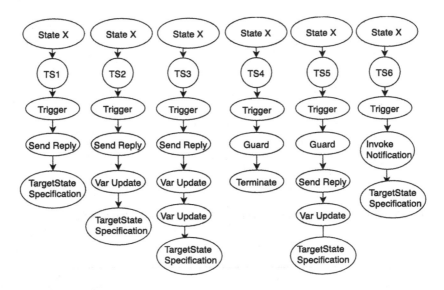

Figure 11.9 Representation of combined behavior of cluster 1 models [43].

updated. As a result, the value "Updated" is sent as a reply. Once the reply is sent, the transition state specifies the same state *Operational* as a target state.

A discussion of the different types of clones, based on the number of occurrences of each type of transition state in the models, is given as follows:

- *Type A Clones.* The elements shown in Fig. 11.7 that are represented at the same height and are part of the same hierarchical cluster can be considered type A clones. Examples of such clones are models 20, 14, and 17 and models 18, 12, and 15. An inspection of a collection of models of this type showed that these models had a single state X with multiple transition states. The behavior of the transition states is as shown in Fig. 11.9. An inspection of models 19, 16, 10, and 13 showed that they each had 18 occurrences of the pattern TS1; one occurrence each of patterns TS2, TS3, and TS4; and eight occurrences of pattern TS5. TS6 however, did not occur in these models.

- *Type B Clones.* We can use the dendrogram as guidance to identify elements that are not exactly the same as but could be considered similar to each other (up to 10% distance, as an intuitive estimate). Model 8, for instance, is similar to the cluster of models 19, 16, 10, and 13; that group is already mentioned above as Type A clones. Model 8 in this case is highly similar to those in the Type A cluster, but contains additionally two occurrences of TS5. This makes it a Type B clone compared to the rest of the models mentioned.

- *Type C Clones.* Again by inspecting the dendrogram, we can consider, for instance, models 27 and 24 as candidate Type C clones to validate. Table 11.1 shows the

Table 11.1 Number of occurrences of transition states in models 24 and 27.

Transition states	Model 24	Model 27
TS1	5	6
TS2	1	1
TS3	1	1
TS4	1	2
TS5	7	4
TS6	2	0

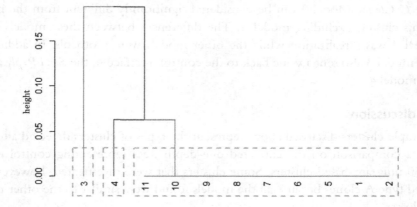

Figure 11.10 A dendrogram representation of cluster 2 with potential clones [43].

Figure 11.11 Representation of behavior of cluster 2 models [43].

number of times each transition state pattern was found in the models; the number of occurrences is slightly different for four out of six transition state patterns. Therefore, we manually label these as Type C clones as well.

We further examined another example cluster to validate the results of SAMOS. Fig. 11.10 shows the resulting dendrogram for Cluster 2. The three types of clones in this cluster are discussed as follows. Note that all the models in this cluster share a common pattern (with minor differences as will be discussed below), as shown in Fig. 11.11.

- *Type A Clones.* All the elements in this cluster excluding models 3 and 4 can be considered type A clones. The models were all protocols, defining state machines with this structure. The action of sending a reply is associated with a control interface defined in the model. In each of these models, it was observed that the value of the reply sent to the control interfaces was *void*.

- *Type B Clones.* The models excluding model 3 and model 4 could be considered similar to model 4, though not exactly the same. Upon investigating these models, it was noted that the difference between the other models and model 4 is in the action *Send Reply*. While the other models sent an empty reply to the control interface, model 4 replied to the control interface with a value. Since this is a small percentage of change between these models, model 4 and the models excluding model 3 can be considered type B clones.
- *Type C Clones.* Model 3 can be considered significantly different from the models in this cluster, excluding model 4. The differences between these models is that model 3 was a realization while the other models were protocols. In addition to this, model 3 also sent a value back to the control interface in the *Send Reply* action, like model 4.

Overall discussion

The example clusters discussed above represent the types of clusters detected after performing a comparison on the extracted one-depth trees representing control models on the 50 unigram-based clusters. Some clusters that were investigated, however, only contained type A clones because all the models found were similar to the other models in that cluster.

While eliminating clones was straightforward for cases in data models, this is not as easy for control models. The presence of duplicates in terms of a sequence of actions might be inevitable if that is the intended behavior of the models. This presents the case for the idea that not all clones can be considered harmful, and some are in fact intended. However, many occurrences of some transition state patterns have been found in the models. The transition state pattern TS1 as seen in the example cluster 1 (Fig. 11.9) was found 18 times each in two inspected models. For such transition states, maybe the language could allow for an easier representation of such a pattern to make it easier for a user to implement this sequence of actions.

According to the domain expert, "*detecting such patterns of control behavior definitely can be used to investigate whether the user could benefit from a more comfortable syntax. Then an evaluation is needed that needs to take into account:*

1. *whether the new syntax requires more time to learn by the user, and*
2. *whether the simplification really simplifies a lot (see below).*

For instance, in the example above, even for TS1, the user will need to specify the trigger somehow. In case of a nonvoid reply, also the reply value will need to be specified. So, TS1 cannot be replaced by one simple keyword. It will always need two or three additional inputs from the user. In this case, we will not likely simplify this pattern. However, the way of thinking to inspect whether we can support the user with simplifying the language is interesting. It will always be a tradeoff between introducing more language concepts vs. writing (slightly) bigger models."

Another suggestion for control models is to investigate the unigram clusters to find the different types of patterns found within the control models. Following this, checking what models do not adhere to these patterns might reveal outliers to investigate, to find unexpected behavior.

A domain expert commented, *"I see the line of reasoning and it brings me to the idea of applying Machine Learning to the collection of models and let the learning algorithm classify the models. Then, investigating the outliers indeed might give some information about models that are erroneous. However, these outliers could also be models describing one single aspect of the system, which would justify the single instance of a pattern. However, I would expect that the erroneous models would also have been identified by other, less costly, means such as verification, validation, review, etc."*

11.7.3 Overview on multiple ASML MDE ecosystems

As introduced in Section 11.4, ASML has a very diverse conglomerate of MDE ecosystems, developed and maintained by different groups and involving different domains in the company's overall operation. While the architects and managers might have a good idea of (parts of) the enterprise-level big picture, we would like to (semiautomatically) investigate the relation among the different ecosystems with respect to the domains.

Objectives

Given the multitude of languages which belong to the various ecosystems, we would like to perform a concept analysis via hierarchical clustering based on the terms used in the metamodels which represent the abstract syntaxes of those languages. Note that we will use the terms metamodel and language interchangeably through our case studies. We have two main subobjectives in this case study. First we would like to get a good overall picture of the enterprise ecosystem and its compartmentalization into meaningful domains and subdomains. It is worthwhile to investigate, e.g., whether different ecosystems occupy distinct or intersecting conceptual spaces. Furthermore, it can be interesting to see what close-proximity metamodel pairs or clusters across different ecosystems imply. Furthermore, we can study whether this information leads to quality improvement opportunities in the ecosystems, such as metamodel refactoring and reuse of language fragments.

Approach

To address the objectives above, we process the 86 metamodels belonging to three ecosystems. Using SAMOS, we extract the element names from the metamodels, using the normalization steps including tokenization and lemmatization. We then compute the VSM over the words, using a tf-idf (with normalized log idf as in [4]) setting also using advanced NLP features such as WordNet checks for semantic relatedness. We then apply hierarchical clustering with average linkage over the cosine distances in the vector space.

Results and discussion

We present our result in the dendrogram depicted in Fig. 11.12. Each leaf in the dendrogram corresponds to a metamodel, and all the metamodels are color-coded with respect to their ecosystems. The colored leaves are also projected into the horizontal bar as a complementary visualization. The joints of the leaves and branches can be traced in the y-axis, which denotes the distance (dissimilarity) of the (groups of) metamodels. For instance, metamodel pairs in the lower parts of the dendrogram (such as *ds_resource* and *resource*) are very similar. By discussing with the language engineers and domain experts for each ecosystem, we gathered a list of remarks that address the objectives above. Next we present a representative summary of those findings, along with key subobjectives of this case study.

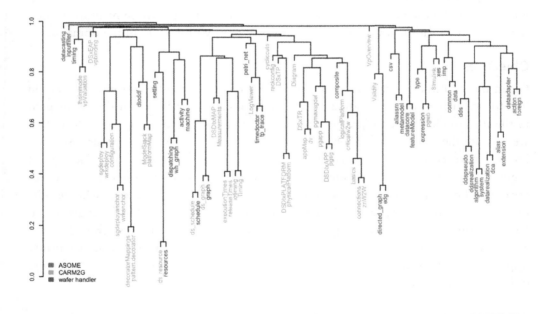

Figure 11.12 Dendrogram depicting the result of clustering the 86 metamodels. Colors denote the ecosystems, while each leaf corresponds to a single metamodel in the color-coded ecosystem [42]. Images from this thesis were reproduced with permission from Technische Universiteit Eindhoven.

Some remarks involving the general overview, domains, subdomains, and proximities across ecosystems would include the following:

- The ecosystems roughly occupy distinct conceptual spaces. As can be seen from the horizontal bar at the bottom, ASOME models are mostly on the right-hand side, while wafer handler (less consistently) is in the middle regions.
- There are however small intersections (i.e., impurities in the colored bar or different colors in the subtrees) among the ecosystems. These are not always surprising or bad because different ecosystems might reuse languages and potentially share subdomains. However, our automated analysis allows having a full overview, in contrast to partial insights of the individual experts.
- Within ecosystems, the domain experts can already detect subdomains, such as platform, deployment, timing, and scheduling for CARM2G and data for ASOME.

We wish to conclude with the following points regarding highly similar metamodels within and across ecosystems:

- Except two metamodels within CARM2G which are highly similar (height < 0.2, i.e., *decoratorMappings* and *pattern.decorator*) – one of which happens to be very small and insignificant, so discarded by the domain experts – no metamodels within a single ecosystem are too similar. This indicates a healthy design, where each language deals with a distinct conceptual subspace. There are still somewhat similar (height < 0.4) pairs, which might lead to a consideration of a within-ecosystem refactoring. Examples for such pairs would include *sgdeployanchor* and *wrkanchor* from CARM2G.
- Across the ecosystems, there is a pattern of similar (height < 0.4) pairs for CARM2G and wafer handler, specifically for *resources, schedule,* and *graph* metamodels. This is apparently due to the fact that wafer handler borrowed these metamodels from CARM2G in early development, while making custom changes as required in time. Our visualization correctly reveals this in a straightforward manner.
- Similarly, somewhat similar couples of metamodels across ASOME vs. the other ecosystems exist, though not as similar as the ones above. Examples are *aliassm* vs. *metamodel*, which partly contain state machine languages, and *expression* vs. *pswb*, which partly contain expression languages.

In summary, according to the feedback we received from the domain experts, such an automated and visual overview of the MDE ecosystems used within a company indeed reveals useful information. This can be used to aid the governance, usage, and maintenance of the ecosystems. However, some additional information, such as dependencies across languages, the corresponding model instantiations and their relations, usage, etc., could be utilized to further augment our study. Furthermore, we currently cannot detect subtle relations among similar languages which use different terminology. The experts exemplified it by various graph description languages, some of which use the terms *node, edge*; others use *task, dependency*. This can potentially be mitigated by using a domain-specific thesaurus, in contrast to just relying on general-purpose WordNet for synonyms.

11.7.4 Cross-DSL clone detection across ecosystems

The concept analysis performed above only deals with the element names, and not the other information in the metamodels such as types, attributes, and the structure. It also treats metamodels as a whole. In this case study, we would like to perform a more precise and fine-grained analysis on the metamodel fragments (i.e., subparts), in order to reveal similar fragments across, as well as within, the different ecosystems and languages.

Objectives

As metamodels across the different ecosystems can have duplicate or highly similar fragments (due to various reasons, e.g., clone-and-own approaches in development or language limitations [5]), we would like to perform clone detection in a more accurate manner, including all the information in the metamodels (not only names). We would like to inspect the clones, their nature (why they occur), and their distribution across the ecosystems. As in the model clone detection case studies, we are also interested to identify potential candidates among these clones which can be used for improving the MDE ecosystems, e.g., in terms of elimination or refactoring.

Approach

We considered the 86 metamodels representing three ecosystems in this study. Using SAMOS, we extracted the one-depth subtrees with full set of model element information from the metamodels, with the EClass scope. Note that we ignored EClasses with no content and supertypes (i.e., zero number of contained elements), assuming they would make less significant cases for refactoring. We then computed the VSM over the subtrees, using the *tree-hung* setting [5]. Finally, we applied the clone detection procedure with reachability clustering over the masked Bray–Curtis distances in the vector space.

Results and discussion

Using SAMOS, we found 9 Type A, 13 exclusively Type B (i.e., discarding Type A clusters), and 55 exclusively Type C clone clusters. Table 11.2 gives some of the interesting clusters, which we will discuss next.

Here we provide a discussion of Type A clones and opportunities for eliminating duplication:

- There are not too many Type A clones overall and they are quite small (size < 3). This indicates little redundancy in general in terms of exact duplication.
- Clusters 1 and 2 show two examples of small clones across different languages, which can be easily refactored and reused.
- Cluster 3 shows an exact clone across ecosystems for the resource language, which we discovered in the previous study to be an evolution/modification from CARM2G to wafer handler.

Table 11.2 Some of the EClass-scope clones in the metamodels (reported using the convention *metamodelName$EClassName*); **t** denotes the clone type (A, B, or C), **s** the average size of the clones in a cluster (with respect to the total number of attributes, operations, etc., for each clone, counting the EClass itself as well), and **eco** the ecosystem of the cluster, where A = ASOME, C = CARM2G, W = wafer handler [42].

id	Cluster	t	s	eco
1	dca$LiteralMapping imp$LiteralMapping	A	3	A
2	criticalw2w$BlockName cycliccuts$BlockName	A	2	C
3	ds_resource$ResourceModel resources$ResourceModel	A	1	CW
4	pgwb$PG_LBoundary pgwb$PG_UBoundary	A	1	C
5	physicalPlatform${CoHost,Host}	A	1	C
...	...	...	...	...
6	xes$Attribute{Boolean,Date,Float,... }Type	B	9	W
7	dca$DDTargetIdentifier imp$DDTargetIdentifier	B	5	A
8	ds_schedule$Sequence schedule$Sequence	B	5	CW
9	VpOverview$NXT19{50Ai,60Bi,70Ci,... }Type	B	3	C
10	machine${AxisPositionMapEntry,AxisPositionsMapEntry}	B	3	W
11	{dca,imp,basics}$NamedElement	B	2	AC
12	ds_resource${WorkerResourceSet,IOWorkerResourceSet}	B	2	C
...	...	...	...	...
13	imp$EntityRealizationRecipe imp$EntityRecipe	C	13	A
14	data$Entity datarealization$EntityRealization	C	8.5	A
15	pgsg${HierarchicalBlockGroup,ServoGroupAbstract}	C	6.5	C
16	vpbinding$Binding vpbinding$Clause	C	5.5	C
17	timing$PertDistribution timing$TriangularDistribution	C	4.5	W
18	setting${Location,Motion,Physical,...}SettingsMapEntry	C	4.2	W
19	Validity$ConstrainingNode Validity$ValidatableNode	C	4	C
20	action$IfAction $action$SwitchAction	C	3.5	A
21	{connections,DSDIxPLATFORM,DSxTR,... }$Connection	C	3.4	C
22	expression$UnaryExpression pgwb$PG_UnaryExpression	C	3	AC
23	connections$ConnectionList logicalPltfm$ConnectionBundle	C	3	C
24	pgmaxsgdef$Pgma{BlockAlias,BlockGroup,Block,... }Ref	C	3	C
...	...	...	...	...

- Due to our NLP settings (notably ignoring stopwords and typo detection compensating for minor changes), SAMOS finds clone clusters such as 4 and 5 as identical. While they are significantly similar and some of these might indicate room for refactoring, the domain experts generally found them to be uninteresting from a maintenance perspective.

As for Type B and C clones and potential refactoring opportunities, we make the following points:

- There is a significant number of Type B and C clones. This indicates that there might be good opportunities to improve the ecosystems.
- Cluster 6 with sizeable (of average size 9) clones shows a clone pattern that we encountered a few more times in this study. According to the domain experts, *xes* is a generated metamodel from xml schemas. There is hence an opportunity to refactor either the xml schema or the generation process in such a way that the commonality, for instance, is abstracted to a superclass.
- Cluster 7, as well as clusters 13 and 14, shows a cloning pattern which happens a few times in the ASOME ecosystem: design vs. realization/implementation. This is a case where clones are *intended*: this pattern is devised to allow the extension of existing language elements for the sake of (a) backwards compatibility and (b) clear (conceptual) separation of concerns, i.e., abstract design vs. client-specific implementations.
- Cluster 8 (and some more clusters omitted in the table for conciseness) shows modified fragments of the metamodels adopted from CARM2G into wafer handler.
- Cluster 9 indicates near-duplicate entities for different machine types at ASML. These could be easily refactored, for example, into enumerations, which solves the cloning problem.
- Cluster 10 is an interesting case: the only difference between these metamodels is the multiplicity of an EReference. Domain experts remarked that this is an *intended* clone for improving the performance while processing the models in real-time in ASML machines.
- Cluster 11 shows small-sized clones with a single attribute *name*, differing only in cardinality – considered as a very minor issue, which does not urge for a refactoring. Cluster 12 similarly indicates small clones, which the experts commented they could refactor, for instance, using generics.
- Clusters 15 and 16 show medium-sized clones with common EAttributes and EOperations defined, so the common parts can be abstracted into superclasses. However, for the latter, the domain experts remarked that actually there is a superclass which is overridden in subclasses. Due to the limitations of EMF (needing to duplicate the EOperation and pointing to the overridden implementation in EAnnotations), cloning here is supposedly inevitable.
- For the rest of the clusters, the experts indicated a varying degree of usefulness (in terms of refactoring), i.e., low-medium (e.g., in cluster 23 – cardinality difference in small EClasses and in cluster 20 – similar control structures which could be refactored into an abstract superclass) and high (e.g., cluster 17 – one statistical distribution being the ontological superclass of the other smoothed distribution).

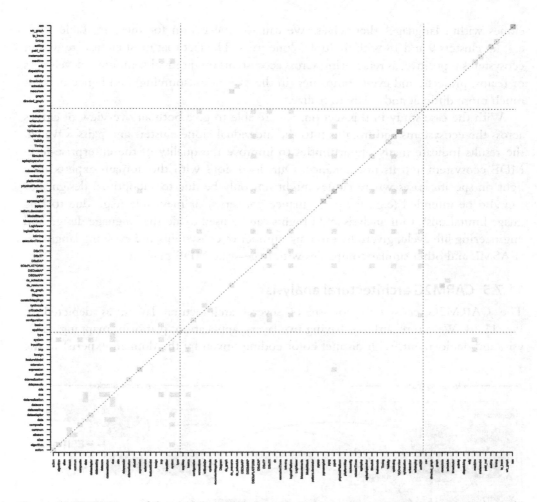

Figure 11.13 Heatmap of the clones across languages and ecosystems. Dashes denote the boundaries of the ecosystems (from left/bottom to right/top: ASOME, CARM2G, wafer handler). Light to dark yellow and red denotes an increasing number of clones [42]. (For interpretation of the colors, the reader is referred to the web version of this chapter.)

Table 11.2 presents the clone occurrences in a flat list. However, we would like to explicitly investigate and visualize the distribution of the clones across languages and ecosystems. To address that, we have constructed the heatmap shown in Fig. 11.13. It is evident from the figure that there are only a few clones across ecosystems. Notable ones include the resource and schedule languages in CARM2G and wafer handler, parts of expression languages across ASOME and CARM2G and some small basic constructs across all three ecosystems (as discussed above in individual clone clusters). Darker yellow and red parts (i.e., high number of clones) are generally on the diagonal, meaning clones within languages themselves. We can see the reason for these in Table 11.2, e.g., in clusters 9 and 18 with multiple clone pairs. The fact that most clones are within ecosystems is positive, as refactoring across ecosystems might involve multiple developers or teams, projects, and even companies (in the case of outsourcing) and hence make it much more difficult and costly to realize.

With the case study in this section, we are able to give both an overview of clones across the ecosystems and insights into the individual clone clusters and pairs. Overall, the results indicate many opportunities to improve the quality of the enterprise-level MDE ecosystem and its maintenance. Our discussions with the domain experts shed light on specific cases where clones might not only be due to suboptimal design, but can also be intended (e.g., for performance concerns) or inevitable (e.g., due to language limitations). Our analysis and insights can be used to aid the language design and engineering life cycle, given the growing number of ecosystems and evolving languages at ASML and other similar companies with large-scale MDE practice.

11.7.5 CARM2G architectural analysis

The CARM2G ecosystem consists of several architectural layers, as depicted in Fig. 11.14. We can regard it as having five layers: application, platform, mapping, analysis, and deployment, with distinct color coding (given by the domain experts) in the

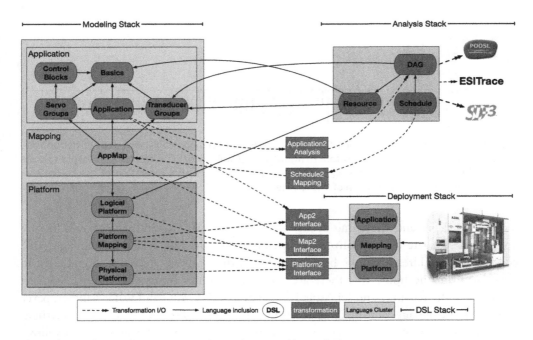

Figure 11.14 CARM2G ecosystem and its architectural layers [42].

figure. As in the previous case study, the relation between the different layers and sublanguages of CARM2G captured in the 41 metamodels is implicit in the domain expertise of the CARM2G developers. We would like to analyze those metamodels and try to automatically infer useful information with respect to the architecture of the ecosystem.

Objectives

By topic modeling the terms (i.e., element names) in the metamodels, we aim to reconstruct architectural partitions (arguably layers) and their relation with the individual metamodels. We formulate the following subobjectives, n being the number of latent topics in the dataset:

1. (unknown n) identifying how many "topics" there are in the dataset, guessing an optimal n;
2. ($n = 5$) assessing the correspondence of the automatically mined topics (i.e., partitions) to the architectural layers given by the domain experts;
3. ($n \neq 5$) identifying additional or redundant partitions or layers by picking different n's.

Approach

As in the first case study, using SAMOS we extracted the element names from the CARM2G metamodels, using the normalization steps including tokenization and lemmatization. After removing regular stopwords in English (such as "of" and "and") and domain-specific stopwords as determined by the domain experts (such as "name" and "type"), we computed a simplistic VSM over the words in the form of a basic frequency matrix (i.e., no idf). We then performed several experiments with Latent Dirichlet Allocation (LDA, see Section 11.9.3 for details) based on Gibbs sampling [21], to infer the topic-term distributions in the dataset. We did not change the default parameters of LDA (due the exploratory nature of this case study and the complexity of the parameter setting [22]); we only kept the number of iterations at a relatively high value of 10,000 to increase the likelihood of convergence to a global maximum.

Results and discussion

Before going into the results involving topic modeling and the individual subobjectives, we would like to present a word cloud for the whole ecosystem, as depicted in Fig. 11.15. According to the domain experts, this is a very nice summary of CARM2G concepts, and it can be used, for instance, to describe and document the ecosystem and to teach it to new language engineers and modelers.

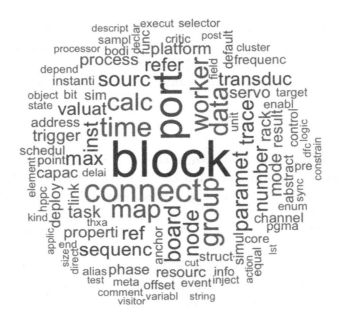

Figure 11.15 Word cloud representation for the whole CARM2G ecosystem [42].

To address the first subobjective, we ran LDA with n from 2 to 50 and analyzed the graphs of several metrics in the ldatuning package[5] to investigate near-optimal (minimized or maximized depending on the metric) values for n, as shown in Fig. 11.16. We can deduce various near-optimal − while aiming for a small n as much as possible − picks for n: $n = 4, 6$ (Deveaux2014), 20 (Griffith2004), 6, 10, 16 (CaoJuan2009), and 15, 17 (Arun2010). Two of these metrics have optimum values close to $n = 5$, as given to us by the domain experts, while others predict a larger number of topics. We proceed with $n = 5$, and we will discuss the implications of picking a lower or higher n later in this section.

After establishing that the number of topics given by the domain experts is (nearly) agreed on by some of the metrics above, we proceeded with topic modeling with $n = 5$. For the second subobjective, we are interested in prominent terms per topic, terms by metamodel, and the distribution of the topics by metamodel. To evaluate the results, we used a subset of 15 metamodels chosen by the domain experts as key representatives of the CARM2G architectural layers (see Fig. 11.14). In Fig. 11.17, we present the results of topic modeling specifically for those key metamodels. The interpretation of the figure is as follows. Each row (i.e., y-axis) represents a topic (labeled with the top five most prominent terms). Each column (i.e., x-axis) represents a key metamodel, as

---

[5] https://cran.r-project.org/package=ldatuning.

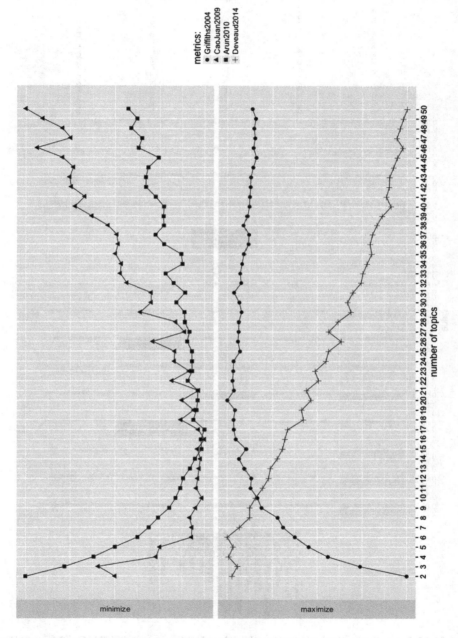

Figure 11.16 Different metric plots for assessing the number of topics [42].

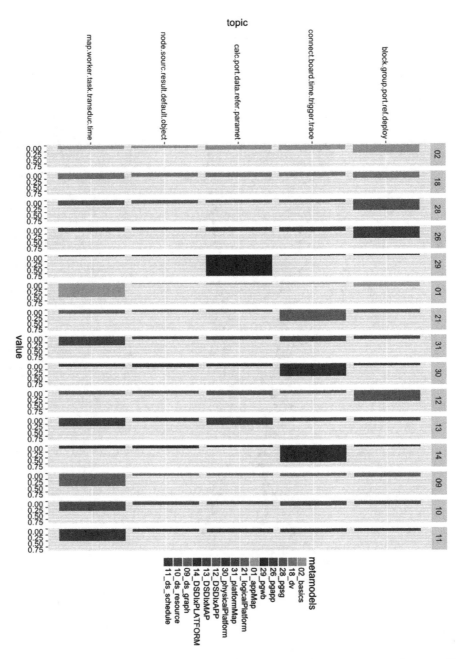

Figure 11.17 Topic distributions per metamodel colored with respect to CARM2G architectural layers: green = application, orange = mapping, blue = platform, purple = interface, red = analysis [42].

also shown in the legend. The bars at each cell of the matrix represent how likely the metamodels are represented by that topic. Each document is associated with a number of topics, hence the probability values in each column for a specific metamodel add up to 1.

Note that we color-coded the metamodels with respect to the architectural layers, green being the application layer, orange mapping, blue platform, purple interface, and red analysis. By inspecting the figure along these color codes, for $n = 5$ we can deduce the following:

Topic-1: The first (top-most) topic roughly represents the application layer, mostly represented in four of the application layer metamodels. Nevertheless, *basics* and *dv* metamodels, being rather generic and common metamodels, have a mixture of the topics; *pg_wb*, which is originally considered in the CARM2G application layer, does not reside in this layer however, and will be discussed below (referred to as L_{wb}). Note that *DSDIxAPP* from the interface layer mostly covers this topic as well.

Topic-2: The second topic can mostly be associated with the platform layer (most related term: *board*). It is found in the platform layer metamodels (except *platformMap*, to be discussed below) and the *DSDIxPlatform* from the interface layer.

Topic-3: In the third topic, *pg_wb* stands out, with almost no association with any other topic. This is due to it being a very large and fundamental language describing general-purpose language building blocks such as statements and expressions.

Topic-4: The fourth topic does not associate with any of the key metamodels, but potentially (a mixture of) some other niche set of languages in the dataset (e.g., variation point languages). We will discuss this further in our experiments with increasing n, referring to it as outlier layer, i.e., L_o.

Topic-5: The final topic has a mixture of mapping and analysis layers; *appMap* naturally is mostly associated with this topic and *platformMap*, *DSDIxMAP* across the other layers as well, all being related. Furthermore, the three metamodels from the analysis layer also consistently reside here.

According to our detailed inspection and the feedback from the domain experts, we argue the following. The most prominent terms per topic give only a limited idea about the topics and layers. However, the partitioning into topics across languages makes a lot of sense. This indeed gives an orthogonal view on the architecture, in terms of the conceptual space. There is still room to change the parameter n for the number of topics, to see whether we can find redundant partitions, and additional (niche) groups of languages besides the standard architectural layers – addressing the final subobjective. Following the different near-optimal estimates as discussed above, we remark on the cases with $n = 3, 4, 6, 10, 16$ in text without giving the figures (due to space limitations).

$n = 4$ We obtain a very similar partitioning as for $n = 5$: roughly the application, platform, and mapping+analysis layers. The topic which did not correspond to any key metamodel disappears. This might indicate a more optimal partitioning than with $n = 5$ if aiming for a high-level layering.

$n = 6$ With larger n, we still get the clear-cut partitions corresponding to platform and mapping+analysis layers; L_o and L_{vb} also remain as is. We see, however, that instead of a single application layer, we have two (with divided probabilities for the related metamodels): one with terms *block, group* and another with *connect, port*. These might partly relate to different aspects of an application description.

$n = 10$ We start getting further decompositions: a mapping+analysis layer into resource with *platformMap, ds_resource* (terms: *worker, map, resource*), and scheduling with *appMap, ds_graph, ds_schedule* (terms: *task, sequence, schedule*). Some of the other partitions, however, start getting a lot fuzzier; platform metamodels for instance are distributed across different topics, *logicalPlatform* is now strongly associated with the application (sub)topic. However, inspecting all the metamodels involved, we discover further topics, sometimes even represented by a single metamodel: system variants (*vpOverview*), variant binding with a visitor pattern (*vpbinding*), deployment and anchor (*configuration, sgdeploy, wrkdeploy, wrkanchor*) intermixed with simulation (*thsimmode*).

$n = 16$ The topics are further diluted, which makes it very difficult to argue about meaningful partitioning compared to the previous run with $n = 10$.

This exploratory study reveals that we can indeed automatically infer valuable architectural information to a certain extent, as a complementary conceptual viewpoint to architectural layering. It can reveal conceptual partitions in an MDE ecosystem for checking architectural conformance, reveal similar groups and subgroups of languages, see the cross-cutting concerns across the languages, etc. The accuracy and reliability of topic modeling on the MDE ecosystems, however, is yet to be quantitatively evaluated and further improved. See Sections 11.8 and 11.10 for threats to validity and potential room for improvements in the future.

11.8. Discussion

We have performed a variety of analyses for the MDE ecosystems at ASML. While we presented discussions for each case study separately, in this section we would like to present an overall discussion for our approach.

For the clone detection studies on models, we have extended SAMOS with partly custom-tailored, domain-specific extraction and comparison methods, particular for the ASOME data and control models. The development of these, with the domain experts in the loop, has indicated that the different nature of the (domain-specific) modeling languages and what the domain experts consider as *relevant* and *irrelevant* pieces of information in the models are crucial for an accurate, intuitive, and actionable clone detection exercise on those models. These additionally lead to implications on the setting and type of clone detection desired. For example, for the control models, the domain experts were interested in structural clones, while not so much for the data models.

As for the accuracy for the model and metamodel clone detection, we have achieved considerable success in general. However, especially for the structural clone detection for control models, which has been a new extension to SAMOS as introduced in this work, our approach possesses certain shortcomings. We will discuss these as threats to validity later in this section.

For both models and metamodel clones, we have participated in discussions with the domain experts on the nature of the clones and actionability for improving the MDE ecosystems. Our discussions reveal that some of those clones are indeed harmful and desirable to eliminate or refactor, while others might be inevitable due to language restrictions or even intended, e.g., for certain design goals, performance criteria, or backwards compatibility. Some of those harmful clones are indeed confirmed by the domain experts to be potential candidates for improvement, e.g., in the form of refactoring or abstraction. On the other hand, other such harmful clones have been identified as difficult or undesirable to refactor. Reasons for these would include deliberate design decisions (e.g., keeping singleton repositories, as reported in Section 11.7.1) or organizational limitations (e.g., language clones across ecosystems maintained by different teams, as reported in Section 11.7.4).

Interestingly, the results of the clone detection in control models might be used not to refactor the models themselves, but to introduce new language concepts, e.g., in the form of syntactic sugar or abstractions. This could increase the modelers' consistency and efficiency. Nevertheless, there can be certain limitations, such as the additional learning time for the new syntax and additional modeling effort in the case of abstractions.

Furthermore, we have discovered another use of model clone detection thanks to our discussions with the domain experts. When the cloning pattern is expected and desirable in a certain set of models, we can investigate the occurrence of those clone fragments in all the expected models. *Outlier* models, i.e., expected to have this pattern but not detected in the corresponding clone clusters, might actually indicate inconsistent design. We believe this to be an interesting additional use of SAMOS, and we hope to investigate this angle of clone detection in our future work.

Our studies on the system of ecosystems, i.e., the languages and their corresponding metamodels, have been shown to be potentially useful for maintaining the growing and evolving system of ecosystems at ASML. A high-level conceptual overview of the enterprise-level ecosystem and finer-grained clone detection on the languages can provide valuable sources of information in an automatic manner, to understand and monitor the ecosystems, while identifying certain shortcomings of those ecosystems, for instance, in the form of duplication and cloning. The architectural analysis we have performed on the CARM2G ecosystem, on the other hand, can provide a complementary conceptual perspective, in terms of automatic architectural reconstruction and conformance checking with respect to the intended layering. The limitations of the architectural study, a newly explored type of analysis in SAMOS, will be elaborated in the next section as threats to validity.

Threats to validity

Thanks to our extension in this work, SAMOS has been adapted for detecting clones in ASOME data and control models. However, there are several threats to validity for our current implementation. Data models have been compared in a structure-agnostic manner (i.e., using unigrams) at a relatively small scope (i.e., structured types and LevelAA; not, e.g., the whole model with a deeper containment hierarchy). For larger scopes we would need to use more powerful settings of SAMOS, capturing structure as well (e.g., subtrees, as done for control models).

On the other hand, clone detection for control models has been done on the *Protocol* scope using a similar structure-agnostic setting of unigrams, followed by another comparison using subtrees. The use of one-depth subtrees allowed us to reduce the computational time for comparison while still including structural information from the models (when compared, for example, to comparing full trees per model fragment). Note that this is still an approximation, and it could lead to certain inaccuracies, in which case maybe fully fledged (and very costly) graph comparison techniques should be employed instead. Obviously there is a tradeoff between the accuracy and the running time (hence the scalability) of the selected techniques.

Another issue arises with the requirement for selectively employing *ordered* comparison and *unordered* comparison for certain parts of the models. In the current implementation of SAMOS, we have it one way or another for the whole process. ASOME Control models prove to be a mixture of both, where order matters for the list of sequential actions and does not matter for the list of states in a state machine. A selective combination of both would be needed for a more accurate representation and comparison in the case of control models.

The comparison of elements for control models using the *No Name* setting is similar to the *blind renaming* approach taken in [19]. In such an approach, the identifiers of all the model elements are blindly renamed to the same name, effectively ignoring the relevance of names for the comparison. This approach allows us to find model elements that have similar structure but different values for elements such as guards or triggers or target state specification. While this improves the recall of the results found, the behavior of the two states as shown in Fig. 11.18 cannot be distinguished. The two cases on the left-hand side of the figure are treated the same, as depicted on the right-hand side.

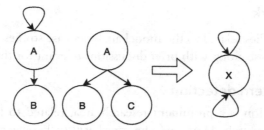

Figure 11.18 Counterexample for blind renaming, where SAMOS (erroneously) would not distinguish between the two cases [43].

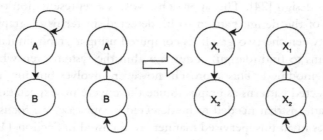

Figure 11.19 Counterexample where consistent renaming would be inaccurate [43].

While the structure is mainly captured by the extracted trees, some structural value is also attached to the names of elements, especially target state specifications. While consistent renaming of model elements might solve this problem, this approach was not taken because the order in which these states are renamed could result in inaccurate comparison results; see Fig. 11.19 for an example.

As for the language analyses presented in this work, several threats to validity exist as well. These would include, for instance, inaccurate NLP for language elements due to the lack of domain-specific dictionaries, cryptic element identifiers, and abbreviations. The topic modeling analysis part, however, is treated in a more exploratory manner in this work, in contrast to the domain analysis and clone detection parts, which have been validated considerably in previous work. The accuracy and reliability of topic modeling used for the architectural analysis is yet to be studied in detail, quantitatively evaluated, and further improved as well. As emphasized in Section 11.9.3, the technique used for topic modeling, namely, LDA, is very sensitive to the parameter settings, especially the number of topics. Hierarchical variants of LDA could be investigated to partly overcome this limitation. More specialized topic modeling approaches for shorter bodies of text (e.g., in social media data) could also be experimented with, as the languages in our case also have significantly less content (in the form of metamodel identifiers) than regular text documents.

11.9. Related work

There are various studies related to the model analytics case studies in this work. In this section we present those along with brief discussions relevant for this work.

11.9.1 Model pattern detection

Model pattern detection is a prominent research area, related to the tasks we are interested in for our research. However, the word *pattern* has been mostly considered synonymous to *design* patterns or *anti*patterns in the literature [23]. One approach uses pattern detection as a means to comprehend the existing design of a system to further improve this design [24]. This approach involves a representation of the system at hand, as well as of the design pattern to be detected, in terms of graphs. Ultimately, the similarity between the two graphs is computed using a graph similarity algorithm. The chapter claims to find (design) patterns within the system even when the pattern has been slightly modified. This approach, however, involves building a collection or catalogue of expected patterns as graphs. Since there were no expectations (by ASML) of the kind of patterns that needed to be detected in our case, we focused on finding, e.g., model clones in an unsupervised manner, as discussed in Section 11.9.2.

11.9.2 Model clone detection

While code clones have been previously explored in abundance and hence can be associated with some standard definition and classifications [25,26], relatively little work has been done in the field of model clone detection, resulting in the lack of such clear definitions. Model clones have been defined as "*unexpected overlaps and duplicates in models*" [27]. Störrle discusses the notion of model clones in depth, defining them as "*a pair of model fragments with a high degree of similarity between each other*" [28]. Model fragments are further defined as model elements closed under the containment relationship (the presence of this relationship between elements implies that the child in the relationship cannot exist independently of its parent).

Quite a few approaches advocate representing and analyzing models with respect to their underlying graphs, for clone detection purposes. One such approach involves representing Simulink models in the form of a labeled model graph [29]. In such graph-based methods, the task of finding clones in the models boils down to finding similar subgraphs within the constructed model graph. To do this, all maximal *clone pairs* are found within the graph (with a specification as to what constitutes a clone pair in their case). The approach of finding these maximal clone pairs is NP-complete and to reduce the running time, [29] the approach is modified to construct a similarity function for two nodes as a measure of their structural similarity. Finally, the detected clone pairs are aggregated using a clustering algorithm to find the resulting *clone classes* in the model. The disadvantage of this approach, however, is that *approximate* clones are not captured.

The work presented by Holthusen et al. [30] compares block-based models by assigning weights to relevant attributes for comparison, such as names, functions of the block, and interfaces. A similarity measure is defined to assign a value for the comparison and this value is stored for every pair of blocks being compared. This approach is taken to find variability in models in the automotive domain. Variations were introduced to a base model to add or remove functionality. By inspecting the similarity values, one could find models similar to a selected base model. SAMOS also uses the idea of computing similarity using a VSM to represent the occurrence of features in each model.

Störrle provides a contradictory notion however, i.e., that for some UML models, the graph structure may not necessarily be the most important aspect of the models to consider for clone detection [28]. He discusses that for some UML models, most of the information worth considering resides in the nodes as opposed to the links between these nodes. Therefore, the approach taken in this chapter defines the similarity of model fragments as the similarity of the nodes in such model fragments instead of the similarity of the graph structure of these model fragments. To construct this measure of similarity, the approach involves using heuristics based on the names of the elements being compared. Such an approach is justified when considering that *most elements that matter to modelers are named* [28]. This approach works for models where structure does not represent much in terms of model behavior. However, when the behavior of the models is represented in terms of structure, this approach cannot be used.

11.9.3 Topic modeling

Topic modeling, an approach in Information Retrieval and Machine Learning domains, involves a set of statistical techniques in text mining to automatically discover prominent concepts or topics in natural language text document collections [21,6]. Topics are typically conceived as collections or distributions of frequently co-occurring words in the corpus, which are assumed to be often semantically related. Topic models are often employed as an effective means to work on unstructured and unlabeled data such as natural language text, to infer some latent structure in the form of topic distributions (over the documents) and term distributions (over the topics).

Topic modeling applications for software engineering

Besides in text mining tasks, topic models are used in other disciplines, such as bioinformatics and computer vision, and recently in software engineering (SE) as well. Various surveys in the SE literature investigate the application of topic modeling to subdomains such as SE [22], mining software repositories (MSRs) [31], and Software Architecture (SA) [32]. The overall goal is to exploit automated techniques to better understand the underlying systems and processes, aid in reconstructing and improving certain parts of them, and eventually increase their quality in a cost-effective manner. A large volume of literature can be found on topic modeling for SE and MSR tasks, such as concept, aspect and feature mining or location from source code, clustering similar SE artifacts, recovering traceability links among heterogeneous sets of SE artifacts/entities (e.g., source code, documentation, requirements), bug localization and prediction, test case prioritization, evolution analysis, and finally clone detection [22,31]. The common denominator of all those approaches is the fact that there exists textual content in all those artifacts. Based on a similar observation of textual content in SE artifacts and the fact that they might also contain architectural information, another set of approaches investigate the use of topic modeling in architecture-related tasks. The exhaustive list of activities to be supported by topic modeling in the mapping study by Bi et al. [32] includes architectural understanding, (automatic) recovery, and documentation on the one hand, and architectural analysis, evaluation, and maintenance on the other hand. The authors in general emphasize the value of those activities, such as architectural understanding for distribution of responsibilities in a software system, architectural analysis for evaluating the conformance in the case of a layered architecture, and so on.

All the topic modeling approaches reported in the three surveys above typically operate on a set of traditional software artifacts, notably source code and documentation. In a recent work, Perez et al. [33] observe this as well, and propose applying feature location directly on the models in model-based product families. They however use it in a very particular setting: for assessing the fitness of model fragments in a query reformulation problem using genetic algorithms. To the best of our knowledge, there are no approaches in the literature which apply topic modeling for SA-related tasks in MDE and DSL ecosystems, in which we are interested in this work.

Latent Dirichlet Allocation

One of the most popular topic modeling techniques, also in SE tasks [22,32,31], is LDA [34]. LDA is a particular probabilistic (Bayesian) variant of topic modeling, which assumes Dirichlet prior distributions on the topics (per document, θ) and words (per topic, ϕ) and fits a generative model on the word occurrences in the corpus. Similar to the VSM setting (see Section 11.3), a collection of documents is transformed into a frequency matrix. Instead of the distance and measurement (as done for clustering in Section 11.3), the matrix is fed to LDA, which identifies the latent variables (topics) hidden in the data. The probability distributions θ and ϕ effectively describe the entire corpus. LDA relies on a set of hyperparameters to be set in advance, notably n being the number of topics, α and β being the parameters of the prior Dirichlet distributions, and additional ones depending on the particular inference technique used.

While the details of the statistical inference process (e.g., computing the posterior distributions using collapsed Gibbs sampling [21] as typically used in SE-related topic modeling tasks) is beyond the scope of this work, from an end-user perspective the output of LDA consists of two matrices: (given the fixed number of topics) one for the probability of each document belonging to various topics (i.e., multiple topics allowed, resulting in a kind of soft clustering) and one for the probability of each term belonging to various topics. The term probabilities can be manually inspected, for instance, to deduce what *"concept"* the topic actually corresponds to, while topic probabilities can be used to get the most prominent topics for the documents and identify document similarities.

The regular application of LDA as described above requires that the number of topics is given in advance, unlike, e.g., some other nonparametric variants such as Hierarchical Dirichlet Process [35]. One can either rely on domain expertise with respect to the corpus such that n is already known, or try to estimate the number using various heuristics. The latter involves running LDA with a range of candidate values and trying to optimize certain metrics: maximize the log likelihood of the inferred model [36] or minimize the topic density [37]. There are advanced techniques aiding or automatizing this estimation process; some notable examples within the SE literature include Panichella et al. [22], based on genetic algorithms, and Grant et al. [38], based on heuristics using vector similarity and source code location.

LDA has a proven track record of successful application in mining problems for natural language text documents. Yet one should be cautious while applying it, especially for other types of artifacts. First of all, there is the nontrivial task of determining the parameters of LDA in advance (such as number of topics, as discussed above); an incorrect choice of parameters [31] and even incorrect order of input [39] can lead to nonoptimal results. The authors in [22] further emphasize the difference between natural language text vs. source code, the latter of which has been recently studied and found to have a higher level of regularity than text in English [40], and claim that topic modeling for source code should be treated differently in order to get better results. For other artifacts, such as models, metamodels, and DSLs, no thorough empirical studies have been conducted regarding their nature yet.

11.10. Conclusion and future work

In this chapter, we have presented our approach for model analytics in an industrial context, with various analyses on ASML's MDE ecosystems. We have used and extended our model analytics framework, SAMOS, to operate on ASML's languages and models. We have elaborated the domain-specific extension of SAMOS, specifically for ASML's ASOME data and control models, to enable clone detection on those models. We have provided extensive case studies, where we performed clone detection on ASML's models, and additionally language-level analyses ranging from cross-DSL conceptual analysis and clone detection to architectural analysis for the CARM2G ecosystem. We have presented our findings along with valuable feedback from domain experts on the nature of cloning in the ecosystems and opportunities such as refactoring to support the maintenance and quality of the ever-growing and evolving ecosystems.

Besides the wide range of analyses presented in this work, there is still a lot of room for improvements and future work. While SAMOS has many combinations of settings and scopes available for model clone detection, not all these combinations were chosen for the case studies (considering the time constraints of our collaborative project with ASML). As future work we could explore different aspects of comparison using the different available settings, such as type-based and idf weighting. Furthermore, as indicated in our discussion on threats to validity, advanced comparison schemes (e.g., selective ordered vs. unordered comparison for different model parts) could be integrated to improve the accuracy of our clone detection. Other directions would include the detection of patterns, e.g., design patterns (as in [24]), or antipatterns. As a useful example application of this, one could create a pattern catalogue and find what models do not adhere to these patterns (i.e., as a potential indication of unexpected behavior in models). Finding structural clones, especially in the control models, is another promising direction for future work. Lastly, the language analyses could be improved to overcome the limitations as addressed in our discussion on the threats to validity, e.g., with more sophisticated NLP and more advanced, fine-tuned topic modeling techniques. Considering the time dimension of the languages, it would also be very interesting to investigate their evolution, in terms of *concept drift* [41] and cloning.

References

[1] R. Schiffelers, Empowering high tech systems engineering using mdse ecosystems (invited talk), in: E. Guerra, M. van den Brand (Eds.), Proc. of the 10th Int. Conf. on Theory and Practice of Model Transformation – Held as Part of STAF 2017, in: Lecture Notes in Computer Science, Springer, 2017, p. XI.

[2] C.J. Kapser, M.W. Godfrey, Supporting the analysis of clones in software systems, Journal of Software Maintenance and Evolution: Research and Practice 18 (2006) 61–82.

[3] C.J. Kapser, M.W. Godfrey, "Cloning considered harmful" considered harmful: patterns of cloning in software, Empirical Software Engineering 13 (2008) 645.

[4] Ö. Babur, L. Cleophas, M. van den Brand, Hierarchical clustering of metamodels for comparative analysis and visualization, in: Proc. of the 12th European Conf. on Modelling Foundations and Applications, 2016, pp. 2–18.

[5] Ö. Babur, L. Cleophas, M. van den Brand, Metamodel clone detection with SAMOS, Journal of Computer Languages 51 (2019) 57–74.

[6] D.M. Blei, J.D. Lafferty, Topic models, in: Text Mining, Chapman and Hall/CRC, 2009, pp. 101–124.

[7] W. Alberts, ASML's MDE going Sirius, https://www.slideshare.net/Obeo_corp/siriuscon2016-asmls-mde-going-sirius, 2016. (Accessed 12 November 2018).

[8] Ö. Babur, L. Cleophas, Using n-grams for the automated clustering of structural models, in: Theory and Practice of Computer Science, Springer International Publishing, Cham, 2017, pp. 510–524.

[9] Ö. Babur, Clone detection for Ecore metamodels using n-grams, in: The 6th International Conference on Model-Driven Engineering and Software Development, ScitePress, 2018, pp. 411–419.

[10] Ö. Babur, L. Cleophas, M. van den Brand, Model analytics for feature models: case studies for S.P.L.O.T. repository, in: Proc. of MODELS 2018 Workshops, Co-Located with the 21st Int. Conf. on Model Driven Engineering Languages and Systems, 2018, pp. 787–792.

[11] R Core Team, R: A Language and Environment for Statistical Computing, R Foundation for Statistical Computing, Vienna, Austria, 2014.

[12] R. Schiffelers, Y. Luo, J. Mengerink, M. van den Brand, Towards automated analysis of model-driven artifacts in industry, in: 6th International Conference on Model-Driven Engineering and Software Development, ScitePress, 2018, pp. 743–751.

[13] R.R.H. Schiffelers, W. Alberts, J.P.M. Voeten, Model-based specification, analysis and synthesis of servo controllers for lithoscanners, in: Proceedings of the 6th International Workshop on Multi-Paradigm Modeling, MPM '12, ACM, New York, NY, USA, 2012, pp. 55–60.

[14] S. Adyanthaya, Robust Multiprocessor Scheduling of Industrial-Scale Mechatronic Control Systems, Ph.D. thesis, Technische Universiteit Eindhoven, Eindhoven, 2016.

[15] B. van der Sanden, M. Reniers, M. Geilen, T. Basten, J. Jacobs, J. Voeten, R. Schiffelers, Modular model-based supervisory controller design for wafer logistics in lithography machines, in: 2015 ACM/IEEE 18th International Conference on Model Driven Engineering Languages and Systems, MODELS, 2015, pp. 416–425.

[16] L. van der Sanden, Performance Analysis and Optimization of Supervisory Controllers, Ph.D. thesis, Department of Electrical Engineering, 2018.

[17] J. Nogueira Bastos, Modular Specification and Design Exploration for Flexible Manufacturing Systems, Ph.D. thesis, Department of Electrical Engineering, 2018.

[18] M.H. Alalfi, J.R. Cordy, T.R. Dean, M. Stephan, A. Stevenson, Models are code too: near-miss clone detection for Simulink models, in: 28th IEEE International Conference on Software Maintenance, 2012, pp. 295–304.

[19] J. Chen, T.R. Dean, M.H. Alalfi, Clone detection in Matlab stateflow models, Software Quality Journal 24 (2016) 917–946.

[20] M. Mondai, C.K. Roy, K.A. Schneider, Micro-clones in evolving software, in: Proc. of the 25th International Conference on Software Analysis, Evolution and Reengineering, IEEE, 2018, pp. 50–60.

[21] M. Steyvers, T. Griffiths, Probabilistic topic models, Handbook of Latent Semantic Analysis 427 (2007) 424–440.

[22] A. Panichella, B. Dit, R. Oliveto, M. Di Penta, D. Poshyvanyk, A. De Lucia, How to effectively use topic models for software engineering tasks? An approach based on genetic algorithms, in: Proceedings of the 2013 International Conference on Software Engineering, IEEE Press, 2013, pp. 522–531.

[23] M. Stephan, J.R. Cordy, Identifying instances of model design patterns and antipatterns using model clone detection, in: Proc. of the 7th International Workshop on Modeling in Software Engineering, IEEE Press, 2015, pp. 48–53.

[24] N. Tsantalis, A. Chatzigeorgiou, G. Stephanides, S.T. Halkidis, Design pattern detection using similarity scoring, IEEE Transactions on Software Engineering 32 (2006) 896–909.

[25] C.K. Roy, J.R. Cordy, R. Koschke, Comparison and evaluation of code clone detection techniques and tools: a qualitative approach, Science of Computer Programming 74 (2009) 470–495.

[26] R. Koschke, Survey of research on software clones, in: Dagstuhl Seminar Proceedings, Schloss Dagstuhl-Leibniz-Zentrum für Informatik, 2007.

[27] D. Rattan, R. Bhatia, M. Singh, Model clone detection based on tree comparison, in: 2012 Annual IEEE India Conference, INDICON, 2012, pp. 1041–1046.

[28] H. Störrle, Towards clone detection in UML domain models, in: Proc. of the Fourth European Conference on Software Architecture: Companion Volume, ACM, 2010, pp. 285–293.

[29] F. Deissenboeck, B. Hummel, E. Jürgens, B. Schätz, S. Wagner, J.-F. Girard, S. Teuchert, Clone detection in automotive model-based development, in: Prof. of the 30th Int. Conf. on Software Engineering, ACM, 2008, pp. 603–612.

[30] S. Holthusen, D. Wille, C. Legat, S. Beddig, I. Schaefer, B. Vogel-Heuser, Family model mining for function block diagrams in automation software, in: Prof. of the 18th Int. Software Product Line Conference: Companion Volume for Workshops, Demonstrations and Tools, vol. 2, ACM, 2014, pp. 36–43.

[31] T.-H. Chen, S.W. Thomas, A.E. Hassan, A survey on the use of topic models when mining software repositories, Empirical Software Engineering 21 (2016) 1843–1919.

[32] T. Bi, P. Liang, A. Tang, C. Yang, A systematic mapping study on text analysis techniques in software architecture, Journal of Systems and Software 144 (2018) 533–558.

[33] F. Pérez, J. Font, L. Arcega, C. Cetina, Automatic query reformulations for feature location in a model-based family of software products, Data & Knowledge Engineering 116 (2018) 159–176.

[34] D.M. Blei, A.Y. Ng, M.I. Jordan, Latent Dirichlet allocation, Journal of Machine Learning Research 3 (2003) 993–1022.

[35] Y.W. Teh, M.I. Jordan, M.J. Beal, D.M. Blei, Hierarchical Dirichlet processes, Journal of the American Statistical Association 101 (2006) 1566–1581.

[36] T.L. Griffiths, M. Steyvers, Finding scientific topics, Proceedings of the National Academy of Sciences 101 (2004) 5228–5235.

[37] J. Cao, T. Xia, J. Li, Y. Zhang, S. Tang, A density-based method for adaptive LDA model selection, Neurocomputing 72 (2009) 1775–1781.

[38] S. Grant, J.R. Cordy, Estimating the optimal number of latent concepts in source code analysis, in: Source Code Analysis and Manipulation (SCAM), 2010 10th IEEE Working Conference on, IEEE, 2010, pp. 65–74.

[39] A. Agrawal, W. Fu, T. Menzies, What is wrong with topic modeling? And how to fix it using search-based software engineering, Information and Software Technology 98 (2018) 74–88.

[40] A. Hindle, E.T. Barr, Z. Su, M. Gabel, P. Devanbu, On the naturalness of software, in: Software Engineering (ICSE), 2012 34th International Conference on, IEEE, 2012, pp. 837–847.

[41] I. Žliobaitė, M. Pechenizkiy, J. Gama, An overview of concept drift applications, in: Big Data Analysis: New Algorithms for a New Society, Springer, 2016, pp. 91–114.

[42] Ö. Babur, Model Analytics and Management, IPA Dissertation Series, Technische Universiteit Eindhoven, 2019.

[43] A. Suresh, Model Analytics for ASML's Data and Control Modeling Languages, Master thesis, Technische Universiteit Eindhoven, 2019.

Index

Printed in the United States
By Bookmasters